MSFC-MAN-503

I0023992

GEORGE C. MARSHALL SPACE FLIGHT CENTER

# SATURN V FLIGHT MANUAL

THIS PUBLICATION REPLACES MSFC-MAN-503 DATED 1 APRIL 1968

NATIONAL AERONAUTICS AND SPACE ADMINISTRATION

NASA

# 1 NOVEMBER 1968

MSFC - Form 454 (Rev October 1967)

**Published by Books Express Publishing**
Copyright © Books Express, 2012
ISBN 978-1-78039-847-1

Books Express publications are available from all good retail and online booksellers. For publishing proposals and direct ordering please contact us at: info@books-express.com

INSERT LATEST CHANGED PAGES.
DESTROY SUPERSEDED PAGES.

# LIST OF EFFECTIVE PAGES

NOTE: The portion of the text affected by the changes is indicated by a vertical line in the outer margins of the page.

TOTAL NUMBER OF PAGES IN THIS PUBLICATION IS 252, CONSISTING OF THE FOLLOWING:

* The asterisk indicates pages changed, added, or deleted by the current change.

NASA

# SATURN V FLIGHT MANUAL
## SA-503

## FOREWORD

This manual was prepared to provide the astronaut with a single source reference as to the characteristics and functions of the SA-503 launch vehicle and the AS-503 manned flight mission. A revision to the manual, incorporating the latest released data on the vehicle and mission, will be released approximately 30 days prior to the scheduled launch date.

The manual provides general mission and performance data, emergency detection system information, a description of each stage and the IU, and a general discussion of ground support facilities, equipment, and mission control. A bibliography identifies additional references if a more comprehensive study is desired.

This manual is for information only and is not a control document. If a conflict should be discovered between the manual and a control document the control document will rule.

Recommended changes or corrections to this manual should be forwarded, in writing, to the Saturn V Systems Engineering Management Office (I-V-E), MSFC, Attention: Mr. H.P. Lloyd; or to the Crew Safety and Procedures Branch (CF-24), MSC, Attention: Mr. D.K. Warren.

Arthur Rudolph

Manager, Saturn V Program
George C. Marshall Space
Flight Center

D.K. Slayton

Director of Flight
Crew Operations
Manned Spacecraft Center

## REVISION NOTE

The manual has been completely revised to incorporate vehicle and mission changes and to take advantage of improvements in presentation which have been developed since the original release. The information in the manual describes the vehicle configuration and mission characteristics as defined for the C Prime mission and was prepared from information available approximately thirty days prior to Oct. 25, 1968.

# TABLE OF CONTENTS

# SECTION I

# GENERAL DESCRIPTION

## TABLE OF CONTENTS

## SATURN V SYSTEM DESCRIPTION

The Saturn V system in its broadest scope includes conceptual development, design, manufacture, transportation, assembly, test, and launch. The primary mission of the Saturn V launch vehicle, three-stage-to-escape boost launch of an Apollo Spacecraft, established the basic concept. This mission includes a suborbital start of the third stage (S-IVB) engine for final boost into earth orbit, and subsequent reignition to provide sufficient velocity for escape missions including the lunar missions.

## LAUNCH VEHICLE DEVELOPMENT

The Saturn launch vehicles are the product of a long evolutionary process stemming from initial studies in 1957 of the Redstone and Jupiter missiles. Early conceptual studies included other proven missiles such as Thor and Titan, and considered pay loads ranging from earth orbiting satellites to manned spacecraft such as Dynasoar, Mercury, Gemini, and eventually Apollo.

The Saturn V launch vehicle evolved from the earlier Saturn vehicles as a result of the decision in 1961 to proceed with the Apollo manned lunar mission. As the Apollo mission definition became clear, conceptual design studies were made, considering such parameters as structural dynamics, staging dynamics, and propulsion dynamics.

Design trade-offs were made in certain areas to optimize the launch vehicle design, based on mission requirements. The best combination of design parameters for liquid propellant vehicles resulted in low accelerations and low dynamic loads. Reliability, performance and weight were among primary factors considered in optimizing the design.

Structural design carefully considered the weight factor. Structural rigidity requirements were dictated largely by two general considerations: flight control dynamics and propellant slosh problems. Gross dimensions (diameter & length) were dictated generally by propellant tankage size.

As propulsion requirements were identified, system characteristics emerged: thrust levels, burning times, propellant types and quantities. From these data, engine requirements and characteristics were identified, and the design and development of the total launch vehicle continued, centered around the propulsion systems.

Some of the principal design ground rules developed during the conceptual phase, which were applied in the final design, are discussed in the following paragraphs.

## VEHICLE DESIGN GROUND RULES

### Safety

Safety criteria are identified by Air Force Eastern Test Range (AFETR) Safety Manual 127-1 and AFETR Regulation 127-9.

Crew safety considerations required the development of an Emergency Detection System (EDS) with equipment located throughout the launch vehicle to detect emergency conditions as they develop. If an emergency condition is detected, this system will either initiate an automatic abort sequence, or display critical data to the flight crew for their analysis and reaction.

Each powered stage is designed with dual redundant range safety equipment which will effect engine cutoff and propellant dispersion in the event of a launch abort after liftoff. Engine cutoff results from closing valves and terminating the flow of fuel and oxidizer. Propellant is dispersed by detonating linear-shaped charges, thereby longitudinally opening the propellant tanks.

### Stage Separation

The separation of the launch vehicle stages in flight required design studies involving consideration of many parameters, such as time of separation, vehicle position, vehicle attitude, single or dual plane separation, and the type, quantity, and location of ordnance.

The launch vehicle stages separate in flight by explosively severing a circumferential separation joint and firing retrorocket motors to decelerate the spent stage. Stage separation is initiated when stage thrust decays to a value equal to or less than 10% of rated thrust. A short coast mode is used to allow separation of the spent stage, and to effect ullage settling of the successive stage prior to engine ignition.

A delayed dual plane separation is employed between the S-IC and S-II stages, while a single plane separation is adequate between the S-II and S-IVB stages.

### Umbilicals

In the design and placement of vehicle plates, consideration was given to such things as size, locations, methods of attachment, release, and retraction.

The number of umbilicals is minimized by the combining of electrical connectors and pneumatic and propellant couplings into common umbilical carriers. Location of the umbilicals depended upon the location of the vehicle plates, which were limited somewhat by the propellant tanking, plumbing, and wiring runs inside the vehicle structure. Umbilical disconnect and retraction systems are redundant for reasons of reliability and safety.

### Electrical Systems

An electrical load analysis of the launch vehicle provided the basic data (voltage, frequency, and power requirements) for design of the electrical system.

Such factors as reliability, weight limitations, and weight distributions dictated requirements to minimize electrical wiring, yet distribute the electrical loads and power sources throughout the launch vehicle. Each stage of the vehicle has its own independent electrical system. No electrical power is transferred between stages; only control signals are routed between stages.

Primary flight power is supplied by wet cell batteries in each stage. The sizes, types, and characteristics are discussed in subsequent sections of this manual. Where alternating current, or direct current with a higher voltage than the batteries is required, inverters and/or converters convert the battery power to the voltages and frequencies needed.

All stages of the launch vehicle are electrically bonded together to provide a unipotential structure, and to minimize current transfer problems in the common side of the power systems.

## MANUFACTURE AND LAUNCH CONCEPTS

The development of the vehicle concept required concurrent efforts in the areas of design, manufacture, transportation, assembly, checkout, and launch.

The size and complexity of the vehicle resulted in the decision to have detail design and manufacture of each of the three stages, the Instrument Unit (IU), and the engines accomplished by separate contractors under the direction of MSFC.

This design/manufacturing approach required the development of production plans and controls, and of transportation and handling systems capable of handling the massive sections.

The assembly, checkout, and launch of the vehicle required the development of an extensive industrial complex at KSC. Some of the basic ground rules which resulted in the KSC complex described in Section VIII are:

1. The vehicle will be assembled and checked out in a protected environment before being moved to the launch site.

2. A final checkout will be performed at the launch site prior to launch.

3. Once the assembly is complete, the vehicle will be transported in the erect position without disconnecting the umbilicals.

4. Automatic checkout equipment will be required.

5. The control center and checkout equipment will be located away from the launch area.

## LAUNCH REQUIREMENTS

Some of the launch requirements which have developed from the application of these ground rules are:

1. Several days prior to the actual launch time, the vehicle is moved to the launch area for prelaunch servicing and checkout. During most of this time, the vehicle systems are sustained by ground support equipment. However, at T-50 seconds, power is transferred to the launch vehicle batteries, and final vehicle systems monitoring is accomplished. In the event of a hold, the launch vehicle can operate on internal power for up to 12 hours before a recycle for batteries would be required.

2. While in the launch area, environmental control within the launch vehicle is provided by environmental control systems in the mobile launcher (ML) and on the pad. The IU also utilizes an equipment cooling system, in which heat is removed by circulation of a methanol-water coolant. During preflight, heat is removed from the coolant by a Ground Support Equipment (GSE) cooling system located on the ML. During flight, heat is removed from the coolant by a water sublimator system.

3. While in transit between assembly area and launch area, or while in the launch area for launch preparations, the assembled launch vehicle must withstand the natural environment. The launch vehicle is designed to withstand 99.9% winds during the strongest wind month, while either free standing or under transport, with the damper system attached. In the event of a nearby explosion of a facility or launch vehicle, the Saturn V will also withstand a peak overpressure of 0.4 psi.

4. To more smoothly control engine ignition, thrust buildup and liftoff of the vehicle, restraining arms provide support and holddown at four points around the base of the S-IC stage. A gradual controlled release is accomplished during the first six inches of vertical motion.

## RELIABILITY AND QUALITY ASSURANCE

The Apollo Program Office, MA, has the overall responsibility for development and implementation of the Apollo reliability and quality assurance (R & QA) program. NASA Centers are responsible for identifying and establishing R & QA requirements, and for implementing an R & QA program to the extent necessary to assure the satisfactory performance of the hardware for which they are responsible. The Apollo R & QA program is defined by the Apollo Program Development Plan, M-D MA 500 and Apollo R & QA Program Plan, NHB 5300-IA.

Crew safety and mission success are the main elements around which the R & QA program is built. The primary criterion governing the design of the Apollo system is that of achieving mission success without unacceptable risk of life or permanent physical disablement of the crew.

It is Apollo program policy to use all currently applicable methods to ensure the reliability and quality of the Apollo/Saturn system. Some of these methods are discussed in subsequent paragraphs.

### Analysis of Mission Profiles

The mission profile is analyzed to determine the type and

scope of demands made on equipment and flight crew during each phase of the mission. This has resulted in the incorporation of design features which will enable the flight crew to detect and react effectively to abnormal circumstances. This permits the flight crew to abort safely if the condition is dangerous or to continue the normal mission in an alternate mode if crew safety is not involved but equipment is not operating properly.

## Failure Effects and Criticality Analyses

The modes of failure for every critical component of each system are identified. The effect of each failure mode on the operation of the system is analyzed, and those parts contributing most to unreliability are identified. These analyses have resulted in the identification of mission compromising, single-point failures, and have aided in the determination of redundancy requirements and/or design changes.

## Design Reviews

A systematic design review of every part, component, subsystem, and system has been performed using comprehensive check lists, failure effects analysis, criticality ratings, and reliability predictions. These techniques have enabled the designer to review the design approach for problems not uncovered in previous analyses. In the R & QA area, the preliminary design review (PDR) and critical design review (CDR) required by the Apollo Program Directive No. 6 represents specialized application of this discipline.

## VEHICLE DEVELOPMENT FLOW

Principal milestones in the hardware and mission phases of the Apollo program are shown in figure 1-1.

## Certification and Review Schedules

Certificates of Flight Worthiness (COFW) function as a certification and review instrument. A COFW is generated for each major piece of flight hardware. The certificate originates at the manufacturing facility, and is shipped with the hardware wherever it goes to provide a time phased historical record of the item's test results, modifications, failures, and repairs.

The program managers pre-flight review (PMPFR) and the program directors flight readiness review (PDFRR) provide a final assessment of launch vehicle, spacecraft, and launch facility readiness at the launch site. During the final reviews, the decision is made as to when deployment of the world wide mission support forces should begin.

## TRANSPORTATION

The Saturn stage transportation system provides reliable and economical transportation for stages and special payloads between manufacturing areas, test areas and KSC. The various modes of transportation encompass land, water, and air routes.

Each stage in the Saturn V system requires a specially designed transporter for accomplishing short distance land moves at manufacturing, test, and launch facilities. These transporters have been designed to be compatible with manufacturing areas, dock facility roll-on/roll-off

requirements, and to satisfy stage protection requirements.

Long distance water transportation for the Saturn V stages is by converted Navy barges and landing ship dock type ocean vessels. Tie-down systems provide restraint during transit. Ocean vessels are capable of ballasting to mate with barges and dock facilities for roll-on/roll-off loading. Docks are located at MSFC, KSC, Michoud, MTF, and Seal Beach, California (near Los Angeles).

Air transportation is effected by use of a modified Boeing B-377 (Super Guppy) aircraft. This system provides quick reaction time for suitable cargo requiring transcontinental shipments. For ease in loading and unloading the aircraft, compatible ground support lift trailers are utilized.

A Saturn transportation summary is presented in figure 1-2.

## LAUNCH VEHICLE DESCRIPTION

## GENERAL ARRANGEMENT

The Saturn V/Apollo general configuration is illustrated in figure 1-3. Also included are tables of engine data, gross vehicle dimensions and weights, ullage and retrorocket data, and stage contractors.

## INTERSTAGE DATA FLOW

In order for the Saturn V launch vehicle and Apollo spacecraft to accomplish their objectives, a continuous flow of data is necessary throughout the vehicle. Data flow is in both directions: from spacecraft to stages, and from stages to the spacecraft. The IU serves as a central data processor, and nearly all data flows through the IU.

Specific data has been categorized and tabulated to reflect, in figure 1-4, the type of data generated, its source and its flow. Each stage interface also includes a confidence loop, wired in series through interstage electrical connectors, which assures the Launch Vehicle Digital Computer (LVDC) in the IU that these connectors are mated satisfactorily.

## RANGE SAFETY AND INSTRUMENTATION

## GENERAL

In view of the hazards inherent in missile/space vehicle programs, certain stringent safety requirements have been established for the Air Force Eastern Test Range (AFETR). Figure 1-5 illustrates the launch azimuth limits and destruct azimuth limits for the Atlantic Missile Range (AMR).

Prime responsibility and authority for overall range safety is vested in the Commander, AFETR, Patrick AFB, Florida. However, under a joint agreement between DOD and NASA, ground safety within the confines of the Kennedy Space Center will be managed by NASA.

To minimize the inherent hazards of the Saturn/Apollo program, a number of safety plans have been developed and implemented in accordance with AFETR regulations.

These plans cover all phases of the Saturn/Apollo program from design, through launch of the vehicle, into orbit.

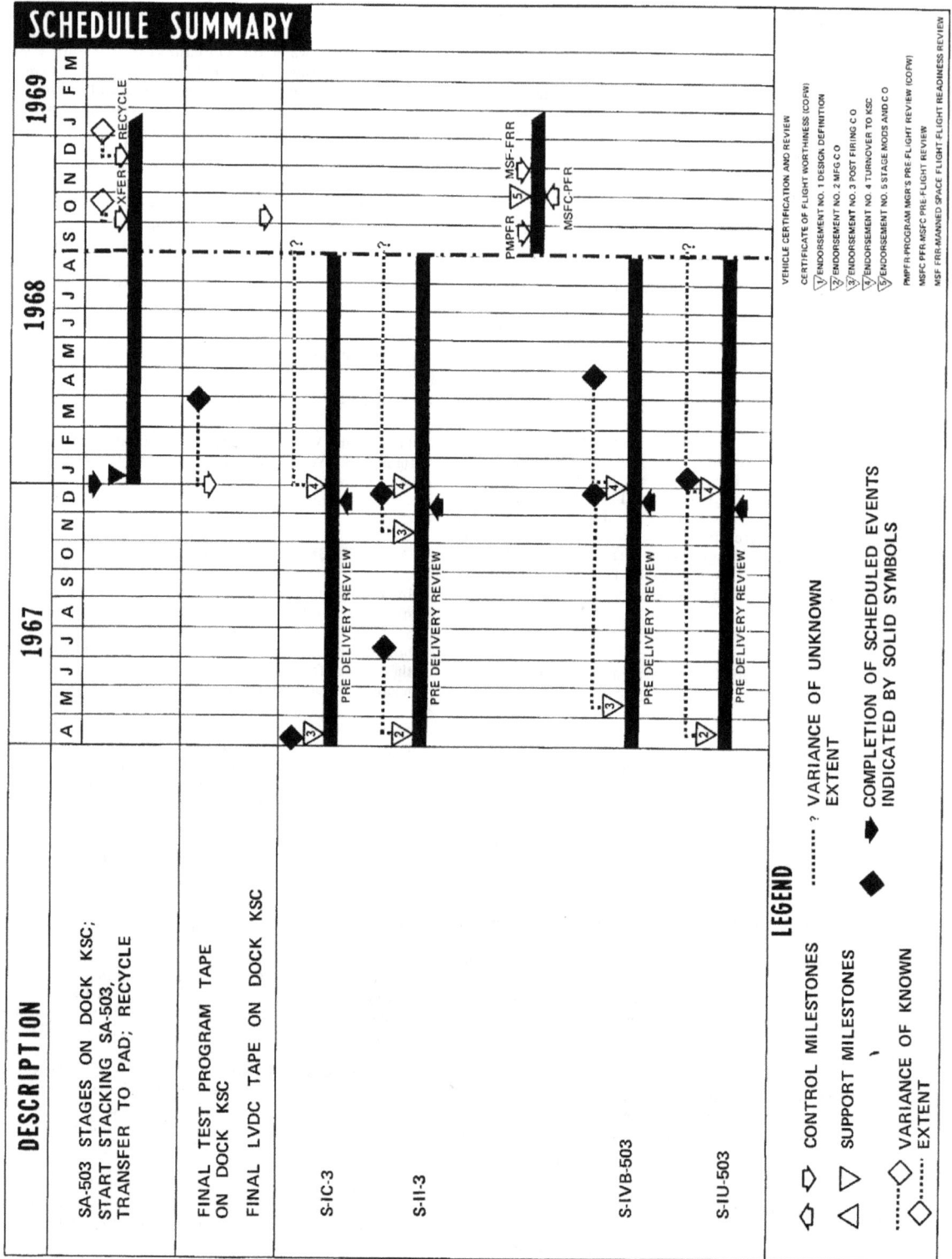

Figure 1-1

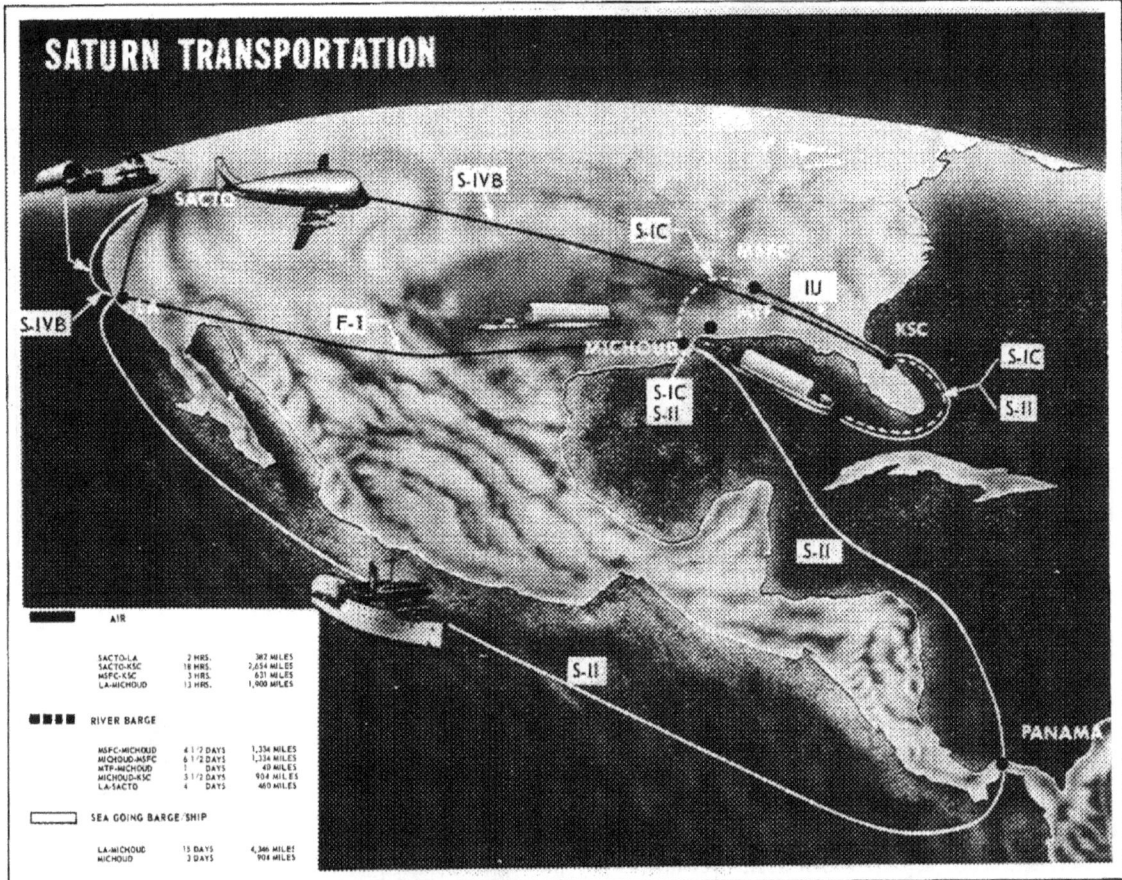

## SATURN TRANSPORTATION

S-IVB

SACTO

S-IC

MSFC

IU

LA

KSC

S-IVB

F-1

MICHOUD

S-IC
S-II

S-IC

S-II

S-II

PANAMA

| AIR | | |
|---|---|---|
| SACTO-LA | 2 HRS. | 382 MILES |
| SACTO-KSC | 18 HRS. | 2,654 MILES |
| MSFC-KSC | 3 HRS. | 651 MILES |
| LA-MICHOUD | 13 HRS. | 1,000 MILES |

| RIVER BARGE | | |
|---|---|---|
| MSFC-MICHOUD | 4 1/2 DAYS | 1,334 MILES |
| MICHOUD-MSFC | 6 1/2 DAYS | 1,334 MILES |
| MTF-MICHOUD | 1 DAYS | 40 MILES |
| MICHOUD-KSC | 3 1/2 DAYS | 904 MILES |
| LA-SACTO | 4 DAYS | 460 MILES |

| SEA GOING BARGE/SHIP | | |
|---|---|---|
| LA-MICHOUD | 15 DAYS | 4,346 MILES |
| MICHOUD | 3 DAYS | 904 MILES |

Figure 1-2

To enhance the development and implementation of the range safety program, two general safety categories have been established: ground safety and flight safety.

## GROUND SAFETY

The ground safety program includes a ground safety plan which calls for the development of safety packages. The major categories covered by these packages are:

1. Vehicle Destruct System. This package includes a system description, circuit descriptions, schematics, ordnance system description, specifications, RF system description, installation, and checkout procedures.

2. Ordnance Devices. This package includes descriptive information on chemical composition and characteristics, mechanical and electrical specifications and drawings, and electrical bridgewire data.

3. Propellants. This package includes descriptive data on chemical composition, quantities of each type, locations in the vehicle, handling procedures, and hazards.

4. High Pressure Systems. This package includes types of gases, vehicle storage locations, pressures, and hazards.

5. Special Precautionary Procedures. This package covers possible unsafe conditions, and includes lightning safeguards, use of complex test equipment, and radiological testing.

Also included under ground safety are provisions for launch area surveillance during launch activities. Surveillance methods include helicopters, search radars, and range security personnel. Automatic plotting boards keep the range safety officer (RSO) informed of any intrusion into the launch danger zones by boats or aircraft.

To further assist the RSO in monitoring launch safety, a considerable amount of ground instrumentation is used. A vertical-wire sky screen provides a visual reference used during the initial phase of the launch to monitor vehicle attitude and position. Television systems photographing the launch vehicle from different angles also provide visual reference. Pulsed and CW tracking radars and real time telemetry data provide an electronic sky screen, which displays on automatic plotting boards, and charts the critical flight trajectory parameters.

# SATURN V LAUNCH VEHICLE

| SOLID ULLAGE ROCKET AND RETROROCKET SUMMARY | | | | |
|---|---|---|---|---|
| STAGE | TYPE | QUANTITY | NOMINAL THRUST AND DURATION | PROPELLANT GRAIN WEIGHT |
| S-IC | RETROROCKET | 8 | 75,800 POUNDS * 0.541 SECONDS | 278.0 POUNDS |
| S-II | ULLAGE | 4 | 23,000 POUNDS † 3.75 SECONDS | 336.0 POUNDS |
| | RETROROCKET | 4 | 34,810 POUNDS ‡ 1.52 SECONDS | 268.2 POUNDS |
| S-IVB | ULLAGE | 2 | 3,390 POUNDS † 3.87 SECONDS | 58.8 POUNDS |

| ENGINE DATA | | | | | |
|---|---|---|---|---|---|
| STAGE | QTY | ENGINE MODEL | NOMINAL THRUST EACH | NOMINAL THRUST TOTAL | BURN TIME |
| S-IC | 5 | F-1 | 1,522,000 | 7,610,000 | 150.7 SEC |
| S-II | 5 | J-2 | 228,000 | 1,140,000 | 367 SEC |
| S-IVB | 1 | J-2 | 203,000 | 203,000 | 156 & 336 SEC |

| STAGE DIMENSIONS | | | STAGE WEIGHTS | |
|---|---|---|---|---|
| | DIAMETER | LENGTH | DRY | AT LAUNCH |
| S-IC Base (including fins) | 63.0 FEET | 138 FEET | 305,100 POUNDS | 4,792,200 POUNDS |
| S-IC Mid-stage | 33.0 FEET | | | |
| S-II Stage | 33.0 FEET | 81.5 FEET | 88,400 POUNDS | 1,034,900 POUNDS |
| S-IVB Stage | 21.7 FEET | 59.3 FEET | 33,142 POUNDS | 262,300 POUNDS |
| Instrument Unit | 21.7 FEET | 3.0 FEET | 4,873 POUNDS | 4,873 POUNDS |

| SATURN V STAGE MANUFACTURERS | |
|---|---|
| STAGE | MANUFACTURER |
| S-IC | THE BOEING COMPANY |
| S-II | NORTH AMERICAN-ROCKWELL |
| S-IVB | McDONNELL - DOUGLAS CORP. |
| S-IU | INTERNATIONAL BUSINESS MACHINE CORP. |

NOTE: THRUST VALUES, WEIGHTS, AND BURN TIMES ARE ALL APPROXIMATIONS.

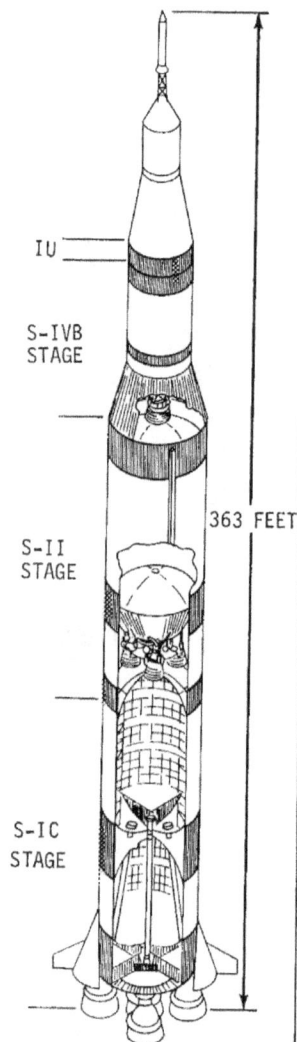

IU

S-IVB STAGE

363 FEET

S-II STAGE

S-IC STAGE

PRE-LAUNCH LAUNCH VEHICLE GROSS WEIGHT ≈ 6,094,073 POUNDS

\* MINIMUM VACUUM THRUST AT 120°F

† AT 170,000 FT. AND 70°F

‡ NOMINAL VACUUM THRUST AT 60°F

Figure 1-3

# STAGE ELECTRICAL INTERFACE FLOW

**SPACECRAFT TO IU**

+28 VDC TO EDS
LV ENGINES CUTOFF TO EDS
ATTITUDE ERROR SIGNAL
Q-BALL PITCH AND YAW
S-IVB ENGINE CUTOFF
AGC COMMAND POWER
S-IVB IGNITION SEQUENCE
   START
AUTO ABORT DEACTIVATE                 (M)
INITIATE S-II/S-IVB                       (M)
   SEPARATION
SPACECRAFT CONTROL                    (M)
   DISCRETE
TRANSLUNAR INJECTION                  (M)
   INHIBIT

   (M) = MANUALLY INITIATED

**IU TO SPACECRAFT**

EDS LIFTOFF
EDS AUTO ABORT
+28 VDC FOR EDS
+28 VDC FOR Q BALL
S-IVB ULLAGE THRUST OK
GUIDANCE REFERENCE RELEASE
AGC LIFTOFF
Q BALL TEMPERATURE SENSING
S-II AND S-IVB FUEL TANK
   PRESSURE                            (V)
LV ATTITUDE REFERENCE
   FAILURE                             (V)
LV RATE EXCESSIVE                       (V)
EDS ABORT REQUEST                       (V)
S-II START/SEPARATION                   (V)
STAGE ENGINES OUT                       (V)

   (V) = VISUALLY DISPLAYED

**S-IVB TO IU**

+28 VDC FOR TIMING
SWITCH SELECTOR ADDRESS
   VERIFICATION
ENGINE ACTUATOR POSITIONS
ATTITUDE CONTROL RATE GYROS
   SIGNALS
ATTITUDE CONTROL ACCELEROMETER
   SIGNALS
LOX TANK PRESSURE
FUEL TANK PRESSURE
RSCR & PD EBW FIRING UNIT
ARM AND ENGINE CUTOFF ON
ENGINE THRUST OK
TELEMETRY SIGNALS

**S-II TO S-IVB**

+28 VDC FOR RETRO-ROCKET
   PRESSURE TRANSDUCER
S-IVB ENGINE START ENABLE

**S-II TO IU**

ENGINE ACTUATOR POSITIONS
+28VDC FOR TIMING
S-IC STAGE SEPARATED
AFT INTERSTAGE SEPARATED
S-II STAGE SEPARATED
S-II ENGINE OUT
S-II PROPELLANT DEPLETION
SWITCH SELECTOR VERIFY
FUEL TANK PRESSURE
ENGINE THRUST OK
LOX TANK PRESSURE

**IU TO STAGES**

STAGE ENGINE ACTUATOR COMMANDS
STAGE ENGINE ACTUATOR MEASURING
   VOLTAGES
+28 VDC FOR SWITCHING AND
   TIMING
STAGE SWITCH SELECTOR SIGNALS
(VERIFY, COMMAND, ADDRESS,
   READ, RESET, ENABLE)
STAGE EDS COMMAND ENGINES OFF
S-IVB ATTITUDE CONTROL SYSTEM
   COMMANDS
TELEMETRY CLOCK AND SYNC.

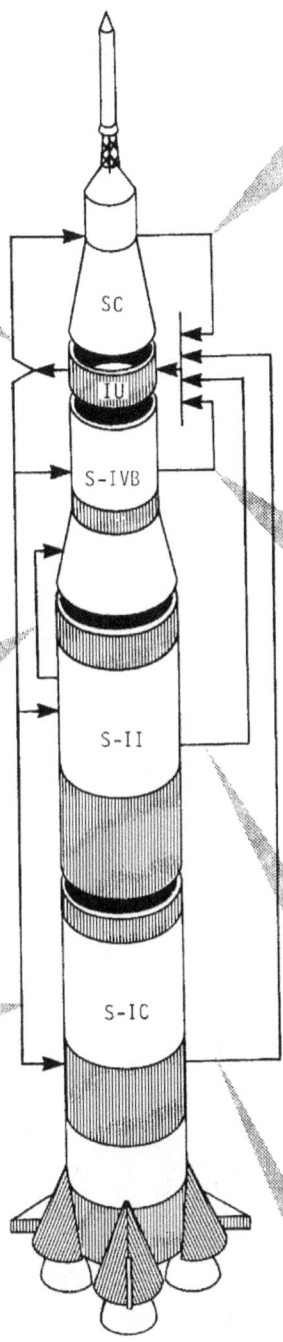

**S-IC TO IU**

ATTITUDE CONTROL ACCELEROMETER
   SIGNALS
ATTITUDE CONTROL RATE GYRO
   SIGNALS
+28 VDC FOR TIMING
ENGINES OUT
OUTBOARD ENGINE CUTOFF
S-II ENGINES START ENABLE
SWITCH SELECTOR ADDRESS
   VERIFY
S-IC THRUST OK

SC

IU

S-IVB

S-II

S-IC

Figure 1-4

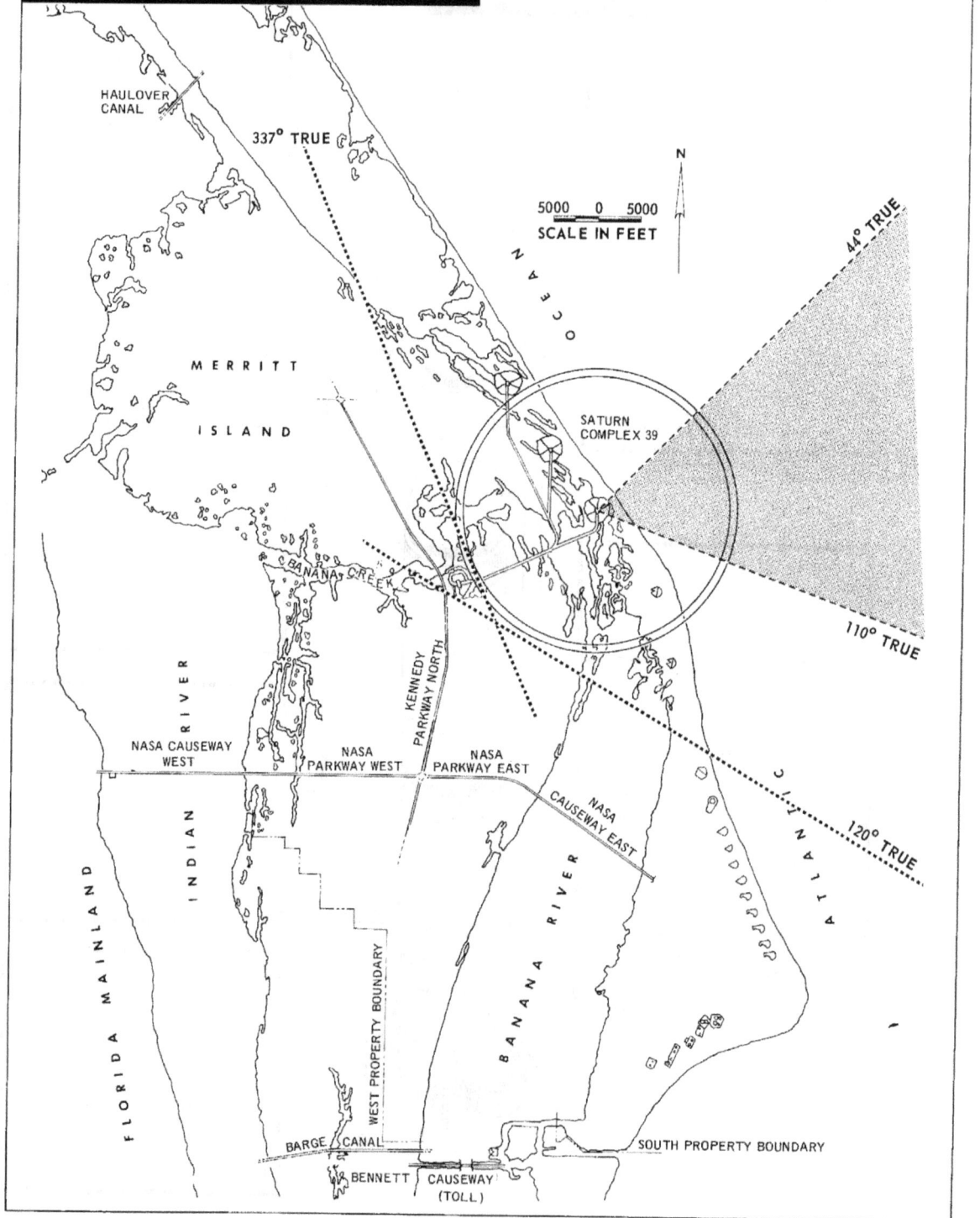

# RANGE SAFETY AZIMUTH LIMITS

Figure 1-5

In the event that the launch vehicle deviates from its planned trajectory, to the degree that it will endanger life or property, the RSO must command destruct by means of the range safety command system. The range safety system is active until the vehicle has achieved earth orbit, after which the destruct system is deactivated (safed) by command from the ground.

## FLIGHT SAFETY

Flight safety planning began during the conceptual phases of the program. One of the requirements of the range safety program is that, during these early phases, basic flight plans be outlined and discussed and, prior to launch, a final flight plan be submitted and approved. As the program develops, the flight planning is modified to meet mission requirements. The flight plan is finalized as soon as mission requirements become firm.

In addition to the normal trajectory data given in the flight plan, other trajectory data is required by the AFETR. This data defines the limits of normality, maximum turning capability of the vehicle velocity vector, instant impact point data, drag data for expended stages and for pieces resulting from destruct action, and location and dispersion characteristics of impacting stages.

In the event the RSO is required to command destruct the launch vehicle, he will do so by manually initiating two separate command messages. These messages are transmitted to the launch vehicle over a UHF radio link. The first message shuts off propellant flow and results in all engines off. As the loss of thrust is monitored by the EDS, the ABORT light is turned on in the Command Module (CM). Upon monitoring a second abort cue, the flight crew will initiate the abort

sequence. The second command from the RSO is for propellant dispersion, and explosively opens all propellant tanks.

Each powered stage of the launch vehicle is equipped with dual redundant command destruct antennae, receivers, decoders, and ordnance to ensure positive reaction to the destruct commands. To augment flight crew safety, the EDS monitors critical flight parameters. Section III provides a more detailed discussion of the EDS.

## LV MONITORING AND CONTROL

All major LV sequences and conditions can be monitored by the flight crew in the Command Module (CM). Normally the LV sequencing is accomplished and controlled automatically by the LVDC in the IU. There are, however, switches and controls in the CM with which the flight crew can assume partial control of the LV and initiate such sequences as engine cutoff, early staging, LVDC update, and translunar injection inhibit. A simplified functional diagram (figure 1-6) illustrates the relationships between CM lights and switches and the LV systems. Many of the lights and switches discussed briefly in subsequent paragraphs are related to the EDS and are discussed in more detail in Section III. Refer to Section III also for an illustration of the CM Main Display Console (MDC).

### MONITORING

#### LV Engine Status Display

The LV engine status display (S-51, figure 3-1) consists of five numbered lights. Each of these numbered lights represents the respective numbered engine on the operating

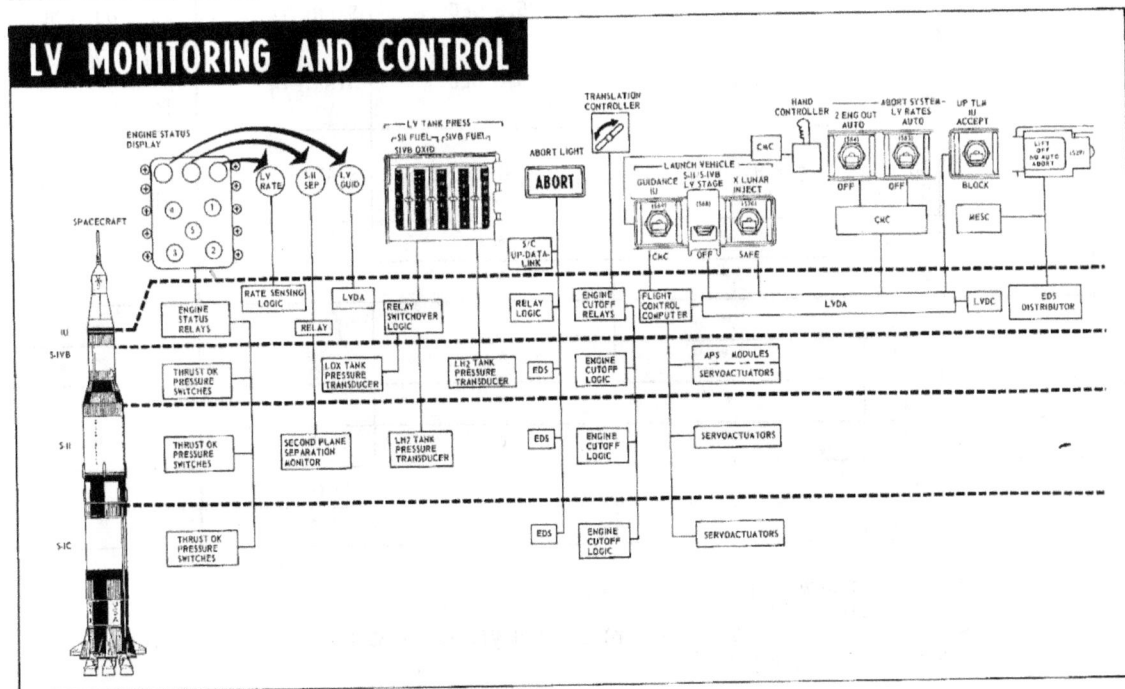

Figure 1-6

stage. (e.g., light number one represents engine number one on the S-IC, S-II, or S-IVB stage; light number two represents engine number two on the S-IC, or S-II stage; etc.).

These lights are controlled by switching logic in the IU. The switching logic monitors THRUST OK pressure switches on each engine of the operating stage and also staging and timing discretes from the LVDC. Figure 1-7 presents a summary of engine status display lights operation and timing.

## LV RATE Light

The LV RATE light (Q-50, figure 3-1) is illuminated any time the LV experiences an excessive pitch, roll, or yaw rate. The light remains energized as long as the excessive rate condition exists. Excessive rates are sensed by the EDS rate gyros in the IU. Refer to LV RATE light paragraph in Section III, and EDS paragraph in Section VII for further discussion.

## S-II SEP Light

The S-II SEP light (Q-51, figure 3-1) serves a dual purpose. Its first function is to illuminate when the S-II start command is issued and then extinguish approximately 30 seconds later

when S-II second plane separation occurs. Its second function occurs later in the mission and is to indicate, by cycling the light on and off, the status of the S-IVB restart sequence. Refer to S-II SEP light paragraph in Section III for further discussion.

## LV GUID Light

The LV GUIDE light (Q-52, figure 3-1) indicates that the IU navigation system stable platform (ST-124-M3) has failed. The light will remain illuminated as long as the failed condition exists. Refer to LV GUID light paragraph in Section III for further discussion.

## ABORT Light

The ABORT light (N-51, figure 3-1) can be illuminated by command from the ground or by the EDS circuitry. However, certain EDS circuits are inhibited during the initial phases of the launch sequence. S-IC multiple engine cutoff is inhibited for 14 seconds after liftoff and multiple engine cutoff automatic abort is inhibited for 30 seconds after liftoff. Refer to ABORT light paragraph of Section III for further discussion.

| NORMAL LV ENGINE STATUS DISPLAY LIGHTS OPERATION | | | | | |
|---|---|---|---|---|---|
| DISPLAY LIGHT | STAGE | ENGINES START COMMAND (LIGHTS ON) | THRUST OK (LIGHTS OFF) | ENGINES OFF - LOSS OF THRUST (LIGHTS ON) | STAGE SEPARATION (LIGHTS OFF) |
| NO. 1 | S-IC | $T_1$ - 08.9 SEC | $T_1$ - 01.5 SEC | START $T_3$ | FIRST PLANE $T_3$ + 00.7 SEC |
| | S-II | $T_3$ + 01.4 SEC | $T_3$ + 05.9 SEC | START $T_4$ | $T_4$ + 00.8 SEC |
| | 1ST BURN S-IVB | $T_4$ + 01.0 SEC | $T_4$ + 05.8 SEC | START $T_5$ | COAST MODE LIGHTS OFF AT $T_5$ + 00.7 SEC |
| | 2ND BURN S-IVB | $T_6$ + 9 MIN 30 SEC | $T_6$ + 9 MIN 40.6 SEC | START $T_7$ | LAUNCH VEHICLE FROM SPACECRAFT (TBD) |
| NO. 2 | S-IC | * | * | * | * |
| | S-II | * | * | * | * |
| NO. 3 | S-IC | * | * | * | * |
| | S-II | * | * | * | * |
| NO. 4 | S-IC | * | * | * | * |
| | S-II | * | * | * | * |
| NO. 5 | S-IC | * | * | START $T_2$ | * |
| | S-II | * | * | * | * |
| * SAME AS LIGHT NUMBER 1 FOR SAME STAGE (TBD) TO BE DETERMINED T-TIMES ARE TIME BASE TIMES. REFER TO SECTION VII FOR TIME BASE DEFINITIONS AND TO SECTION II FOR RELATED FLIGHT TIMES | | | | | |

Figure 1-7

## LIFTOFF - NO AUTO ABORT Lights

The LIFTOFF - NO AUTO ABORT lights (V-50, figure 3-1) are two independent lights in an integral light/switch assembly. The LIFTOFF light is illuminated to indicate IU umbilical disconnect and LV release command. The LIFTOFF light is turned off at S-IC CECO. The NO AUTO ABORT light is also turned on at IU umbilical disconnect and remains on to indicate no auto abort capability during the early phase of S-IC burn. The NO AUTO ABORT light should go off to indicate the auto abort inhibit timers have run out. Refer to LIFTOFF - NO AUTO ABORT paragraph of Section III for additional coverage.

## LV TANK PRESS Gauges

The LV TANK PRESS Gauges (X-46, figure 3-1) indicate ullage pressures in the S-II and S-IVB oxidizer tank. Prior to S-II/S-IVB separation the two left-hand pointers indicate the pressure in the S-II fuel tank. Subsequent to S-II/S-IVB separation these same two pointers indicate S-IVB oxidizer pressure. The two right-hand pointers indicate S-IVB fuel tank pressure until LV/spacecraft separation.

## CONTROL

Through the use of switches and hand controls the flight crew can assume partial control of the LV during certain periods of flight.

## TRANSLATIONAL CONTROLLER

The TRANSLATIONAL CONTROLLER (not shown in figure 3-1) can be used to accomplish several functions. A manual abort sequence is initiated when the T-handle is rotated counter-clockwise and held in that position for at least 3 seconds. Returning the T-handle to neutral before the 3 seconds expires results only in an engine cutoff signal rather than a full abort sequence. Clockwise rotation of the T-handle transfers control of the spacecraft from the CMC to the stability control system. The T-handle can also provide translation control of the CSM along one or more axes. Refer to TRANSLATIONAL CONTROLLER paragraph in Section III for additional discussion.

## GUIDANCE Switch

The normal position of the GUIDANCE switch (X-58, figure 3-1) is IU. In this position guidance of the LV is controlled by the program in the LVDC. Placing the GUIDANCE switch in the CMC position is effective only during time base 5 ($T_5$), the first portion of the coast mode. At all other times the switch function is blocked by the program in the LVDC. When placed in the CMC position, during $T_5$, the CMC will generate attitude error signals for the launch vehicle flight control computer. Inputs from the hand controller modulate these attitude error signals. If the switch is in the CMC position when $T_6$ begins, control of the launch vehicle is returned to the LVDC automatically. This precludes a restart of the S-IVB with an incorrect launch vehicle attitude.

## S-II/S-IVB Switch

The S-II/S-IVB switch (X-60, figure 3-1) is a dual function switch. Its normal position is OFF. It is used to manually initiate the staging sequence of the S-II from the S-IVB by placing the switch in the LV STAGE position. The switch is

effective for this purpose from $T_3$ + 1.4 seconds up to the start of $T_4$. This switch is also used to manually initiate cutoff of the S-IVB by placing the switch in the LV STAGE position. It is effective for this purpose from $T_4$ + 1.4 seconds up to the S-IVB final cutoff. If the switch is used to initiate staging it must be reset to OFF in order that it can be used later for S-IVB cutoff.

## XLUNAR Switch

Normal position of the XLUNAR switch (X-62, figure 3-1) is INJECT. In this position the LVDC will sequence the S-IVB systems through the steps necessary to accomplish a restart and inject the spacecraft into the mission trajectory. The XLUNAR switch can be used during certain periods of time to inhibit restart of the S-IVB. This translunar injection inhibit (TLI) is a temporary action the first time it is used but if a second injection opportunity is inhibited it becomes final and no restart of the S-IVB can be subsequently accomplished. Timing of these options is under control of the LVDC and is as follows:

1. If the XLUNAR switch is placed in the SAFE position prior to the start of $T_6$ the LVDC will accept the signal upon starting $T_6$ and inhibit the restart sequence.

2. If the XLUNAR switch is placed in the SAFE position after $T_6$ is started, but prior to $T_6$ + 41.0 seconds, the LVDC will accept the signal at $T_6$ + 41.0 seconds and inhibit $O_2H_2$ burner ignition and the remainder of the restart sequence.

3. If the XLUNAR switch is placed in the SAFE position after $O_2H_2$ burner ignition ($T_6$ + 41.3 seconds), but before $T_6$ + 5 minutes 41.3 seconds, the LVDC will accept the signal at $T_6$ + 5 minutes 41.3 seconds and inhibit ambient repressurization.

4. If the XLUNAR switch is placed in the SAFE position after ambient repressurization ($T_6$ + 8 minutes 17.3 seconds), but before TLI commit at $T_6$ + 9 minutes 20 seconds, the LVDC will accept the signal within two seconds and inhibit S-IVB ignition.

5. If the XLUNAR switch is placed in the SAFE position after TLI commit the LVDC will not accept the signal and S-IVB restart will occur.

Refer to OPERATION SEQUENCE paragraph of section VII for definition of time bases.

## UP TLM-IU Switch

Normal position of the UP TLM-IU switch (not shown in figure 3-1) is BLOCK. The switch will be in BLOCK all through boost flight. After injection into waiting orbit the switch may be moved to the ACCEPT position to allow command data to enter the onboard computers. Command up-data from ground control can be for up-dating the guidance program, commanding a telemetry routine, or for several other purposes. Refer to COMMAND COMMUNICATIONS SYSTEMS paragraph in Section VII and the COMMAND SYSTEM paragraph in Section IX for additional discussion.

## ABORT SYSTEM - LV RATES Switch

The normal position of the LV RATES switch (R-63, figure 3-1) is AUTO. In this position the EDS rate gyros in the IU monitor vehicle rates in all three axes and send rate signals to the EDS and LVDC. The flight crew can inhibit an EDS automatic abort due to excessive rates by placing the switch in the OFF position.

## ABORT SYSTEM - 2 ENG OUT Switch

The normal position of the 2 ENG OUT switch (R-60, figure 3-1) is AUTO. In this position the EDS will initiate an automatic abort when it monitors a two engine out condition. The flight crew can inhibit an automatic abort due to a two engine out condition by placing the switch in the OFF position.

## PERCEPTIBLE PRELAUNCH EVENTS

Prelaunch events which occur subsequent to astronaut loading (T-3 hours 40 minutes), and which may be felt or heard by the flight crew inside the spacecraft, are identified in figure 1-8. Other events, not shown, combine to create a relatively low and constant background. This background noise includes the sounds of environmental control, propellant replenishment, control pressure gas supplies, propellant boiloff and low pressure, low volume purges.

Significant noises and vibrations may be caused by the starting or stopping of an operation or they may result from turbulent flow of gases or liquids. Figure 1-8 illustrates those events most likely to be heard of or felt above the background noise or vibration.

Nearly all items shown in figure 1-8 are noise producers and each will have individual characteristics resulting from such things as proximity, volume, and timing. For example, pressurizing the supply spheres in the IU will be more noticeable than pressurizing the supply and purge spheres on the S-IC. Yet with both items starting at the same time, each sound will add to the other to make the total sound. When pressurization of the IU supply sphere ends at T-31 minutes, it is likely that the sound of the pressurization of the S-IC supply and purge will blend with the cold helium sphere pressurization sound and be indiscernible as such in the CM. In this manner, the sounds of the remainder of the items illustrated will rise and fall, join and separate, to form the sound of the Saturn V. At approximately T-6.1 seconds all sounds are hidden by the ignition of the engines on the S-IC.

## POGO

One of the major anomalies of the AS-502 flight of April 4, 1968, was the POGO phenomenon. This phenomenon produced an undesirable longitudinal oscillation in the space vehicle caused by a regenerative feedback of vehicle motion to the propellant feed system. A thrust oscillation buildup, along with a structural response buildup, resulted from a closed loop dynamic effect involving the coupling of the longitudinal vibration of the vehicle structure, the fluid vibration in the propellant ducts, and the hydraulic characteristics of the engine. Pressure fluctuations at the pump inlets, caused by movement of the propellant ducts and pumps relative to the fluid in the ducts, produced lagging fluctuations of engine thrust. Space vehicle instability resulted from a tuning of the propulsion and vehicle structural systems. The onset and eventual cessation of the instability were caused by the change in the propulsive system and vehicle resonant frequencies with time.

The AS-502 space vehicle instability occurred during the latter part of the S-IC burn period. The buildup of longitudinal amplitudes started at about T + 110 seconds, reached a peak at about T + 126 seconds, and decayed to a negligible level by T + 140 seconds. The buildup was determined to be a result of the coalescence of the first longitudinal frequency of the vehicle with the first lox line frequency (see figure 1-9).

A stability analysis of the AS-503 S-IC stage flight indicated that the AS-503 space vehicle (without POGO suppression modification) would be unstable from approximately T + 90 seconds to center engine cutoff.

## ACCEPTABILITY CRITERIA

Criteria were established to evaluate the acceptability of the proposed solutions to the POGO phenomenon. The proposed solutions were required to meet the criteria listed below:

1. Pass a preliminary screening as a technically effective solution without major modifications or schedule impact on AS-503.

2. Provide acceptable gain margins and phase margins for all modes under nominal/tolerance-conditions.

3. Provide system reliability to include system criticality, single point failures, and fail safe potential.

4. Have no adverse impact on other space vehicle systems.

5. Use flight qualified hardware.

6. Have retrofit capability without schedule slippage.

7. Have no adverse impact on crew safety and operational or program requirements.

8. Increase confidence by tests and analyses.

## SOLUTIONS CONSIDERED

Ten possible solutions were initially identified by the POGO working group. For each possible solution (listed in figure 1-10) a preliminary screening was made to determine the technical effectiveness, degree of modification required, and schedule impact. The screening showed that only two of the possible solutions were acceptable for detailed testing and/or analytical treatment. These two solutions were the helium-charged lox prevalve accumulator and the helium injection at the top of the lox suction line (solutions 1 and 2 on figure 1-10). A complete evaluation of each solution was not performed since evaluation of a proposed solution ceased whenever a "No" result was obtained for any criterion.

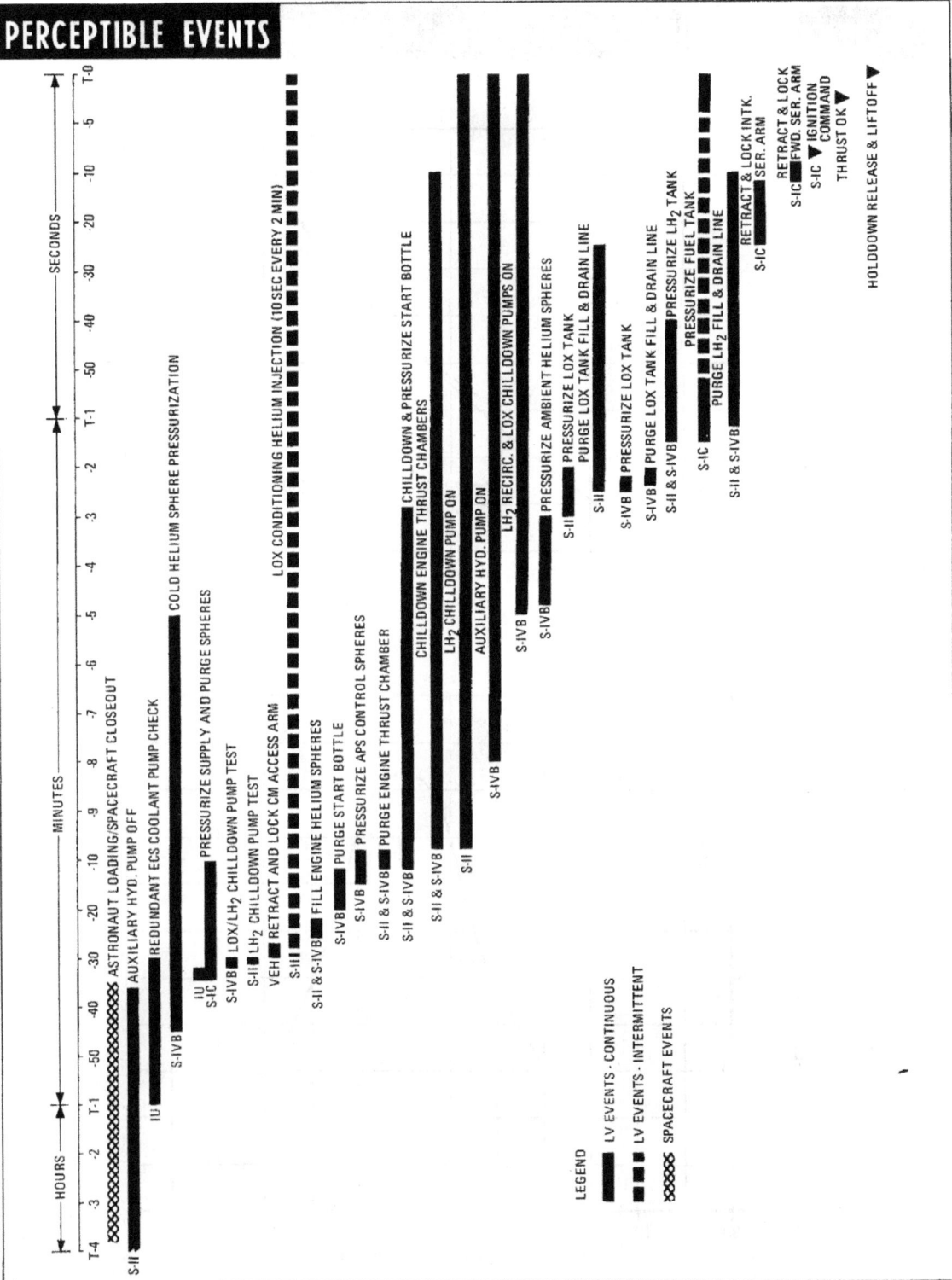

# PERCEPTIBLE EVENTS

Figure 1-8

# AS-502 LONGITUDINAL OSCILLATION TRENDS 110 TO 140 SECONDS

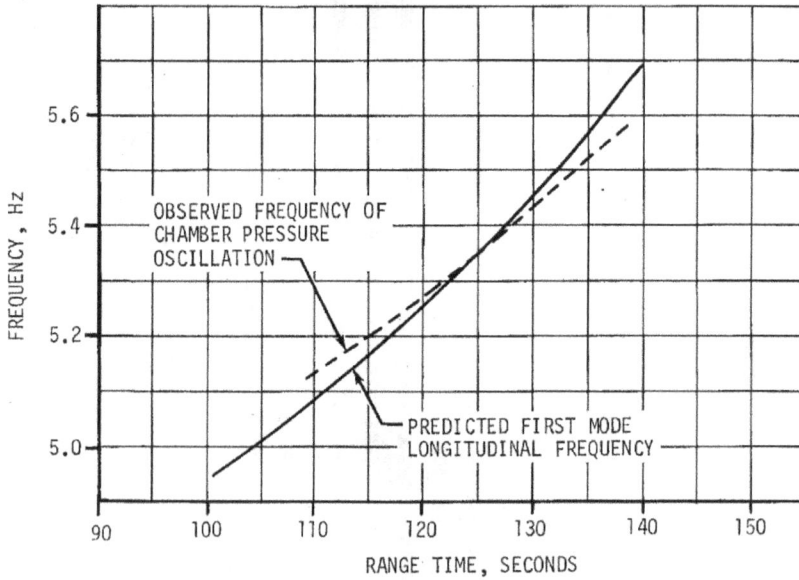

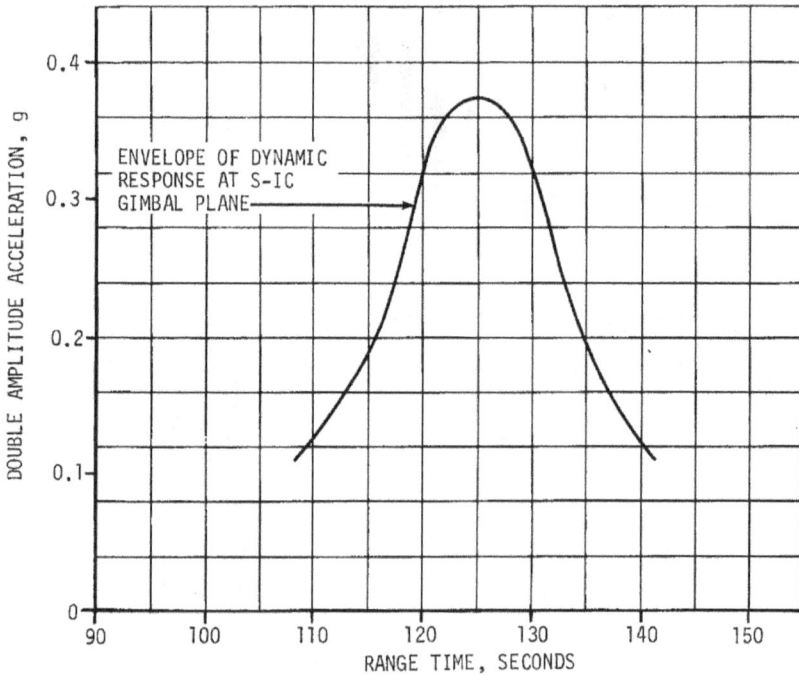

Figure 1-9

## ANALYSIS AND TESTING

A comprehensive testing program was conducted to provide the necessary data to verify the analytical model, to demonstrate the capability of the proposed solution, and to assure solution compatibility with the propulsion system. The program included evaluation of propulsion system dynamics, definition of the operating modes, and extensive static test firings.

The results of the analytical and experimental programs demonstrated the compatibility of both prime POGO solution candidates. With either of the proposed solutions the performance loss was negligible, main chamber and gas generator combustion was stable, and the thrust decay was not affected.

The use of helium-charged lox prevalves as accumulators (surge chambers) in each of the five propellant feed systems was selected as the POGO solution. (See section IV under POGO Suppression for system description.) The selected solution was considered to be superior to the helium injection solution in seven of twelve factors considered and adequate in all twelve factors. This fix has been further verified by the S-IC-6 static firing test.

### EVALUATION OF SOLUTION VERSUS ACCEPTANCE CRITERIA

| SOLUTION | PASSES PRELIMINARY TECHNICAL CRITERIA ANALYSIS/TESTS | NO SCHEDULE IMPACT ON AS-503 | ACCEPTABLE STABILITY GAIN MARGINS | RELIABLE | NO ADVERSE IMPACT ON OTHER SPACE VEHICLE SYSTEMS | USES FLIGHT QUALIFIED HARDWARE | HAS RETROFIT CAPABILITY | NO CREW SAFETY/OPERATIONAL IMPACT | GROUND TESTS INCREASE CONFIDENCE | REMARKS |
|---|---|---|---|---|---|---|---|---|---|---|
| 1. HELIUM CHARGED PREVALVE ACCUMULATOR | YES | YES | YES | YES | YES | YES | YES | YES | YES | FINAL SOLUTION |
| 2. HELIUM INJECTION AT TOP OF LOX SUCTION LINE | YES | YES | NO | ** YES | YES | YES | YES | YES | YES | UNSTABLE IN 2ND MODE |
| 3. LOX PREVALVE ACCUMULATOR CHARGED WITH GOX | YES | YES | * | * | * | * | * | * | NO | GOX CONDENSES |
| 4. LOSSY LOX SUCTION LINE | NO | * | * | * | * | * | * | * | * | NEW DESIGN |
| 5. SUCTION LINE RESTRICTION | YES | NO | * | * | * | * | * | * | * | EXTENSIVE NEW ENGINE TESTING |
| 6. HELIUM INJECTION AT 90 IN. POINT OF LOX SUCTION LINE | NO | * | * | * | * | * | * | * | * | SMALL FREQUENCY CHANGE |
| 7. DISCHARGE LINE ACCUMULATORS | YES | NO | * | * | * | * | * | * | * | MAJOR MODIFICATION |
| 8. SUCTION LINE ACCUMULATORS | YES | NO | * | * | * | * | * | * | * | MAJOR MODIFICATION |
| 9. GAS INJECTION AT LOX PUMP INLET | NO | * | * | * | * | * | * | * | * | SMALL FREQ. CHG./ GOX CONDENSES |
| 10. REDUCED LOX TANK PRESSURE | NO | * | * | * | * | * | * | * | * | INEFFECTIVE |

* UNDETERMINED     ** SYSTEM DOES NOT FAIL SAFE SINCE THE LOSS OF HELIUM FLOW IN ANY LINE WOULD TERMINATE ANY STABILITY GAIN ASSOCIATED WITH THAT LINE

Figure 1-10

# SECTION II

# PERFORMANCE

## TABLE OF CONTENTS

## INTRODUCTION

Saturn V launch vehicle performance characteristics, under the constraints established by environment and mission requirements, are described in this section. Mission profile, variables, requirements and constraints are described in Section X.

## FLIGHT SEQUENCE

The SA-503 vehicle will be launched from Launch Complex 39 (LC-39) at the Kennedy Space Center. The flight sequence phases described in the following paragraphs cover the AS-503 C Prime basic mission as well as the Option 1 mission. In those phases generally applicable to both basic and optional missions, significant differences between missions are noted. A typical sequence of critical launch vehicle events is contained in figure 2-1.

## LAUNCH AND BOOST TO EARTH PARKING ORBIT

The vehicle rises nearly vertically from the pad, for approximately 450 feet, to clear the tower. During this period, a yaw maneuver is executed to provide tower clearance in the event of adverse wind conditions, deviation from nominal flight and/or engine failure. (See figure 2-1 for start and stop times for this and other maneuvers and events). After clearing the tower, a tilt and roll maneuver is initiated to achieve the flight attitude and proper orientation for the selected flight azimuth. For the AS-503 C Prime basic mission the flight azimuth is 72 degrees. For the Option 1 mission the flight azimuth varies between 72 and 108 degrees, depending on time and date of launch. From the end of the tilt maneuver to tilt-arrest, the vehicle flies a pitch program (biased for winds of the launch month) to provide a near zero-lift (gravity-turn) trajectory. Tilt-arrest freezes the pitch attitude to dampen out pitch rates prior to S-IC/S-II separation. The pitch attitude remains constant until initiation of the Iterative Guidance Mode (IGM) which occurs about five seconds after launch escape tower (LET) jettison during the S-II stage flight. Figure 2-2 shows the pitch attitude profile from first motion to earth parking orbit (EPO). Mach 1 is achieved approximately one minute after first motion. Maximum dynamic pressure of approximately 740 pounds per square foot is encountered at 1 minute 16 seconds after first motion. S-IC center engine cutoff occurs at 2 minutes 5.6 seconds after first motion, to limit the vehicle acceleration to a nominal 3.98 g. The S-IC outboard engines are cutoff at 2 minutes 31 seconds after first motion.

A time interval of 4.4 seconds elapses between S-IC cutoff and the time the J-2 engines of the S-II stage reach the 90%

operating thrust level. During this period, ullage rockets are fired to seat the S-II propellant, the S-IC/S-II separation occurs, and the retrorockets back the S-IC stage away from the flight vehicle. Threshold for engine status light OFF is 65% thrust. The S-II aft interstage and the LET are jettisoned 30.5 and 36.2 seconds, respectively, after S-IC cutoff.

During the S-II burn, two oxidizer-to-fuel mixture ratio (MR) shifts are programmed by the flight software. The MR, 5.0 at engine start, is shifted to 5.5 at 2.5 seconds after S-II mainstage (90% thrust level) and remains at this value until shifted to 4.5 at 4 minutes 40 seconds after mainstage. The 4.5 MR yields a reduced thrust at increased specific impulse. The S-II engines are cutoff simultaneously by sensors in either the lox or $LH_2$ tanks.

An interval of 6.5 seconds elapses between S-II cutoff and the time the S-IVB J-2 engine attains 90% operating thrust level (mainstage). During this coast period, the S-IVB ullage rockets are fired to seat the stage propellant, the S-II/S-IVB separation occurs, and retrorockets back the S-II stage away from the flight vehicle. The threshold for engine status light OFF is 65% thrust.

An MR of 5.0 is programmed for J-2 engine start and entire first burn. The S-IVB first burn inserts the vehicle into a 100 nautical mile (NMI) altitude, nearly circular, EPO.

## CIRCULAR EARTH PARKING ORBIT (BASIC MISSION)

During the first revolution in EPO, the Command and Service Module (CSM) is separated from the launch vehicle. In earth orbit, spacecraft validation operations specified by the mission are performed, after which the Command Module (CM) is separated from the Service Module and returned to earth.

The S-IVB stage with the Instrument Unit (IU) and Lunar Test Article-B (LTA-B) is restarted during the second revolution in EPO (first injection opportunity) for injection into a typical lunar trajectory.

## CIRCULAR EARTH PARKING ORBIT (OPTION 1 MISSION)

At first S-IVB engine cutoff, the 70-pound thrust auxiliary propulsion system (APS) engines are started and operated for approximately 88 seconds. The $LH_2$ continuous vents open approximately 49 seconds after insertion. This venting provides a continuous low-level thrust to keep the S-IVB propellant seated against the aft bulkheads.

The normal vehicle attitude in parking orbit has Position I pointed toward the center of the earth (astronauts heads are down), the vehicle longitudinal axis in the inertial orbital plane and perpendicular to the radius vector, and the nose ahead. A maneuver to the appropriate attitude for landmark sighting is performed between 45 minutes and 1.5 hours after insertion. The maneuver to the sighting attitude consists of a 180-degree roll, followed by a pitch maneuver, that places the vehicle longitudinal body axis 20 degrees below the local horizon. This attitude, referenced to the local horizon, is maintained until the return maneuver to the normal coast attitude.

While in EPO, spacecraft and launch vehicle systems are checked out and verified for translunar injection.

## TRANSLUNAR INJECTION BOOST

The translunar injection boost is part of an ordered flight sequence that begins at initiation of the preignition sequence. The flight computer signals the beginning of the preignition sequence when it determines that the vehicle position satisfies a predesignated geometrical relationship with the target vector. At this time the computer resets to Time Base 6. If a translunar injection inhibit signal from the CM is not sensed (not applicable to unmanned TLI of basic mission) the computer issues the signals that lead to S-IVB reignition. These signals include start helium heater ($O_2H_2$ burner), close $LH_2$ tank continuous vent valves, ignite APS ullage engines, restart S-IVB J-2 engine and cutoff ullage engines.

During the preignition sequence, thrust from the continuous $LH_2$ vent keeps the propellants seated until $O_2H_2$ burner ignition. The vent is then closed to enable the burner to pressurize both the lox and $LH_2$ propellant tanks. The burner is a pressure-fed system and the nominal burner thrust magnitude continues to increase as the pressure in the propellant tanks increases. The burner is part of a dual repressurization system; if the burner fails to operate properly, ambient helium is provided to complete tank repressurization.

For the basic mission, the unmanned S-IVB/IU/LTA-B is placed into a typical lunar transfer trajectory during the second revolution in EPO (first injection opportunity, Pacific window). Since the flight azimuth for this mission is fixed at 72 degrees, there is little likelihood that a translunar trajectory will be achieved.

For the Option 1 mission, the space vehicle at translunar injection is placed in a free-return lunar trajectory (see figure 10-2, Section X). The CSM, after separation from the S-IVB/IU/LTA-B, coasts in the free-return trajectory, passing retrograde, about the moon on a return path to earth. During the pass around the moon, the choice between returning directly to earth or deboosting into lunar parking orbit and completing several revolutions about the moon is made. Following lunar orbit, the CSM boosts out of lunar orbit into a transearth trajectory. The CM is separated from the Service Module during the approach to earth and the CM proceeds to a Pacific Ocean landing.

Two opportunities for translunar injection are provided. For first injection opportunity, S-IVB ignition occurs after approximately 1.5 revolutions in EPO (Pacific window). The second opportunity occurs after 2.5 revolutions in EPO.

## SLINGSHOT MODE

Approximately 20 minutes after translunar injection, the CSM is separated from the launch vehicle. To minimize the probability of contact between the CSM and the S-IVB/IU/LTA-B, the trajectory of S-IVB/IU/LTA-B is altered by initiating the slingshot mode. The slingshot mode is achieved by maneuvering the S-IVB/IU/LTA-B to a predetermined attitude (see figure 10-3, in Section X) and dumping residual propellant through the J-2 engines, to obtain a decrease in velocity of approximately 80 feet per second (retrograde dump). This $\Delta V$ is designed to perturb the trajectory so that the influence of the moon's gravitational field increases the velocity of the S-IVB/IU/LTA-B sufficiently to place it in solar orbit (see figure 10-2, Section X). Following the retrograde dump of propellants, the S-IVB stage is "safed" by dumping the remaining propellants and high pressure gas bottles through the nonpropulsive vents which are latched open. The slingshot mode will be simulated in the basic mission.

## FLIGHT PERFORMANCE

The typical flight performance data presented herein are based on launch vehicle operational trajectory studies. These studies were based on the requirements and constraints imposed by the AS-503 C Prime Option 1 mission.

## FLIGHT PERFORMANCE PARAMETERS

Flight performance parameters for a lunar orbit mission are presented graphically in figures 2-2 through 2-20. These parameters are shown for nominal cases for the earth parking orbit and translunar injection boost phases. Parameters shown include pitch attitude angle, vehicle weight, axial force, aerodynamic pressure, longitudinal acceleration, inertial velocity, altitude, range, angle of attack, aerodynamic heating indicator, inertial path angle and inertial heading angle.

## FLIGHT PERFORMANCE AND FLIGHT GEOMETRY RESERVES (OPTION 1 MISSION)

Required propellant reserves for the AS-503 C Prime Option 1 mission are comprised of two components: Flight Performance Reserves (FPR) and Flight Geometry Reserves (FGR). The FPR is defined as the root-sum-square (RSS) combination of negative launch vehicle weight dispersions at TLI due to 3-sigma launch vehicle subsystems and environmental perturbations. The FGR is defined as the reserve propellant required to guarantee the launch vehicle capability to establish a lunar flyby trajectory at any earth-moon geometry. The total reserves required to provide 99.865% assurance that the launch vehicle will complete its primary mission objective is the algebraic sum of the FPR and the FGR.

The FPR for the Option 1 mission is estimated to be 2,390 pounds. The FGR varies with date of launch. For the eight December 1968 launch days, the FGR ranges from 201 to 810 pounds.

The nominal payload capability is 95,557 pounds and the required payload is 87,700 pounds. This leaves a nominal propellant reserve of 7,857 pounds. The difference between the available propellant reserve and the required propellant reserve (FGR plus FPR) is the propellant margin. For payload critical lunar missions, the payload could be increased by an amount equal to the propellant margin. The propellant margins for the December 1968 launch days vary from +4,657 to +5,266 pounds thus assuring launch vehicle capability for injecting the required payload.

## CRITICAL EVENT SEQUENCE

| TIME FROM FIRST MOTION (HR:MIN:SEC) | TIME FROM REFERENCE (HR:MIN:SEC) | EVENT | TIME FROM FIRST MOTION (HR:MIN:SEC) | TIME FROM REFERENCE (HR:MIN:SEC) | EVENT |
|---|---|---|---|---|---|
| -0:00:17.3 | $T_1$ - 0:00:17.6 | Guidance Reference Release (GRR) | 0:12:49.8 | $T_5$ + 0:01:28.1 | S-IVB APS Ullage Engine No. 2 Cutoff |
| 0:00:00.0 | $T_1$ - 0:00:00.4 | First Motion | 0:13:01.9 | $T_5$ + 0:01:40.2 | Begin Orbital Navigation |
| 0:00:00.4 | $T_1$ + 0:00:00.0 | Liftoff | 2:41:00.6 | $T_6$ + 0:00:00.0 | Begin S-IVB Restart Preparations |
| 0:00:01.4 | $T_1$ + 0:00:01.0 | Begin Tower Clearance Yaw Maneuver | 2:41:41.6 | $T_6$ + 0:00:41.0 | LVDC Check for XLUNAR INJECT Signal |
| 0:00:09.4 | $T_1$ + 0:00:09.0 | End Yaw Maneuver | 2:41:41.9 | $T_6$ + 0:00:41.3 | $O_2H_2$ Burner (Helium Heater) On |
| 0:00:10.0 | $T_1$ + 0:00:09.6 | Pitch and Roll Initiation | 2:41:42.8 | $T_6$ + 0:00:42.2 | LH$_2$ Continuous Vent Closed |
| 0:00:29.0 | $T_1$ + 0:00:28.6 | End Roll Maneuver | 2:46:41.9 | $T_6$ + 0:05:41.3 | LVDC Check for XLUNAR INJECT Signal |
| 0:01:00.3 | $T_1$ + 0:00:59.9 | Mach 1 | 2:49:16.9 | $T_6$ + 0:08:16.3 | S-IVB APS Ullage Engine No. 1 Ignition |
| 0:01:16.0 | $T_1$ + 0:01:15.6 | Maximum Dynamic Pressure | 2:49:17.0 | $T_6$ + 0:08:16.4 | S-IVB APS Ullage Engine No. 2 Ignition |
| 0:02:05.6 | $T_2$ + 0:00:00.0 | S-IC Center Engine Cutoff | 2:49:17.9 | $T_6$ + 0:08:17.3 | Ambient Repressurization |
| 0:02:27.0 | $T_2$ + 0:00:21.4 | Begin Tilt Arrest | 2:49:21.9 | $T_6$ + 0:08:21.3 | Helium Heater Off |
| 0:02:31.0 | $T_3$ + 0:00:00.0 | S-IC Outboard Engine Cutoff | 2:50:20.6 | $T_6$ + 0:09:20.0 | Translunar Injection Commit |
| 0:02:31.5 | $T_3$ + 0:00:00.5 | S-II Ullage Rocket Ignition | 2:50:30.6 | $T_6$ + 0:09:30.0 | S-IVB Engine Restart Sequence |
| 0:02:31.7 | $T_3$ + 0:00:00.7 | Signal to Separation Devices and S-IC Retrorockets | 2:50:33.6 | $T_6$ + 0:09:33.0 | S-IVB APS Ullage Engine No. 1 Cutoff |
| 0:02:31.8 | $T_3$ + 0:00:00.8 | S-IC/S-II First Plane Separation Complete | 2:50:33.7 | $T_6$ + 0:09:33.1 | S-IVB APS Ullage Engine No. 2 Cutoff |
| 0:02:32.4 | $T_3$ + 0:00:01.4 | S-II Engine Start Sequence Initiated | 2:50:38.4 | $T_6$ + 0:09:37.8 | S-IVB Ignition, Second Burn (Start Tank Discharge Valve Opens) |
| 0:02:33.4 | $T_3$ + 0:00:02.4 | S-II Ignition (Start Tank Discharge Valve Opens) | 2:50:40.9 | $T_6$ + 0:09:40.3 | S-IVB at 90% Thrust |
| 0:02:35.4 | $T_3$ + 0:00:04.4 | S-II Engines at 90% Thrust | 2:50:43.4 | $T_6$ + 0:09:42.8 | S-IVB MR(5.0) Shift |
| 0:02:36.0 | $T_3$ + 0:00:05.0 | S-II Ullage Thrust Cutoff | 2:55:54.9 | $T_7$ + 0:00:00.0 | S-IVB Engine Cutoff, Second Burn |
| 0:02:37.9 | $T_3$ + 0:00:06.9 | S-II First MR(5.5) Shift | 2:55:55.2 | $T_7$ + 0:00:00.3 | LH$_2$ Continuous and Non-propulsive Vents Open |
| 0:03:01.5 | $T_3$ + 0:00:30.5 | S-II Aft Interstage Drop (Second Plane Separation) | 2:55:55.5 | $T_7$ + 0:00:00.6 | Lox Nonpropulsive Vent Open |
| 0:03:07.2 | $T_3$ + 0:00:36.2 | LET Jettison | 2:56:04.7 | $T_7$ + 0:00:09.8 | Translunar Injection |
| 0:03:12.0 | $T_3$ + 0:00:40.9 | Initiate IGM | 2:56:14.9 | $T_7$ + 0:00:20.0 | Begin Orbital Guidance |
| 0:07:15.4 | $T_3$ + 0:04:44.4 | S-II Second MR(4.5) Shift | 2:57:15.5 | $T_7$ + 0:01:30.6 | Lox Nonpropulsive Vent Closed |
| 0:07:32.4 | $T_3$ + 0:05:01.4 | S-II Fuel Tank Pressurization Flowrate Step | 2:57:34.9 | $T_7$ + 0:01:40.0 | Begin Orbital Navigation |
| 0:08:29.0 | $T_3$ + 0:06:58.0 | Begin Chi Freeze | 3:10:54.9 | $T_7$ + 0:15:00.0 | LH$_2$ Continuous and Nonpropulsive Vents Closed Maneuver Space Vehicle to CSM Separation Attitude |
| 0:08:40.1 | $T_4$ + 0:00:00.0 | S-II Engine Cutoff | 3:15:54.9 | $T_7$ + 0:20:00.0 | CSM Separation |
| 0:08:40.8 | $T_4$ + 0:00:00.7 | S-IVB Ullage Ignition | 3:55:54.9 | $T_7$ + 1:00:00.0 | LH$_2$ Nonpropulsive Vent Open |
| 0:08:40.9 | $T_4$ + 0:00:00.8 | Signal to Separation Devices and S-II Retrorockets | 4:10:54.9 | $T_7$ + 1:15:00.0 | LH$_2$ Nonpropulsive Vent Closed |
| 0:08:41.0 | $T_4$ + 0:00:00.9 | S-II/S-IVB Separation Complete | 4:44:54.9 | $T_7$ + 1:49:00.0 | Maneuver S-IVB/IU/LTA-B to Slingshot Attitude |
| 0:08:41.1 | $T_4$ + 0:00:01.0 | S-IVB Engine Start Sequence, First Burn | 4:55:54.9 | $T_7$ + 2:00:00.0 | LH$_2$ Continuous Vent Open |
| 0:08:44.1 | $T_4$ + 0:00:04.0 | S-IVB Ignition (Start Tank Discharge Valve Opens) | 5:07:54.9 | $T_7$ + 2:12:00.0 | Start Lox Dump for Slingshot $\Delta V$ |
| 0:08:46.6 | $T_4$ + 0:00:06.5 | S-IVB Engine at 90% Thrust | 5:12:54.9 | $T_7$ + 2:17:00.0 | Complete Lox Dump Sequence |
| 0:08:47.6 | $T_4$ + 0:00:07.5 | End Chi Freeze | 5:12:58.1 | $T_7$ + 2:17:03.2 | Lox Nonpropulsive Vent Open |
| 0:08:48.7 | $T_4$ + 0:00:08.6 | S-IVB Ullage Thrust Termination | 5:12:58.3 | $T_7$ + 2:17:03.4 | LH$_2$ Nonpropulsive Vent Open |
| 0:08:53.0 | $T_4$ + 0:00:12.9 | S-IVB Ullage Case Jettison | | | |
| 0:11:13.0 | $T_4$ + 0:02:32.0 | Begin Chi Freeze | | | |
| 0:11:21.7 | $T_5$ + 0:00:00.0 | S-IVB Velocity Cutoff, First Burn | | | |
| 0:11:22.0 | $T_5$ + 0:00:00.3 | S-IVB APS Ullage Engine No. 1 Ignition | | | |
| 0:11:22.1 | $T_5$ + 0:00:00.4 | S-IVB APS Ullage Engine No. 2 Ignition | | | |
| 0:11:31.5 | $T_5$ + 0:00:09.8 | Parking Orbit Insertion | | | |
| 0:11:41.9 | $T_5$ + 0:00:20.2 | Begin Orbital Guidance | | | |
| 0:12:20.7 | $T_5$ + 0:00:59.0 | LH$_2$ Continuous Vent Open | | | |
| 0:12:49.7 | $T_5$ + 0:01:28.0 | S-IVB APS Ullage Engine No. 1 Cutoff | | | |

Figure 2-1

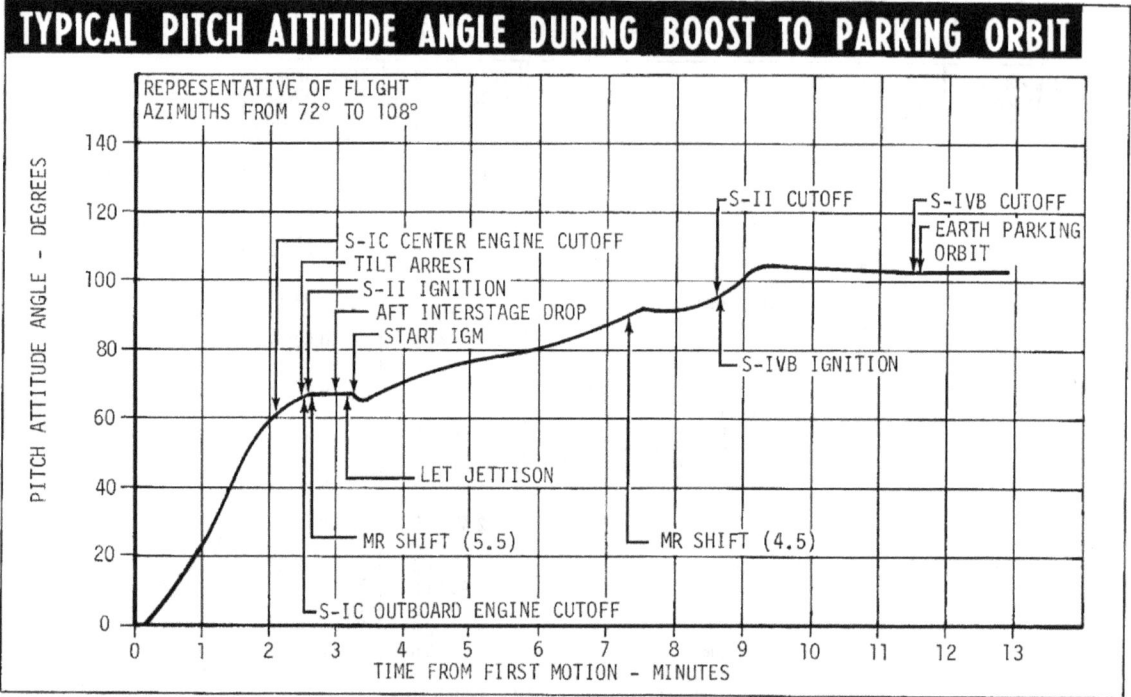

Figure 2-2

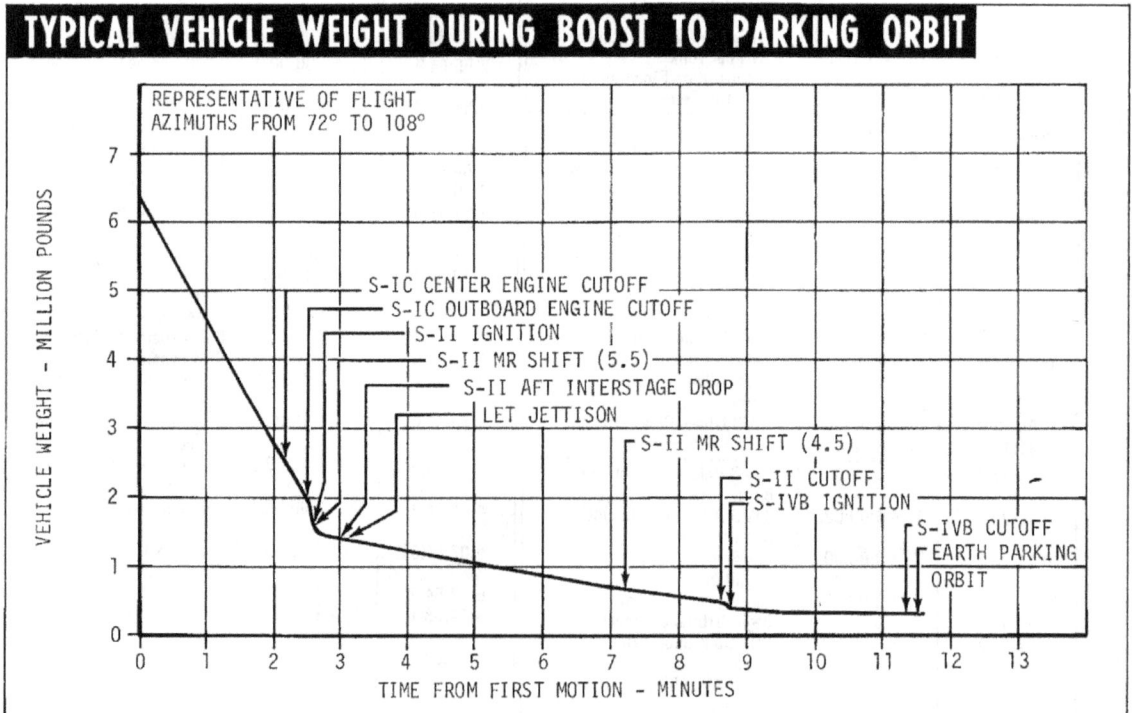

Figure 2-3

## TYPICAL AERODYNAMIC AXIAL FORCE DURING S-IC AND EARLY S-II FLIGHT

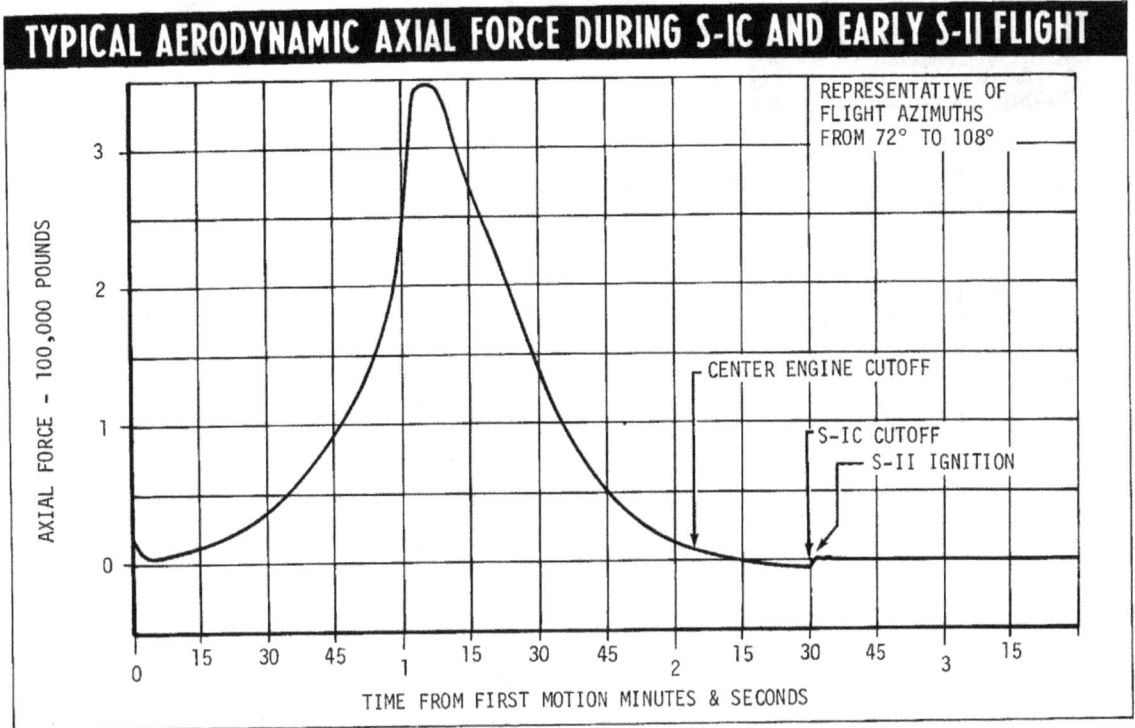

Figure 2-4

## TYPICAL AERODYNAMIC PRESSURE DURING S-IC AND EARLY S-II FLIGHT

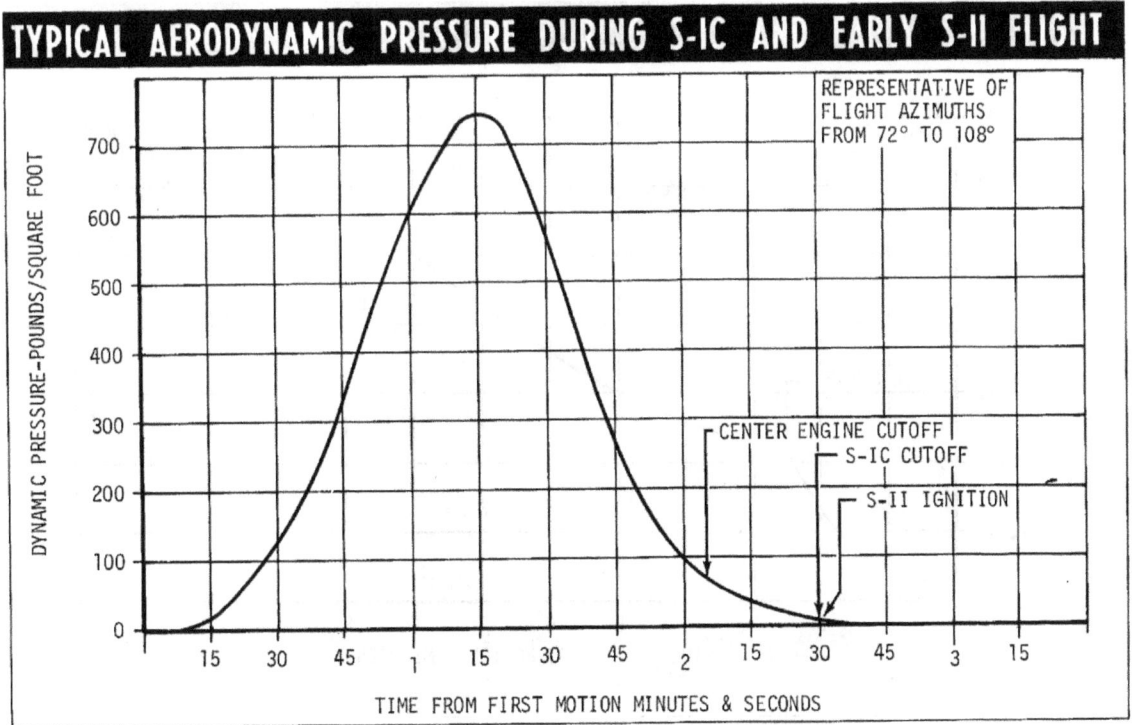

Figure 2-5

## TYPICAL LONGITUDINAL ACCELERATION DURING BOOST TO PARKING ORBIT

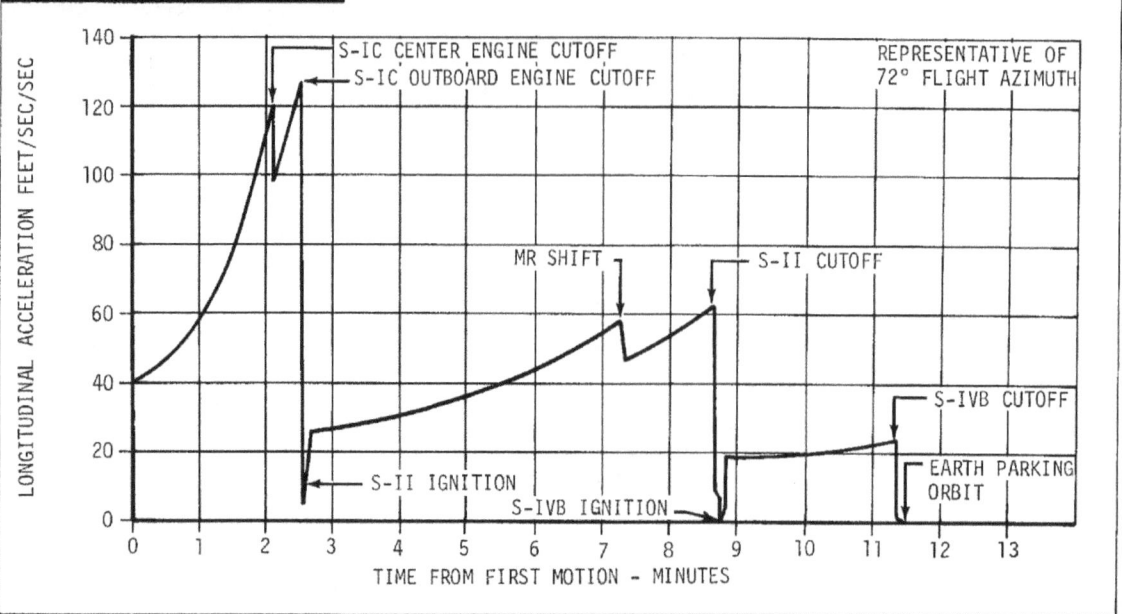

Figure 2-6

## TYPICAL INERTIAL VELOCITY DURING BOOST TO PARKING ORBIT

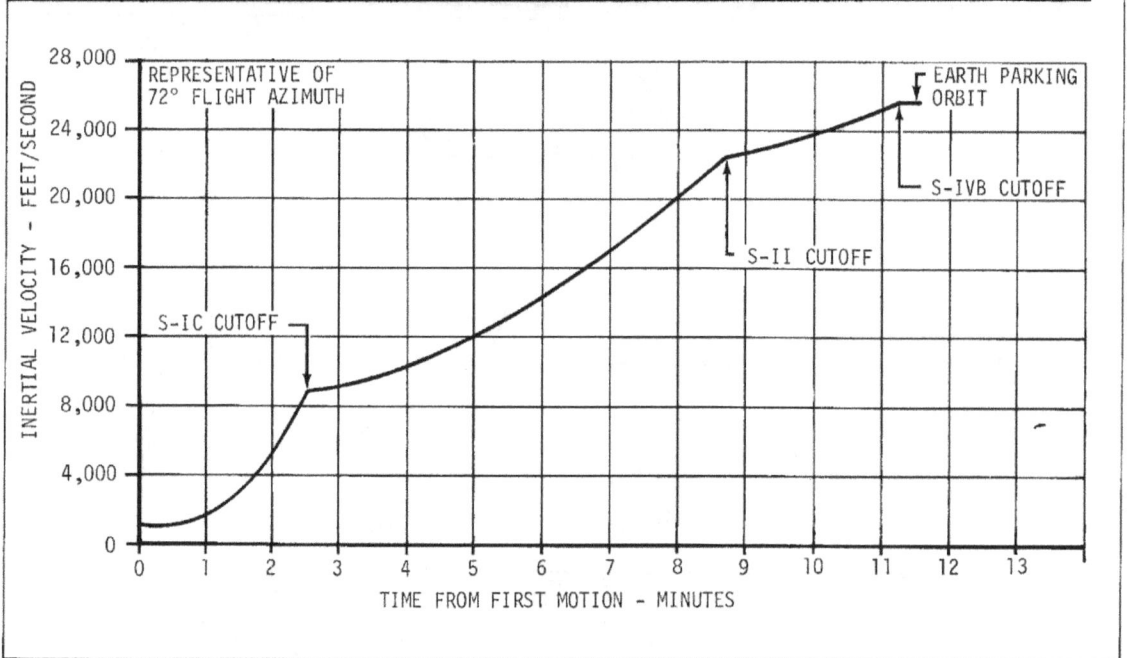

Figure 2-7

# TYPICAL ALTITUDE DURING BOOST TO PARKING ORBIT

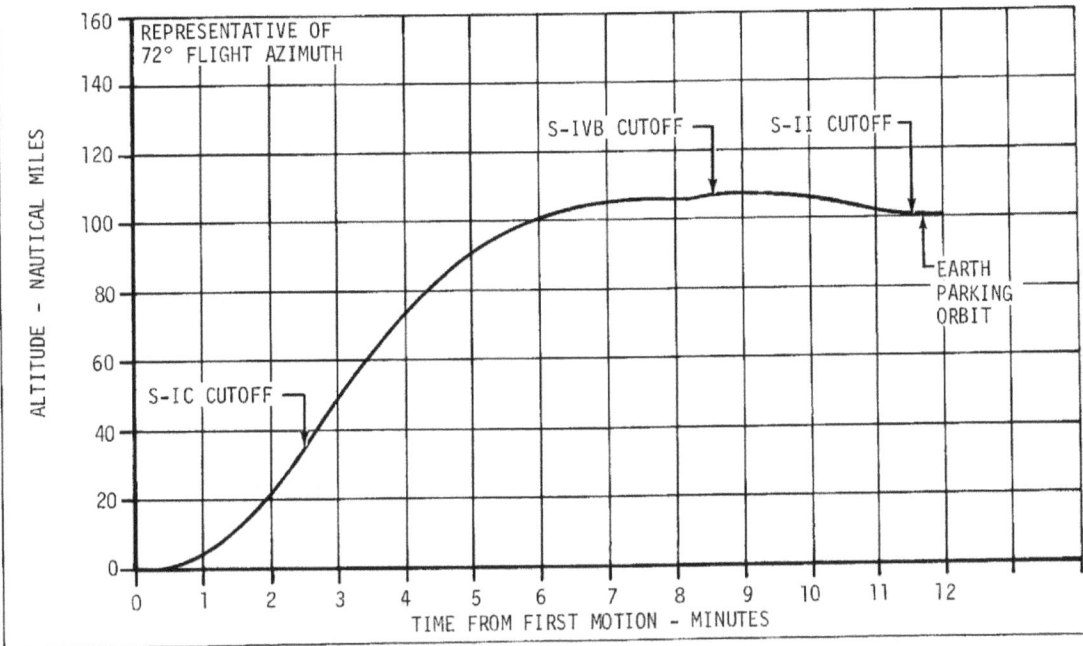

REPRESENTATIVE OF
72° FLIGHT AZIMUTH

S-IVB CUTOFF

S-II CUTOFF

EARTH
PARKING
ORBIT

S-IC CUTOFF

ALTITUDE - NAUTICAL MILES

TIME FROM FIRST MOTION - MINUTES

Figure 2-8

# TYPICAL RANGE DURING BOOST TO PARKING ORBIT

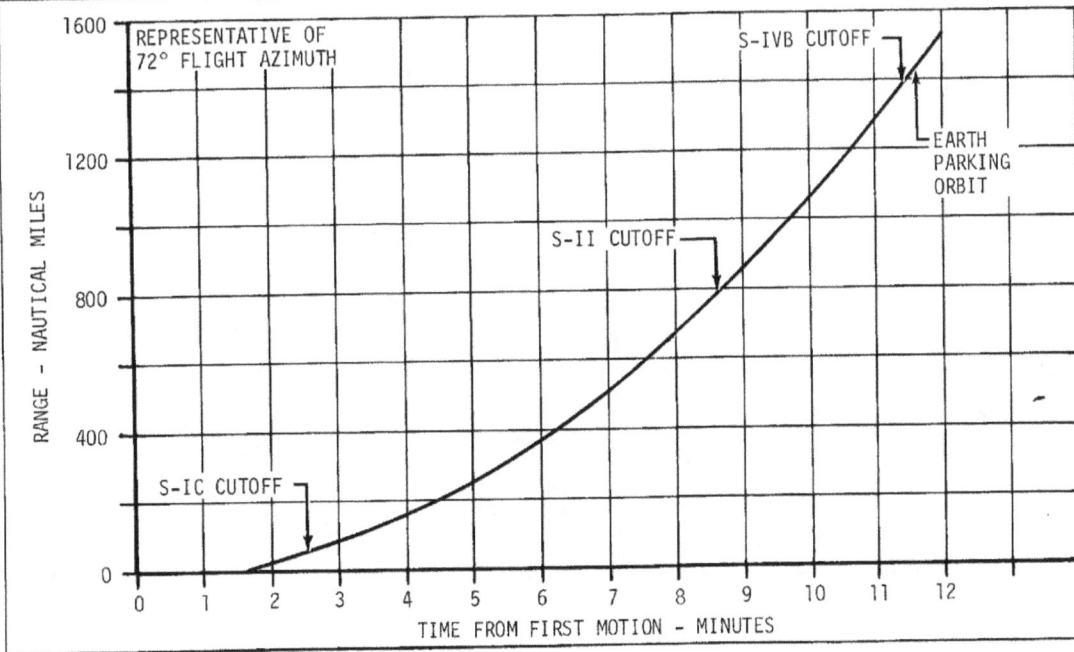

REPRESENTATIVE OF
72° FLIGHT AZIMUTH

S-IVB CUTOFF

EARTH
PARKING
ORBIT

S-II CUTOFF

S-IC CUTOFF

RANGE - NAUTICAL MILES

TIME FROM FIRST MOTION - MINUTES

Figure 2-9

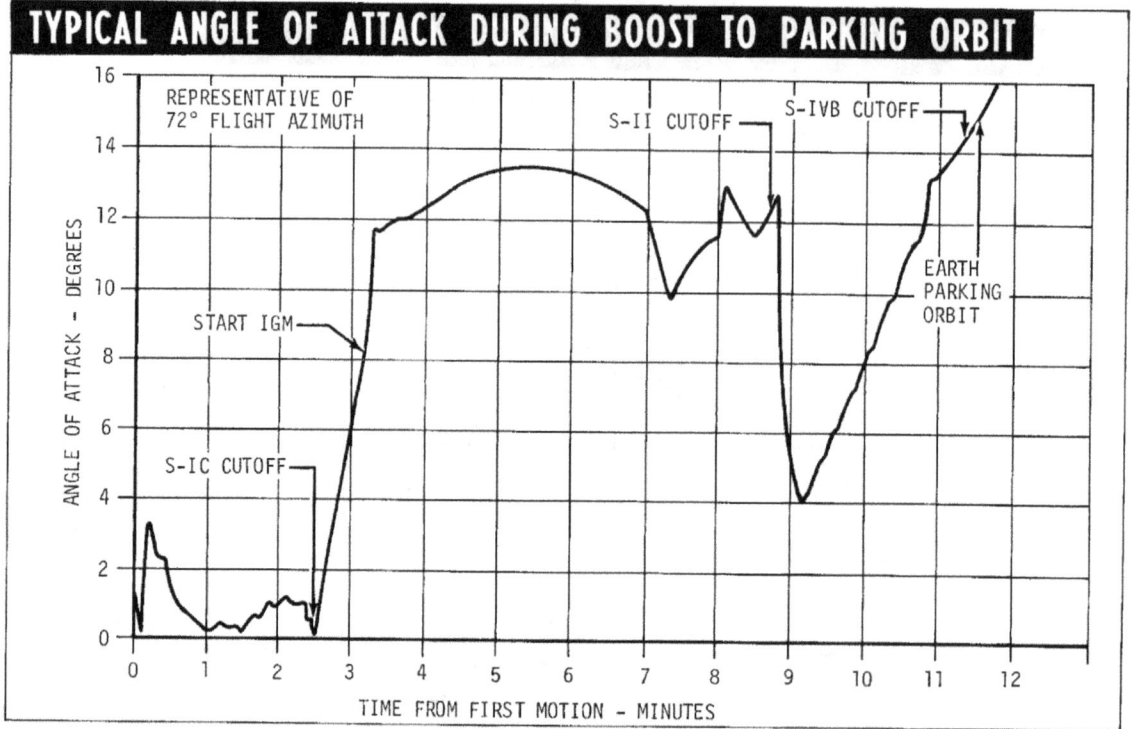

Figure 2-10

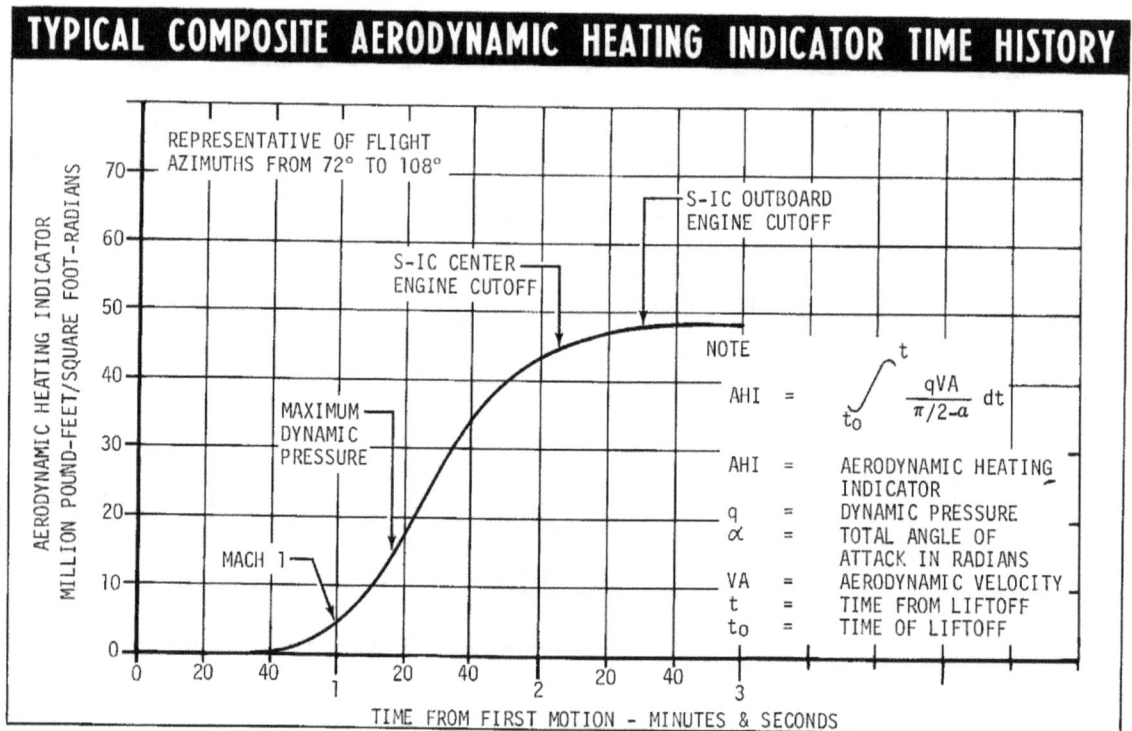

Figure 2-11

# TYPICAL INERTIAL PATH ANGLE DURING BOOST TO PARKING ORBIT

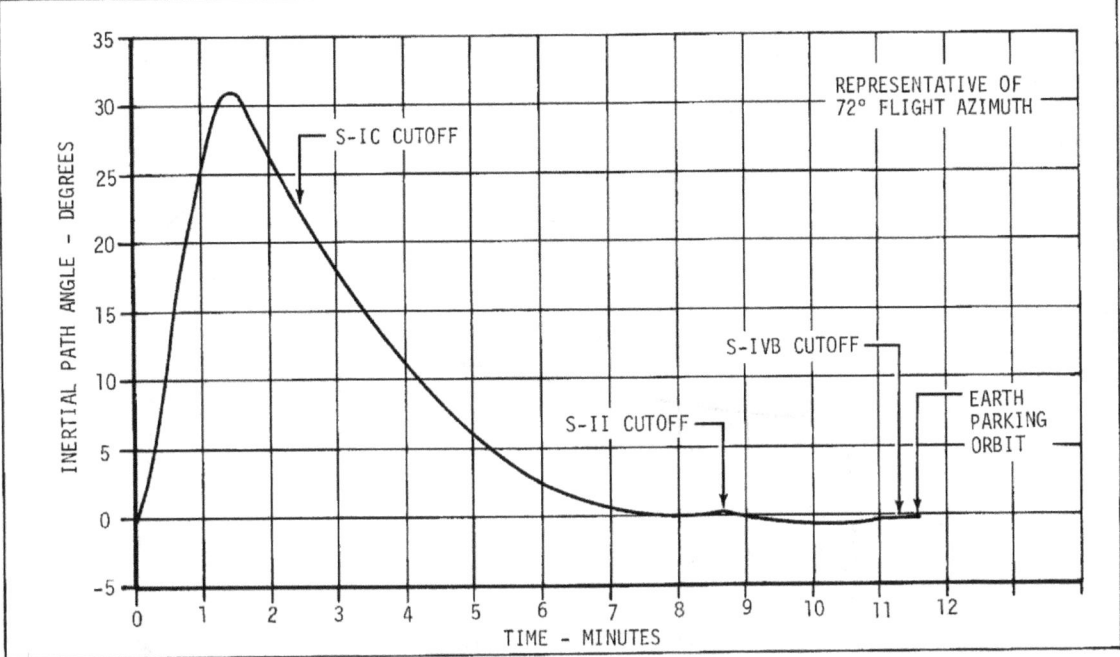

Figure 2-12

# TYPICAL INERTIAL HEADING ANGLE DURING BOOST TO PARKING ORBIT

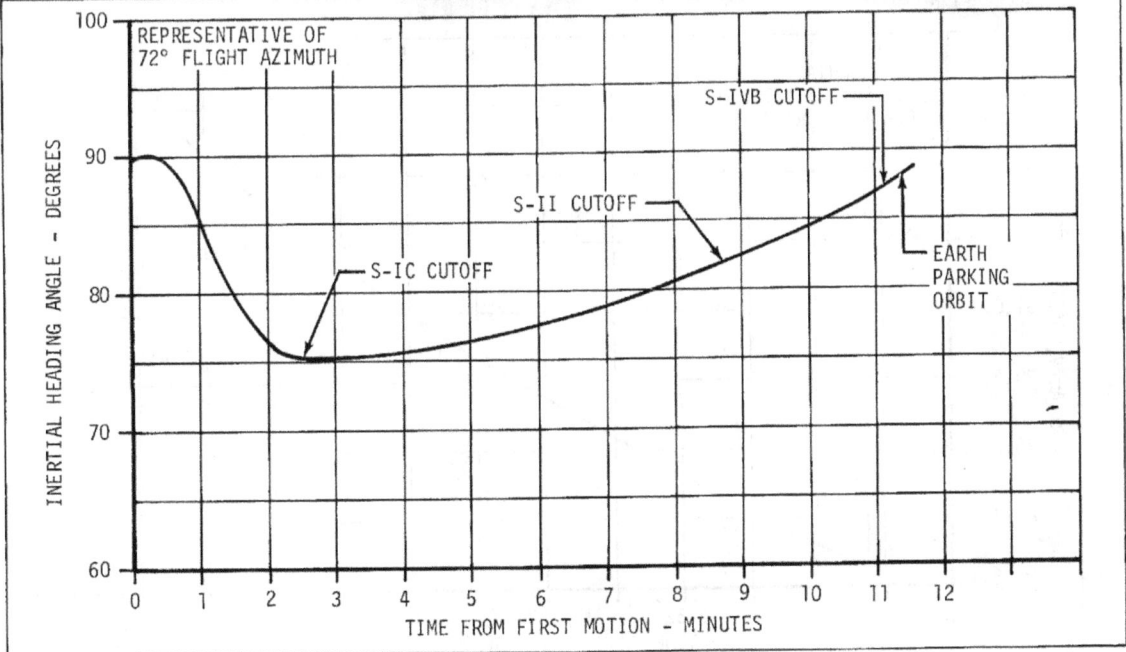

Figure 2-13

## TYPICAL LONGITUDINAL ACCELERATION DURING TLI BOOST

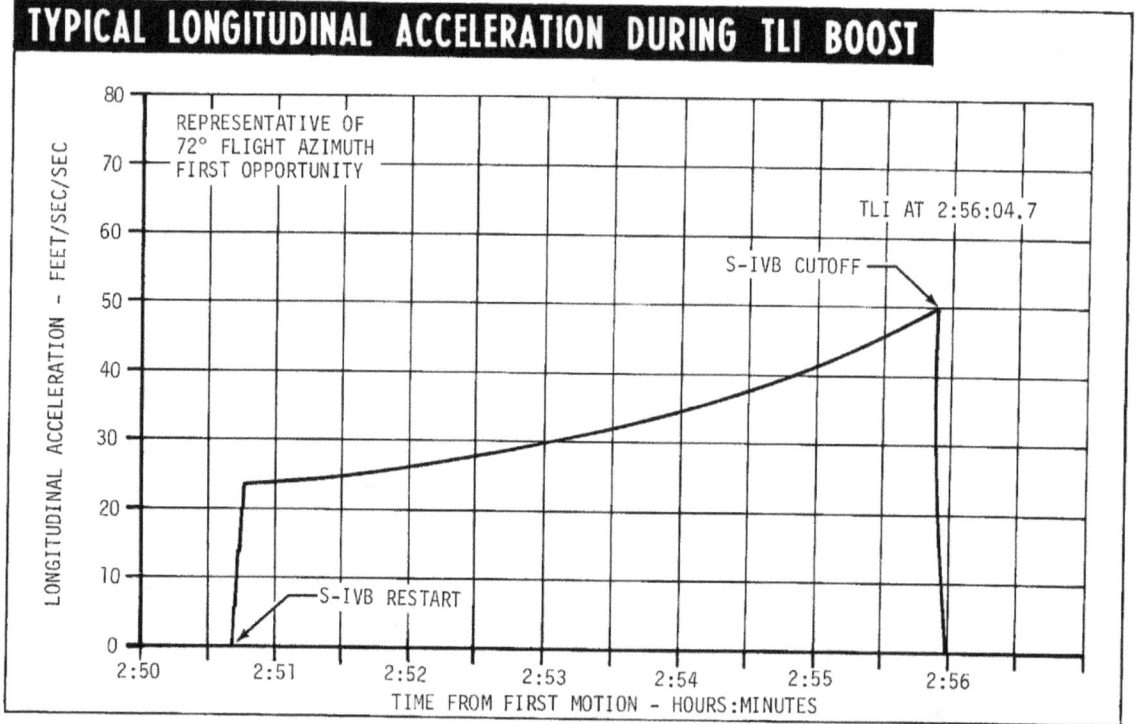

Figure 2-14

## TYPICAL INERTIAL VELOCITY DURING TLI BOOST

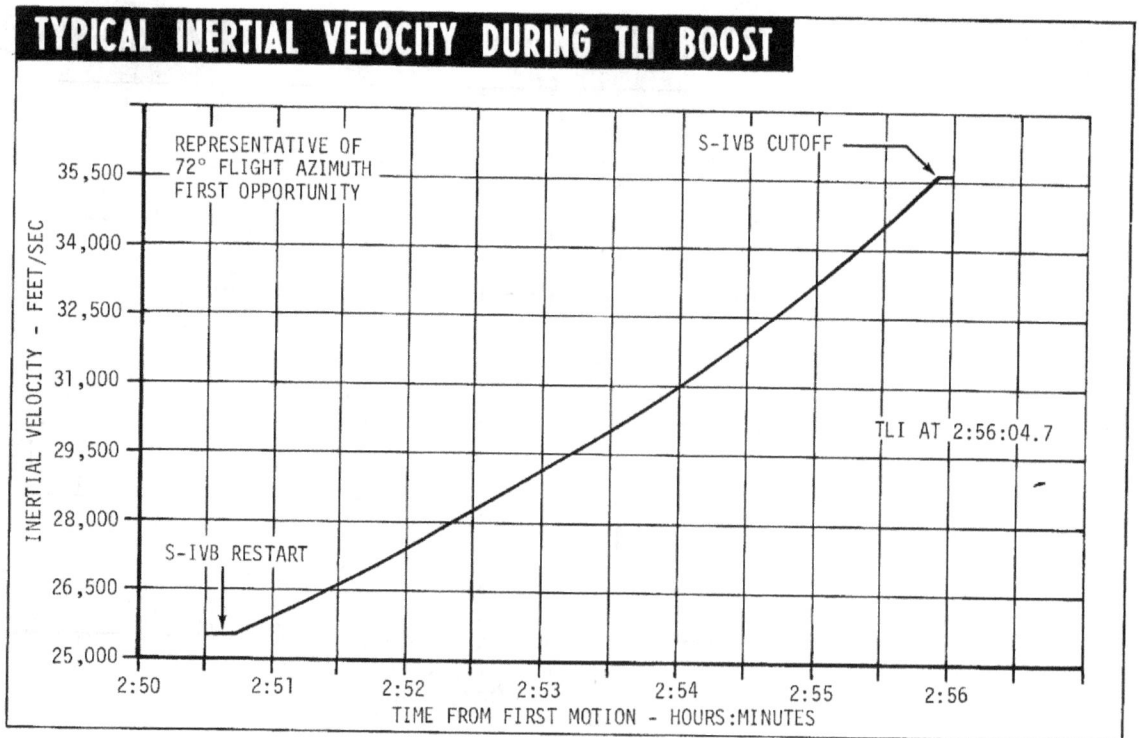

Figure 2-15

## TYPICAL ALTITUDE DURING TLI BOOST

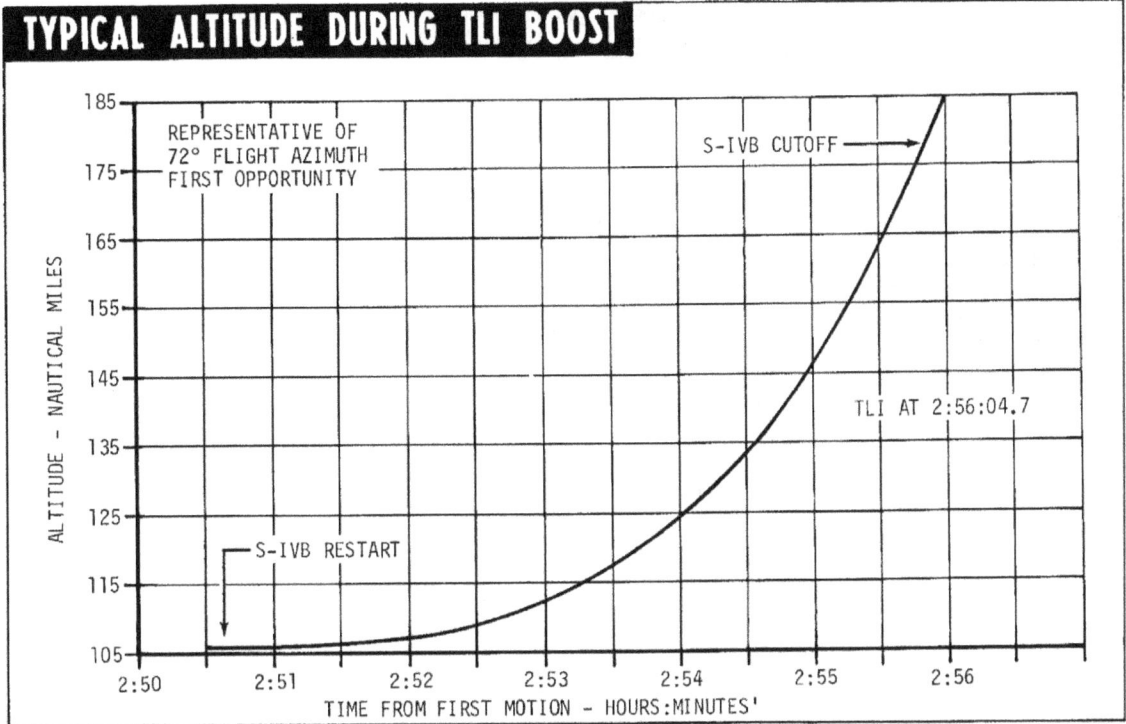

Figure 2-16

## TYPICAL RANGE FROM CAPE KENNEDY DURING TLI BOOST

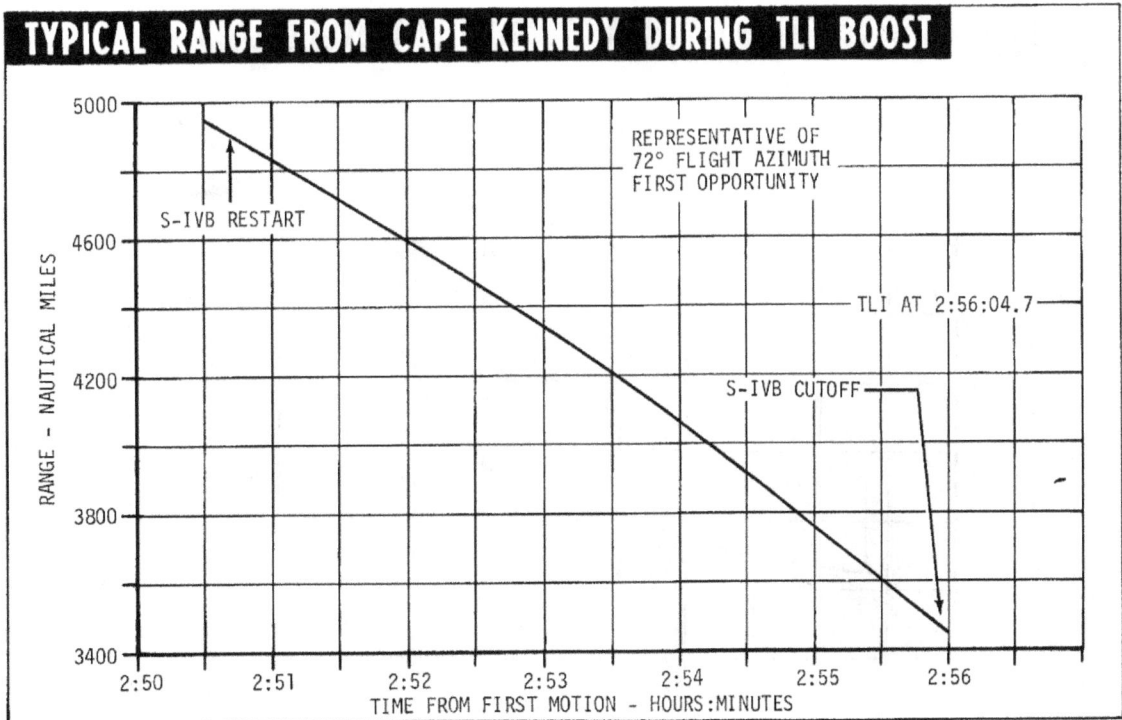

Figure 2-17

## TYPICAL INERTIAL PATH ANGLE DURING TLI BOOST

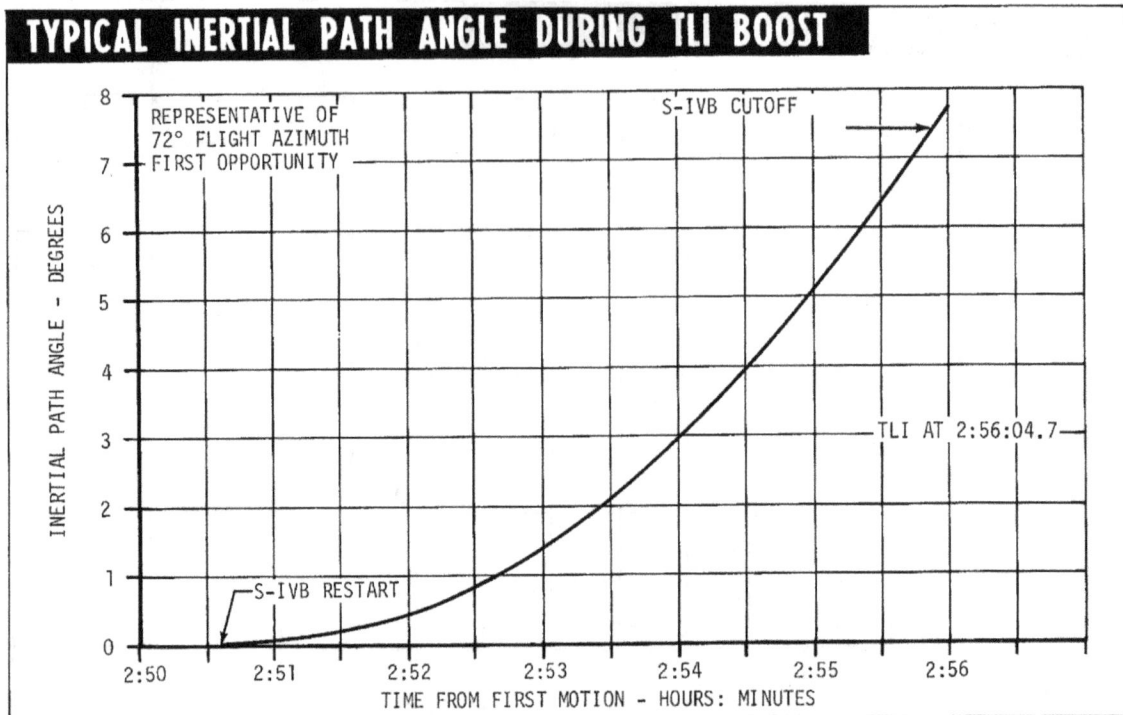

Figure 2-18

## TYPICAL INERTIAL HEADING ANGLE DURING TLI BOOST

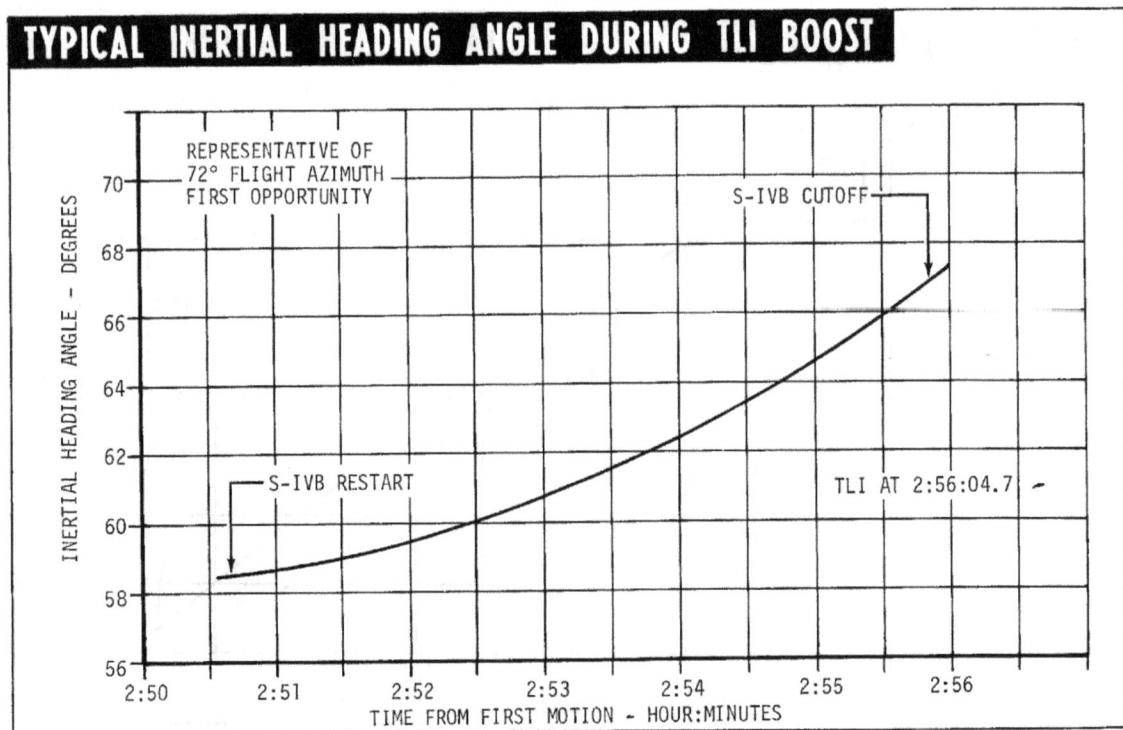

Figure 2-19

# TYPICAL ANGLE OF ATTACK DURING TLI BOOST

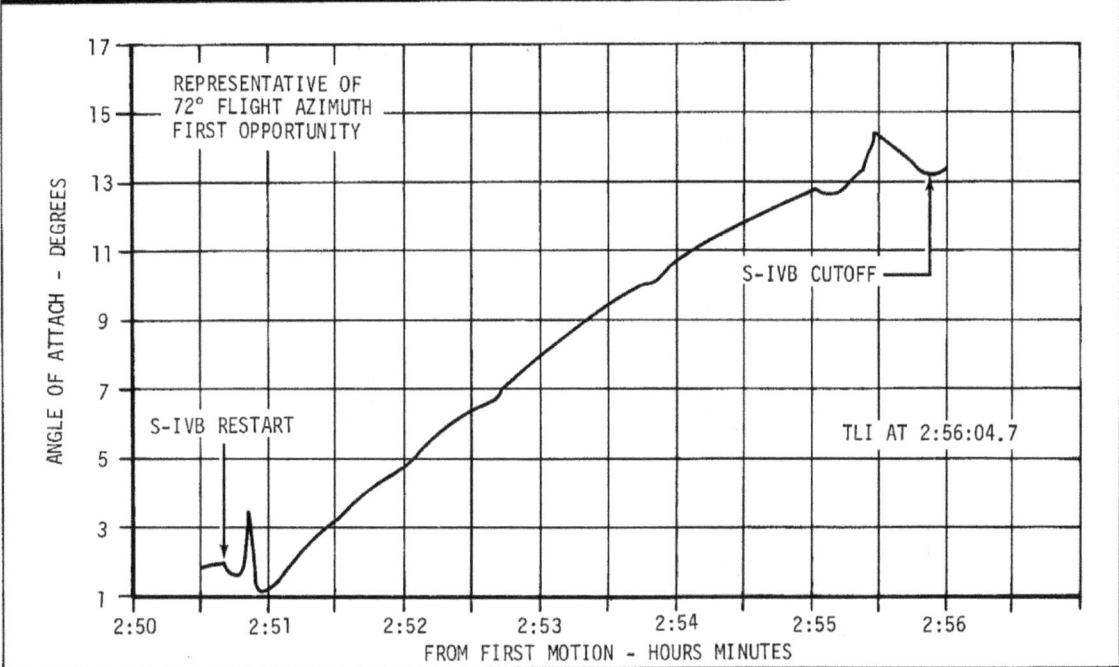

Figure 2-20

## PROPULSION PERFORMANCE

The typical propulsion performance data presented herein are based on flight simulations, stage and engine configuration, and static test firing data.

## PROPELLANT LOADING

A propellant weight summary for each stage is tabulated in figures 2-21 through 2-23. The tables break down propellant use into such categories as usable, unusable, trapped, buildup and holddown, mainstage, thrust decay, and fuel bias.

## ENGINE PERFORMANCE

Stage thrust versus time history for the three stages are graphically presented in figures 2-24 through 2-26.

The thrust profile for the S-IC stage (figure 2-24) shows the thrust increase from the sea level value of approximately 7,648,000 pounds to approximately 9,160,000 pounds at center engine cutoff, where the vehicle has attained an altitude of approximately 142,500 feet. At center engine cutoff, vehicle thrust drops to approximately 7,240,000 pounds.

The S-II stage thrust profile (figure 2-25) is slightly perturbed by the aft interstage drop and launch escape tower jettison. A significant drop in thrust level is noted at the MR shift from 5.5 to 4.5, where the thrust drops from 1,140,000 pounds to 900,000 pounds. Thrust is slightly affected subsequent to the MR shift by the fuel tank pressurization flowrate step which increases tank pressurization to maximum.

The S-IVB stage thrust profiles for first and second burns are shown in figure 2-26. The thrust level for first burn is approximately 203,000 pounds attained with a 5.0 MR. (A predicted thrust level of 230,000 pounds is attainable with a 5.5 MR). The S-IVB second burn is started at a 4.5 MR and is shifted to 5.0 MR 2.5 seconds after 90% thrust is reached. The thrust level for the second burn is approximately 203,000 pounds.

| S-IC STAGE PROPELLANT WEIGHT SUMMARY | | |
|---|---|---|
| | LOX (POUNDS) | RP-1 (POUNDS) |
| CONSUMED PROPELLANT | 3,099,262 | 1,328,084 |
| BUILDUP AND HOLDDOWN | 67,011 | 18,472 |
| MAINSTAGE | 3,017,773 | 1,300,430 |
| THRUST DECAY | 5,468 | 3,434 |
| TAILOFF | 1,635 | 415 |
| FUEL BIAS | --------- | 5,333 |
| PRESSURIZATION | 7,375 | --------- |
| RESIDUAL PROPELLANT | 33,499 | 23,282 |
| TANKS | 2,160 | 9,950 |
| SUCTION LINES | 28,849 | 6,449 |
| INTERCONNECT LINES | 330 | --------- |
| ENGINES | 2,160 | 6,585 |
| ENGINE CONTROL SYSTEMS | --------- | 298 |
| TOTAL | 3,132,761 | 1,351,366 |

Figure 2-21

| S-II STAGE PROPELLANT WEIGHT SUMMARY | | |
|---|---|---|
| | LOX (POUNDS) | LH2 (POUNDS) |
| USABLE PROPELLANT | 785,247 | 150,487 |
|   MAINSTAGE | 783,487 | 148,032 |
|   BIAS | ------- | 1,714 |
|   THRUST BUILDUP | 1,573 | 626 |
|   THRUST DECAY | 187 | 115 |
| | | |
| UNUSABLE PROPELLANT | 8,868 | 4,174 |
|   TRAPPED: | 5,991 | 2,849 |
|     ENGINE | 490 | 48 |
|     LINE | 843 | 186 |
|     RECIRCULATION | 230 | 10 |
|     INITIAL ULLAGE MASS | 421 | 116 |
|     TANK AND SUMP | 4,007 | 2,489 |
|   PRESSURIZATION GAS | 2,877 | 1,209 |
|   VENTED | ------- | 116 |
| | | |
| TOTAL | 794,115 | 154,661 |

Figure 2-22

| S-IVB STAGE PROPELLANT WEIGHT SUMMARY | | |
|---|---|---|
| | LOX (POUNDS) | LH2 (POUNDS) |
| USABLE PROPELLANT | 190,103 | 38,531 |
|   USABLE | 190,103 | 38,110 |
|   (INCLUDES MAINSTAGE | | |
|   FLIGHT PERFORMANCE | | |
|   AND FLIGHT GEOMETRY | | |
|   RESERVES) | | |
| | | |
|   RESIDUAL FUEL BIAS | ------ | 421 |
| UNUSABLE PROPELLANT | 1,057 | 4,819 |
|   ORBITAL | 112 | 3,467 |
|   *FUEL LEAD | ------- | 55 |
|   SUBSYSTEMS | 10 | 302 |
|   ENGINE TRAPPED | 108 | 10 |
|   LINES TRAPPED | 259 | 38 |
|   TANK UNAVAILABLE | 40 | 685 |
|   *BUILDUP TRANSIENTS | 438 | 212 |
|   *DECAY TRANSIENTS | 90 | 50 |
| | | |
| *FOR FIRST AND SECOND BURNS | | |
| TOTAL | 191,160 | 43,350 |

Figure 2-23

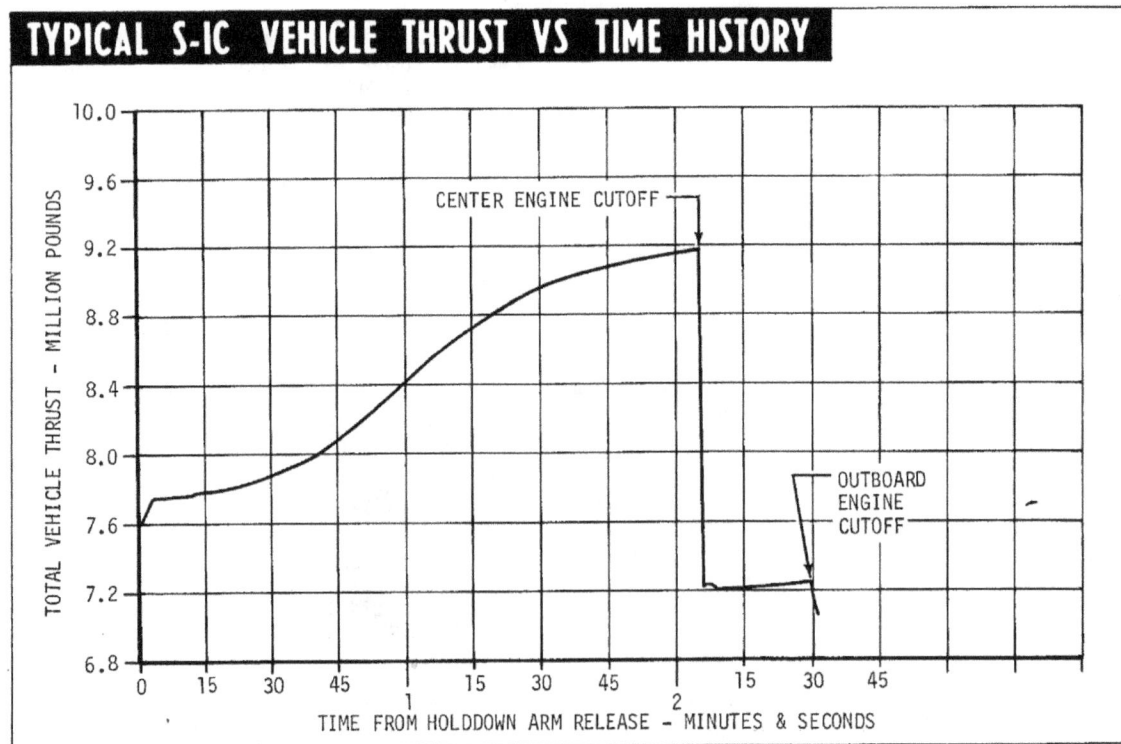

TYPICAL S-IC VEHICLE THRUST VS TIME HISTORY

Figure 2-24

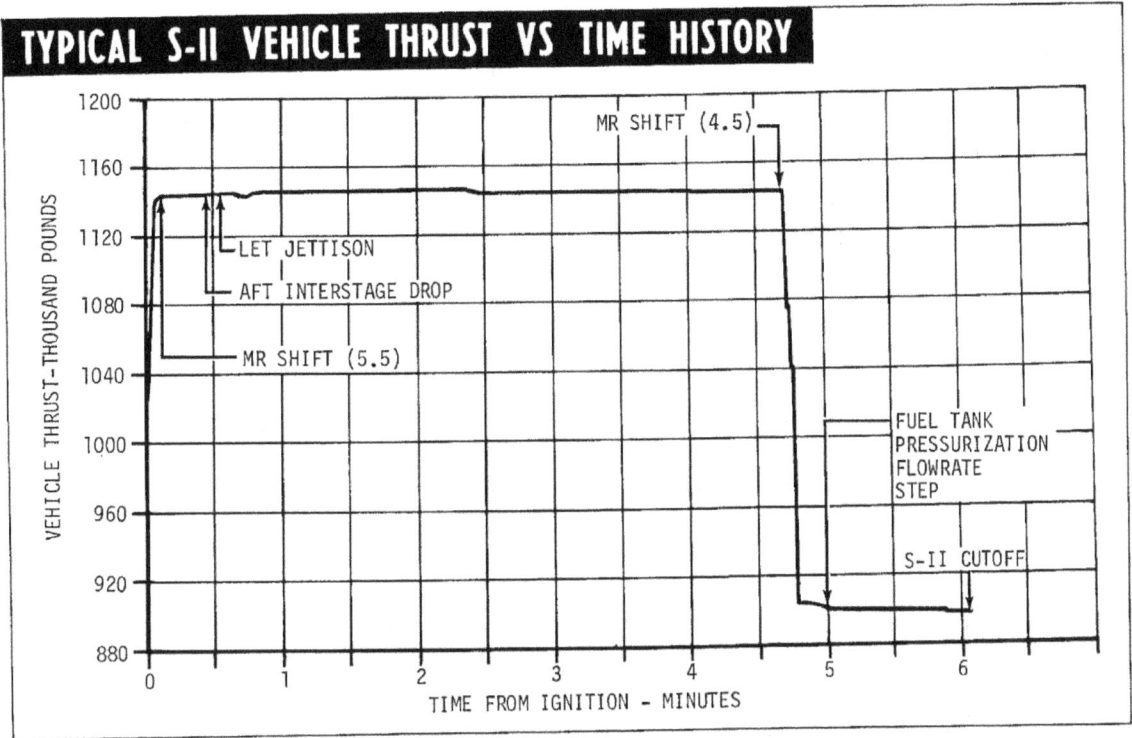

Figure 2-25

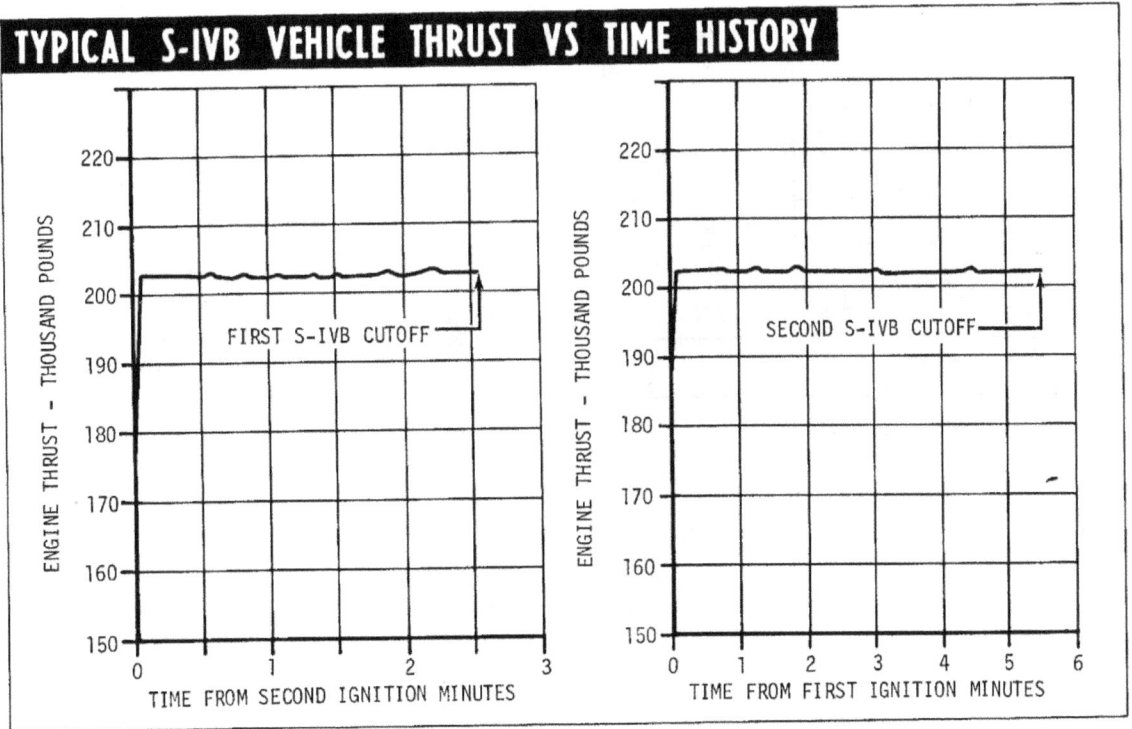

Figure 2-26

## FLIGHT LOADS

Flight loads are dependent on the flight trajectory, associated flight parameters, and wind conditions. These factors are discussed in the following paragraphs.

### WIND CRITERIA

Winds have a significant effect on Saturn V launch vehicle flight loads. Wind criteria used in defining design flight loads for the Saturn V launch vehicle was a scalar wind profile constructed from 95 percentile windiest month speed with 99 percentile shear and a 29.53 feet per second gust. A criteria revision reduces the criteria conservatism for a gust in conjunction with wind shear.

| MINIMUM LOAD INDICATORS FOR DESIGN CRITERIA WINDS AND DIRECTIONAL WINDS | | | | HIGH-Q | CECO |
|---|---|---|---|---|---|
| CONDITION | | | | HIGH-Q | CECO |
| DESIGN WINDS | TENSION | L.I. ** | | 1.27 | 2.14 |
| | | STATION | | 1848 FWD | 2387 AFT |
| | COMP * | L.I. | | 1.33 | 1.42 |
| | | STATION | | 602 AFT | 1541 |
| DIRECTIONAL WINDS | NON-BIASED TRAJECTORY | TENSION | L.I. | 1.34 | SAME AS DESIGN WINDS |
| | | | STATION | 2387 AFT | |
| | | COMP | L.I. | 1.44 | |
| | | | STATION | 602 AFT | |
| | BIASED TRAJECTORY | TENSION | L.I. | 1.53 | SAME AS DESIGN WINDS |
| | | | STATION | 2387 AFT | |
| | | COMP | L.I. | 1.50 | |
| | | | STATION | 602 AFT | |
| *   COMPRESSION | | | | | |
| **   LOAD INDICATOR | | | | | |

Figure 2-27

Trajectory wind biasing reduces flight loads. Wind biased trajectories are used for launch months with predictable wind speed magnitude and direction. Figure 2-27 shows the effect on load indicators of a biased trajectory, non-biased trajectory and design wind conditions.

Variation in bending moment with altitude for the biased trajectory is shown in figure 2-28 for a typical station.

The variation in peak bending moment with azimuth is shown for 95 percentile directional winds and a wind biased trajectory in figure 2-29.

Axial load distribution is the same for all wind conditions. This is illustrated by figure 2-30 which shows the axial load distribution at center engine cutoff (CECO).

### ENGINE OUT CONDITIONS

Engine-out conditions, if they should occur, will effect the vehicle loads. The time at which the malfunction occurs, which engine malfunctions, peak wind speed and azimuth orientation of the wind, are all independent variables which combine to produce load conditions. Each combination of engine-out time, peak wind velocity, wind azimuth, and altitude at which the maximum wind shear occurs, produces a unique trajectory. Vehicle responses such as dynamic pressure, altitude, Mach number, angle-of-attack, engine gimbal angles, yaw and attitude angle time histories vary with the prime conditions. A typical bending moment distribution for a single S-IC control engine-out is shown in figure 2-31. Structure test programs indicate a positive structural margin exists for this malfunction flight condition.

Studies indicate that the immediate structural dynamic transients at engine-out will not cause structural failure. However, certain combinations of engine failure and wind direction and magnitude may result in a divergent control condition which could cause loss of the vehicle.

The "Chi-Freeze" schedule is incorporated into the vehicle guidance program as an alternate to reduce the effect of loss in thrust from an S-IC engine. (Freeze initiation and freeze duration are dependent upon the time at which the loss in thrust occurs.) This schedule holds the pitch attitude command constant, thereby providing a higher altitude trajectory. The higher altitude trajectory minimizes the payload losses into orbit. It also improves the vehicle engine-out dynamic response by providing a lower-velocity entry into the maximum aerodynamic region.

A single control engine-out during S-II powered flight does not produce load conditions which are critical.

## TYPICAL ENVELOPE OF MAXIMUM BENDING MOMENT AT STATION 1156 - NOMINAL FLIGHT

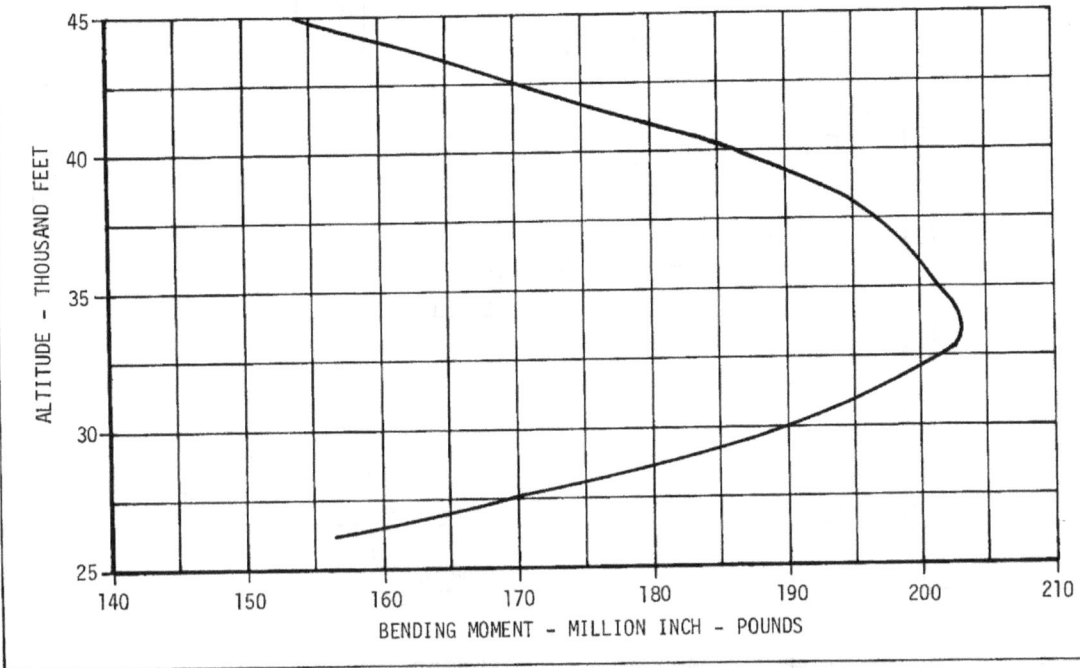

Figure 2-28

## TYPICAL VARIATION OF PEAK BENDING MOMENT WITH WIND AZIMUTH

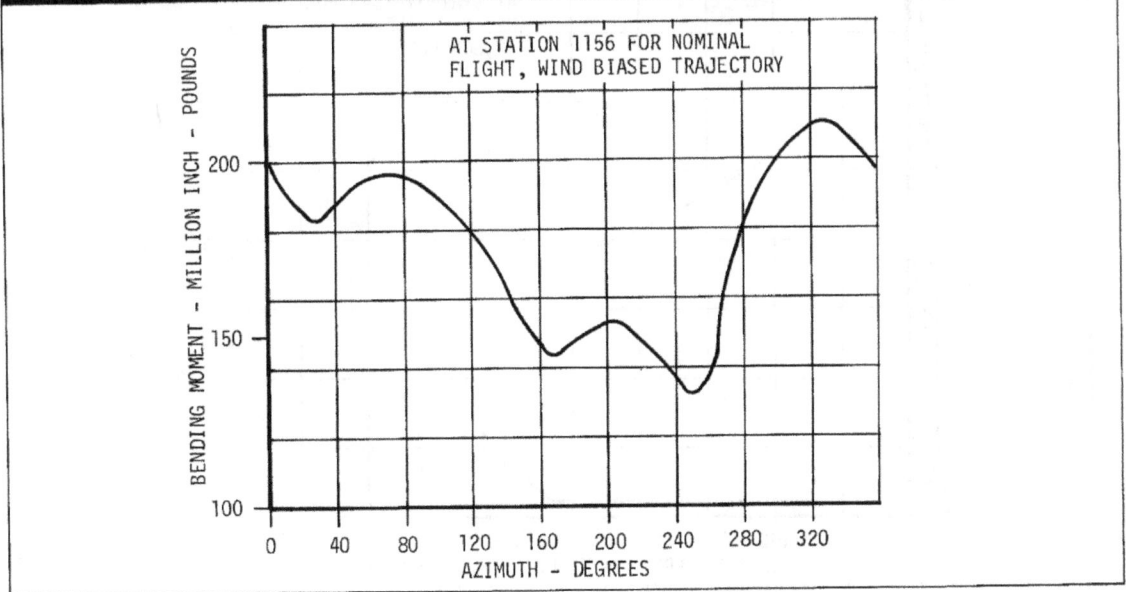

Figure 2-29

# TYPICAL AXIAL LOAD AT CENTER ENGINE CUTOFF

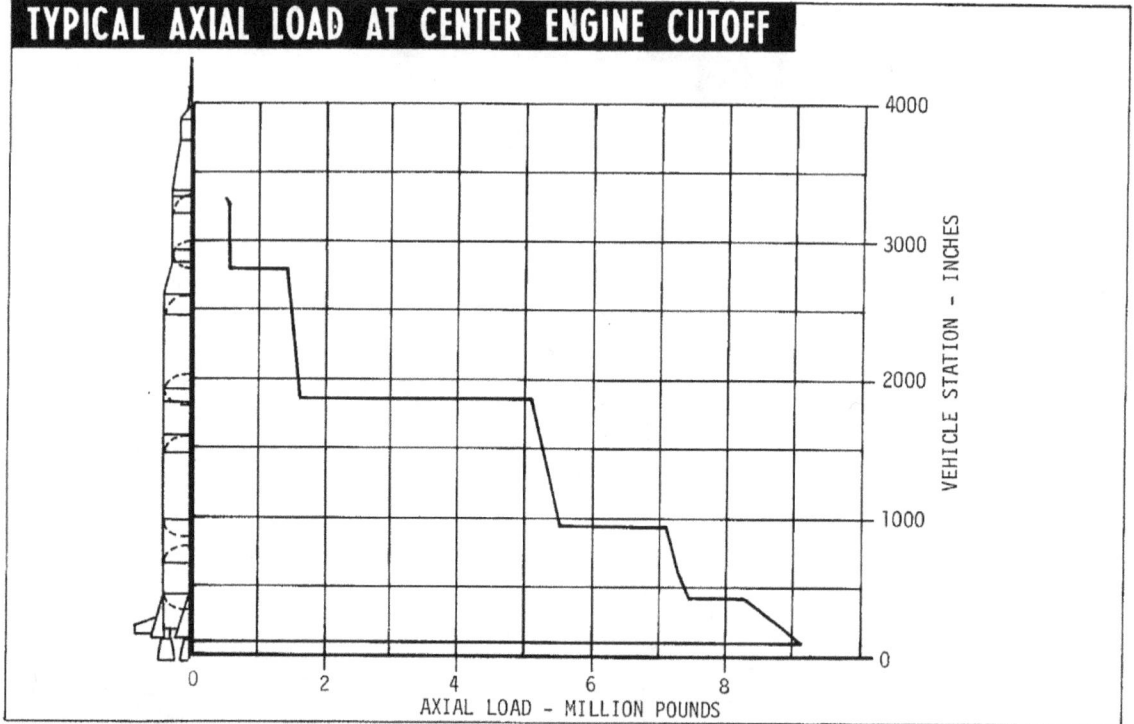

Figure 2-30

# TYPICAL BENDING MOMENT AT MINIMUM FACTOR OF SAFETY

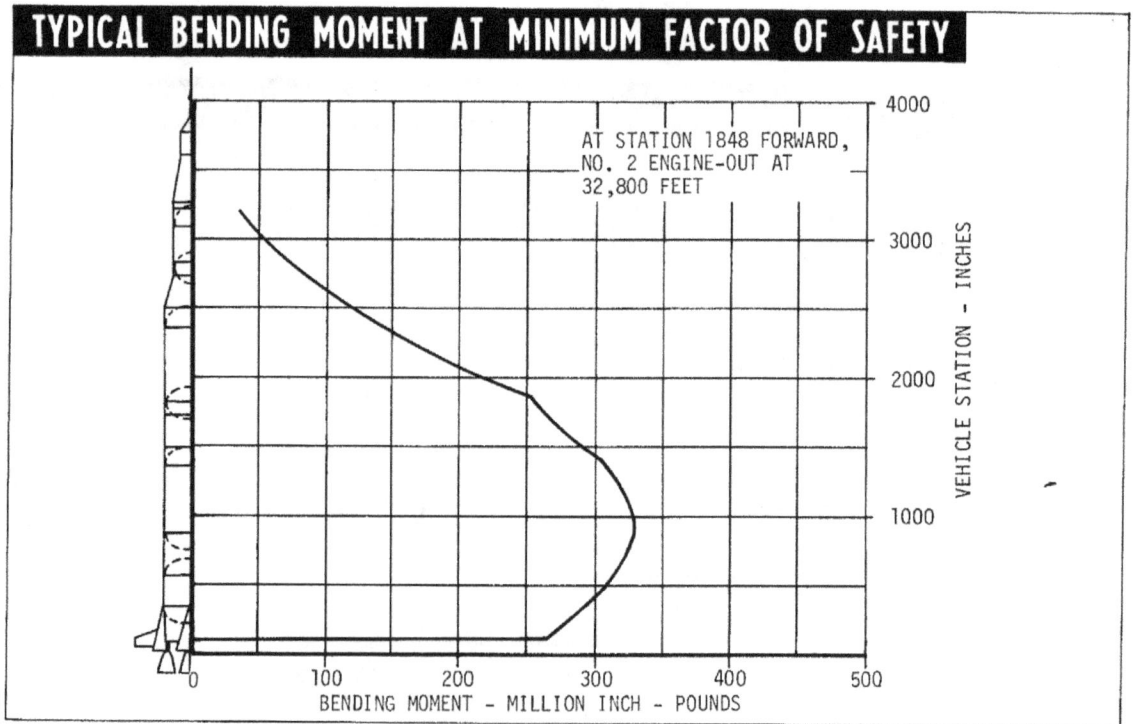

Figure 2-31

# SECTION III
# EMERGENCY DETECTION AND PROCEDURES

## TABLE OF CONTENTS

## EMERGENCY DETECTION SYSTEM

The displays implemented for Emergency Detection System (EDS) monitoring were selected to present as near as possible those parameters which represent the failures leading to vehicle abort. Whenever possible, the parameter was selected so that it would display total subsystem operation. Manual abort parameters have been implemented with redundant sensing and display to provide highly reliable indications to the crewmen. Automatic abort parameters have been implemented triple redundant, voted two-out-of-three, to protect against single point hardware or sensing failures and inadvertent abort.

The types of displays have been designed to provide onboard detection capability for rapid rate malfunctions which may require abort. Pilot abort action must, in all cases, be based on two separate but related abort cues. These cues may be derived from the EDS displays, ground information, physiological cues, or any combination of two valid cues. In the event of a discrepancy between onboard and ground based instrumentation, onboard data will be used. The EDS displays and controls are shown in figure 3-1. As each is discussed it is identified by use of the grid designators listed on the border of the figure.

## EMERGENCY DETECTION SYSTEM DISPLAYS

### FLIGHT DIRECTOR ATTITUDE INDICATOR

The flight director attitude indicator (FDAI) (Q-45, figure 3-1) provides indications of launch vehicle attitude, attitude rates, and attitude errors, except that attitude errors will not be displayed during S-II and S-IVB flight.

Excessive pitch, roll, or yaw indications provide a single cue that an abort is required. Additional abort cues will be provided by the FDAI combining rates, error, or total attitude. Second cues will also be provided by the LV RATE light (Q-50, figure 3-1), LV GUID light (Q-52), physiological, and MCC ground reports.

The FDAI will be used to monitor normal launch vehicle guidance and control events. The roll and pitch programs are initiated simultaneously at + 10 seconds. The roll program is terminated when flight azimuth is reached, and the pitch program continues to tilt-arrest. IGM initiate will occur approximately five seconds after LET jettison during the S-II stage flight.

### LV ENGINES LIGHTS

Each of the five LV ENGINES lights (S-51, figure 3-1) represents the respective numbered engine on the operating stage. (e.g., light number one represents engine number one

on the S-IC, S-II, or S-IVB stage; light number two represents engine number two on the S-IC, or S-II stage; etc.).

These lights are controlled by switching logic in the IU. The switching logic monitors thrust ok pressure switches on each engine of the operating stage and also staging discretes. A light ON indicates its corresponding engine is operating below a nominal thrust level (90% on F-1 engines and 65% on J-2 engines). During staging all lights are turned OFF momentarily to indicate physical separation has occured.

### LV RATE LIGHT

The LV RATE light (Q-50, figure 3-1), when ON, is the primary cue from the launch vehicle that preset overrate settings have been exceeded. It is a single cue for abort, while secondary cues will be provided by FDAI indications, physiological cues, or ground information.

Automatic LV rate aborts are enabled automatically at liftoff, with EDS AUTO (N-60) and LV RATES AUTO (R-63) switches enabled in SC, and are active until deactivated by the crew. EDS auto abort deactivation times will be governed by mission rules. The automatic LV rate abort capability is also deactivated by the launch vehicle sequencer prior to center engine cutoff and is not active during S-II or S-IVB flight.

The automatic abort overrate settings are constant throughout first stage flight. The overrate settings are:

| | | | |
|---|---|---|---|
| Pitch and Yaw | 4.0 (+ 0.5) degrees/sec. ignition | | Liftoff to S-II |
| | 9.2 (+ 0.8) degrees/sec. cutoff | | S-II ignition to S-IVB |
| Roll | 20.0 (+0.5) degrees/sec. cutoff | | Liftoff to S-IVB |

The LV RATE light will illuminate at any time during first, second, or third stage flight if the LV rates exceed these values.

The LV RATE light will also light in response to a separate signal to provide a second cue for a LV platform failure while in the "Max q" region. This circuit is only active during the time the overrate portion of the auto abort system is armed (liftoff to approximately T+2 minutes).

NOTE

The LV RATE light may blink on and off during normal staging.

### LV GUID LIGHT

The LV platform (ST-124M-3) is interrogated every 25 milliseconds for the correct attitude. If an excessive attitude discrepancy is found during three consecutive checks on the fine resolvers, three more checks will be performed on the coarse resolvers. (The LV can continue the mission on the coarse resolvers alone.) If the coarse resolvers fail to control within 15 degrees per second rate of attitude change in any plane, commands sent to the flight control system to change LV attitude will be inhibited and the flight control system will hold the last acceptable command.

# MAIN DISPLAY CONSOLE EDS PANELS

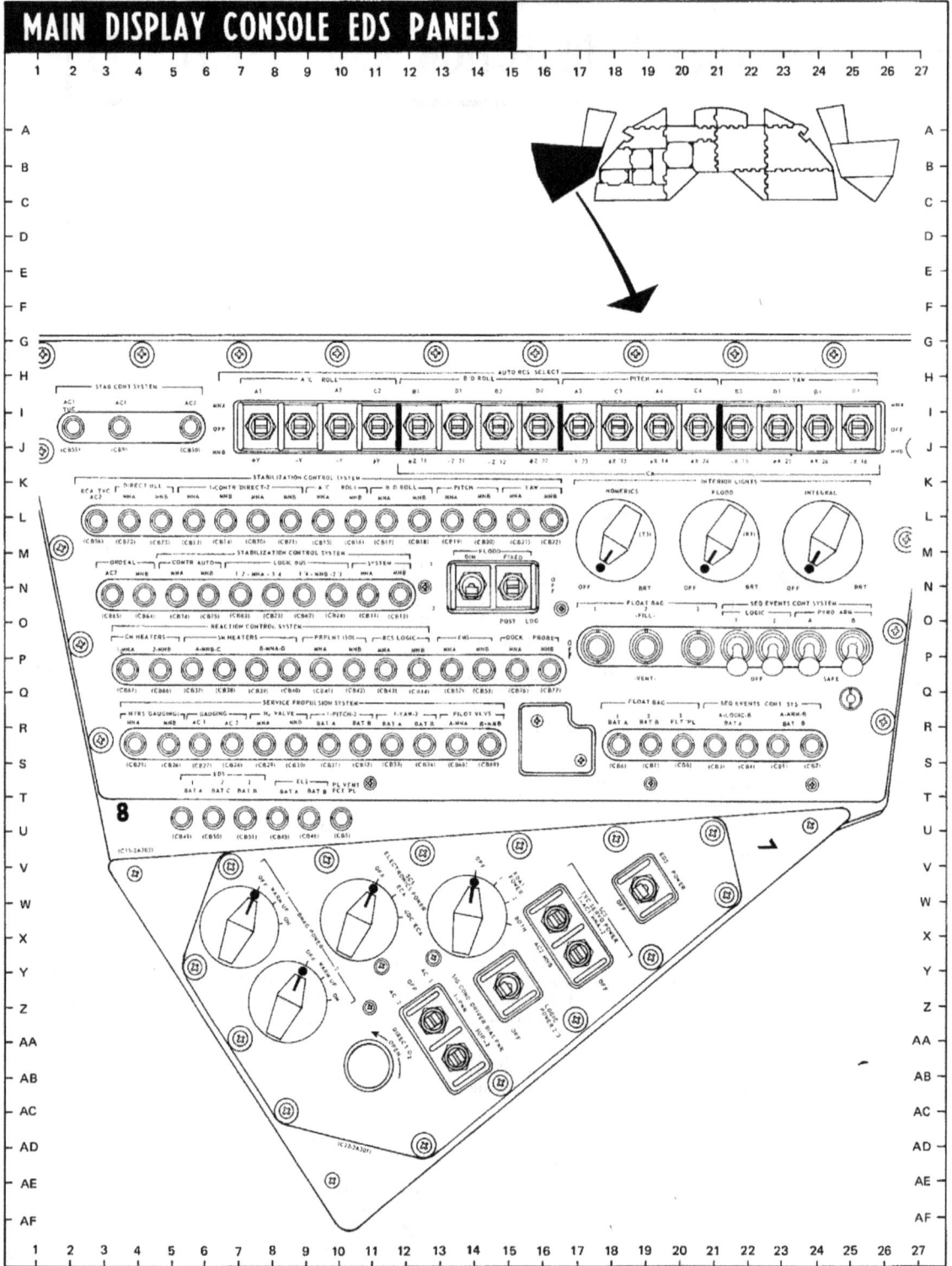

Figure 3-1 (Sheet 1 of 3)

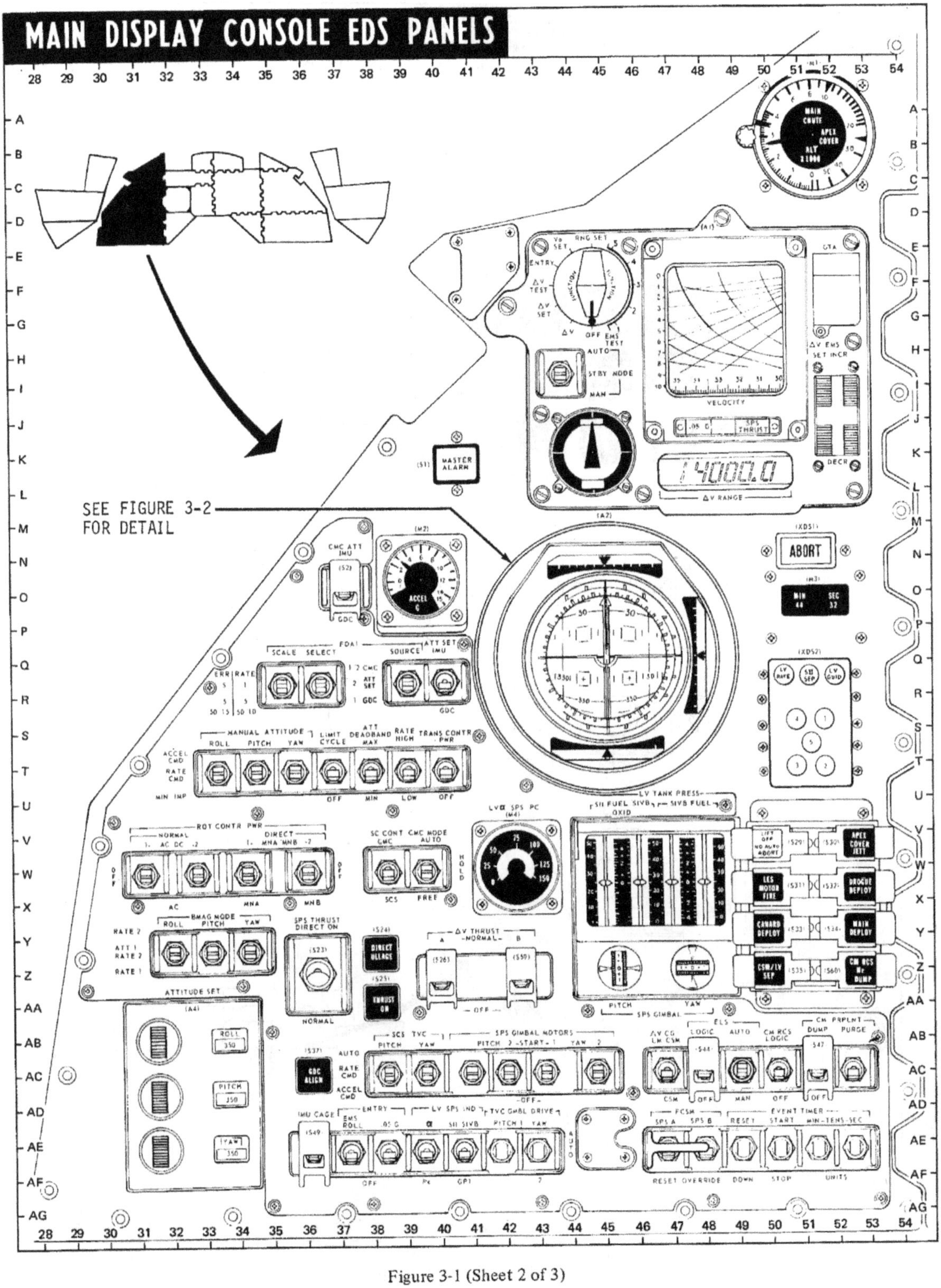

# MAIN DISPLAY CONSOLE EDS PANELS

SEE FIGURE 3-2
FOR DETAIL

Figure 3-1 (Sheet 2 of 3)

# MAIN DISPLAY CONSOLE EDS PANELS

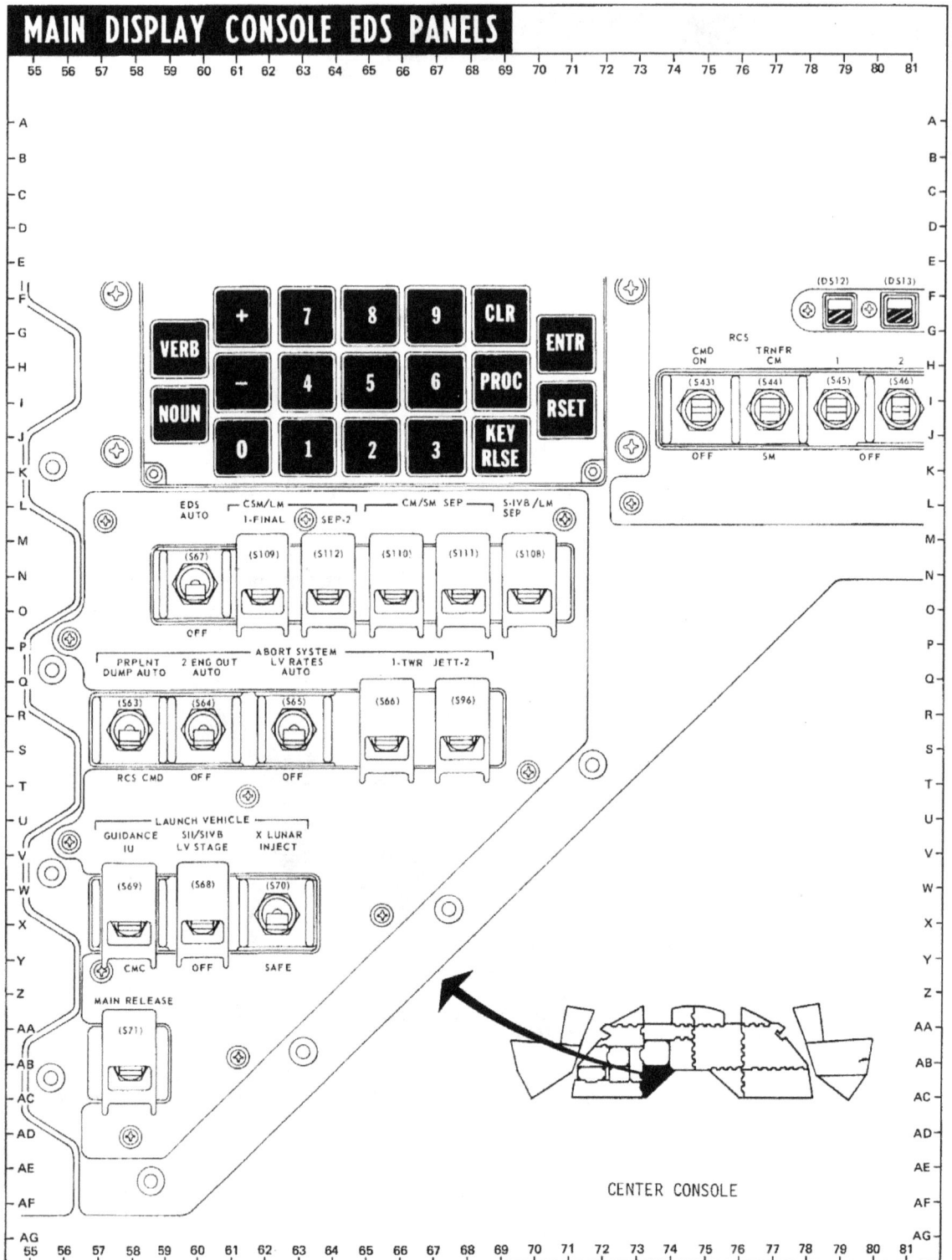

Figure 3-1 (Sheet 3 of 3)

# FLIGHT DIRECTOR ATTITUDE INDICATOR

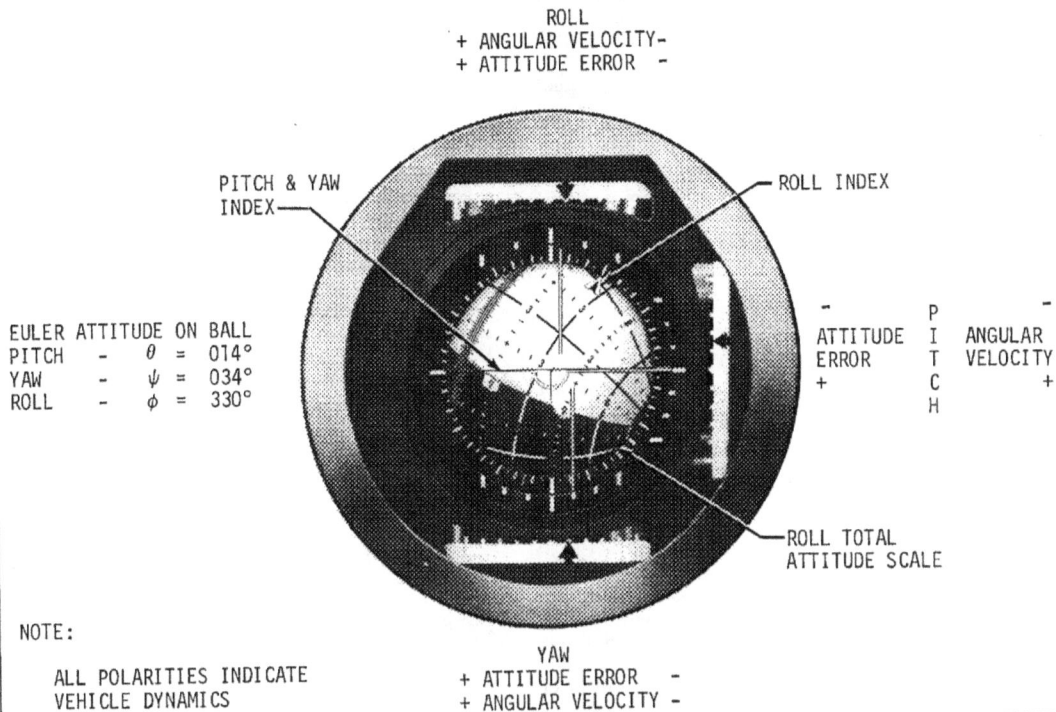

ROLL
+ ANGULAR VELOCITY-
+ ATTITUDE ERROR -

PITCH & YAW
INDEX

ROLL INDEX

EULER ATTITUDE ON BALL
PITCH  -  $\theta$ =  014°
YAW    -  $\psi$ =  034°
ROLL   -  $\phi$ =  330°

ATTITUDE   I   ANGULAR
ERROR      T   VELOCITY
+          C     +
           H

P

ROLL TOTAL
ATTITUDE SCALE

NOTE:

ALL POLARITIES INDICATE
VEHICLE DYNAMICS

YAW
+ ATTITUDE ERROR -
+ ANGULAR VELOCITY -

Figure 3-2

A signal is sent from the LVDA to activate the LV GUID light (Q-52, figure 3-1) in the CM at the same time the flight control commands are inhibited. It is a single cue for abort. Second cues will be provided by the LV RATE light (only when the Auto Abort system is on) and by the FDAI (Q-45), angle of attack ($q\alpha$) (W-42) meter and/or ground information.

## LIFTOFF/NO AUTO ABORT LIGHTS

The LIFTOFF and NO AUTO ABORT lights (V-50, figure 3-1) are independent indications contained in one switch/light assembly.

The LIFTOFF light ON indicates that vehicle release has been commanded and that the IU umbilical has ejected. The S/C digital event timer is started by the same function. The LIFTOFF light is turned OFF at S-IC CECO.

The NO AUTO ABORT light ON indicates that one or both of the spacecraft sequencers did not enable automatic abort capability at liftoff. Automatic abort capability can be enabled by pressing the switch/light pushbutton. If the light remains ON, then the crew must be prepared to back up the automatic abort manually. The NO AUTO ABORT light is also turned OFF at S-IC inboard engine cutoff.

## WARNING

If the NO AUTO ABORT pushbutton is depressed at T-0 and a pad shutdown should occur, a pad abort will result.

### ABORT LIGHT

The ABORT light (N-51, figure 3-1) may be illuminated by ground command from the Flight Director, the Mission Control Center (MCC) Booster Systems Engineer, the Flight Dynamics Officer, the Complex 39 Launch Operations Manager (until tower clearance at +10 seconds), or in conjunction with range safety booster engine cutoff and destruct action. The ABORT light ON constitutes one abort cue. An RF voice abort request constitutes one abort cue.

NOTE

Pilot abort action is required prior to receipt of an ABORT light or a voice command for a large percentage of the time critical launch vehicle malfunctions, particularly at liftoff and staging.

### ANGLE OF ATTACK METER

The angle of attack ($q\alpha$) meter (W-42, figure 3-1) is time shared with service propulsion system (SPS) chamber pressure. the $q\alpha$ display is a pitch and yaw vector summed angle-of-attack/dynamic pressure product ($q\alpha$). It is expressed in percentage of total pressure for predicted launch vehicle breakup (abort limit equals 100%). It is effective as an abort parameter only during the high q flight region from +50 seconds to +1 minute 40 seconds.

Except as stated above, during ascent, the $q\alpha$ meter provides trend information on launch vehicle flight performance and provides a secondary cue for slow-rate guidance and control malfunctions. Primary cues for guidance and control malfunctions will be provided by the FDAI, physiological cues, and/or MCC callout.

Nominal angle of attack meter indications should not exceed 25%. Expected values based on actual winds aloft will be provided by MCC prior to launch.

### ACCELEROMETER

The accelerometer (N-39, figure 3-1) indicates longitudinal acceleration/deceleration. It provides a secondary cue for certain engine failures and is a gross indication of launch vehicle performance. The accelerometer also provides a readout of G-forces during reentry.

### S-II SEP LIGHT

With S-IC/S-II staging, the S-II SEP light (Q-51, figure 3-1) will illuminate. The light will go out approximately 30 seconds later when the interstage structure is jettisoned. A severe overheating problem will occur if the structure is not jettisoned at the nominal time. Under the worst conditions, abort limits will be reached within 25 seconds from nominal jettison time. Confirmation from Mission Control of interstage failure to jettison serves as the second abort cue.

During the earth orbit phase of the mission the S-II SEP light is again used. It is turned ON to indicate the beginning of restart preparations at $T_6$+0.1 seconds. It is turned OFF at $T_6$+41.6 seconds to indicate $O_2H_2$ burner ignition. It is turned ON again at $T_6$+8 minutes 40 seconds and OFF again at $T_6$+9 minutes 20 seconds to indicate translunar injection commit (engine start-10 seconds).

### ALTIMETER

Due to dynamic pressure, static source location, and instrument error the altimeter (A-51, figure 3-1) is not considered to be an accurate instrument during the launch phase.

The primary function of the altimeter is to provide an adjustable (set for barometric pressure on launch date) reference for parachute deployment for pad/near LES aborts. However, the aerodynamic shape of the CM coupled with the static source location produces errors up to 1300 feet. Therefore, the main parachutes must be deployed at an indicated 3800 feet (depends on launch day setting) to ensure deployment of 2500 feet true altitude.

### EVENT TIMER

The event timer (AE-51, figure 3-1) is critical because it is the primary cue for the transition of abort modes, manual sequenced events, monitoring roll and pitch program, staging, and S-IVB insertion cutoff. The event timer is started by the

liftoff command which enables automatic aborts. The command pilot should be prepared to manually back up its start to assure timer operation.

The event timer is reset to zero automatically with abort initiation.

### MASTER ALARM LIGHT

There are three MASTER ALARM lights, one on Main Display Panel 1 (K-41, figure 3-1) one on Main Display Panel 3 and one in the lower Equipment Bay. The three MASTER ALARM lights ON alert the flight crew to critical spacecraft failures or out-of-tolerance conditions identified in the caution and warning light array. After extinguishing the alarm lights, action should be initiated to correct the failed or out-of-tolerance subsystem. If crew remedial action does not correct the affected subsystem, then an abort decision must be made based on other contingencies. Secondary abort cues will come from subsystem displays, ground verification, and physiological indications.

NOTE

The Commander's MASTER ALARM light (K-41) will not illuminate during the launch phase, but the other two MASTER ALARM lights can illuminate and their alarm tone can sound.

### EMERGENCY DETECTION SYSTEM CONTROL

### EDS AUTO SWITCH

The EDS AUTO switch (N-60, figure 3-1) is the master switch for EDS initiated automatic abort. When placed in the AUTO position (prior to liftoff), an automatic abort will be initiated if:

1. a LV structural failure occurs between the IU and the CSM,

2. two or more S-IC engines drop below 90% of rated thrust,

3. LV rates exceed 4 degrees per second in pitch or yaw or 20 degrees per second in roll.

The two engine out and LV rate portions of the auto abort system can be manually disabled, individually, by the crew. However, they are automatically disabled by the LV sequencer prior to center engine cutoff.

### PRPLNT SWITCH

The PRPLNT switch (R-58, figure 3-1) is normally in the DUMP AUTO position prior to liftoff in order to automatically dump the CM reaction control system (RCS) propellants, and fire the pitch control (PC) motor if an abort is initiated during the first 42 seconds of the mission. The propellant dump and PC motor are inhibited by the SC sequencer at 42 seconds. The switch in the RCS CMD position will inhibit propellant dump and PC motor firing at any time.

## TOWER JETTISON SWITCHES

Either of two redundant Tower Jettison switches (R-66, figure 3-1) can be used to fire the explosive bolts and the tower jettison motor. The appropriate relays are also de-energized so that if an abort is commanded, the SPS abort sequence and not the LES sequence will occur. The switches are momentary to the TWR JETT position. Both switches should be activated to ensure that redundant signals are initiated.

No other automatic functions will occur upon activation of the Tower Jettison switches.

## CM/SM SEP SWITCHES

The two CM/SM SEP switches (N-66, figure 3-1) are redundant, momentary ON, guarded switches, spring loaded to the OFF position. They are used by the Command Pilot to accomplish CM/SM separation when required. Both switches should be activated to ensure that redundant signals are initiated.

These switches can also be used to initiate an LES abort in case of a failure in either the EDS or the translational controller. All normal post-abort events will then proceed automatically. However, the CANARD DEPLOY (Y-50) pushbutton should be depressed 11 seconds after abort initiation, because canard deployment and subsequent events will not occur if the failure was in the EDS instead of the translational controller. If the CANARD DEPLOY pushbutton is depressed, all automatic functions from that point on will proceed normally.

## S-II/S-IVB LV STAGE SWITCH

S-II/S-IVB LV STAGE switch (X-60, figure 3-1) activation applies a signal to the LVDA to initiate the S-II/S-IVB staging sequence. This capability is provided to allow the crew to manually upstage from a slowly diverging failure of the S-II stage. If the S-IVB does not have a sufficient velocity increment ($\Delta V$) available for orbit insertion, an abort can be performed at predetermined locations.

The S-II/S-IVB LV STAGE switch can also be used to manually initiate cutoff of the S-IVB stage.

## S-IVB/LM SEP SWITCH

Activation on the S-IVB/LM SEP switch (N-70, figure 3-1) will cause separation, by ordnance, of tension ties securing the LM legs to the SLA and start two parallel time delays of 0.03 seconds each. At expiration of the time delay, the LM tension tie firing circuit is deadfaced and the umbilical guillotine is fired.

## CSM/LM FINAL SEP SWITCHES

The LM docking ring is simultaneously jettisoned with the TWR JETT command during an LES abort. During a normal entry or an SPS abort, the ring must be jettisoned by actuation of either CSM/LM FINAL SEP switch (N-63, figure 3-1). Failure to jettison the ring could possibly hamper normal earth landing system (ELS) functions.

## ELS AUTO/MAN SWITCH

The ELS AUTO/MAN switch (AC-49, figure 3-1) is used in conjunction with either the automatically initiated ELS sequence (post-abort) or with the manually initiated sequence (ELS LOGIC ON). The switch is normally in the AUTO position as it will inhibit all ELS functions when in the MAN position.

## ELS LOGIC SWITCH

The ELS LOGIC switch (AC-48, figure 3-1) is a switch, guarded to OFF, which should only be activated during normal reentry or following an SPS abort, and then only below 45,000 feet altitude. If the switch is activated at any time below 30,000 feet (pressure altitude), the landing sequence will commence, i.e., LES and apex cover jettison and drogue deployment. If activated below 10,000 feet altitude, the main chutes will also deploy.

ELS LOGIC is automatically enabled following any manual or auto EDS initiated LES abort. It should be manually backed up if time permits.

### WARNING

Activation of ELS LOGIC switch below 30,000 feet altitude will initiate landing sequence, i.e., LES and apex cover jettison, and drogue deployment.

## CM RCS PRESS

Any time the CM is to be separated from the SM, the CM RCS must be pressurized. The normal sequence of events for an abort or normal CM/SM SEP is to automatically deadface the umbilicals, pressurize the CM RCS, and then separate the CM/SM. However, if the automatic pressurization fails, the CM RCS can be pressurized by the use of the guarded switch (not shown) located on panel 2.

## MAIN RELEASE

The MAIN RELEASE switch (AB-57, figure 3-1) is guarded to the down position. It is moved to the up position to manually release the main chutes after the Command Module has landed. No automatic backup is provided. This switch is armed by the ELS LOGIC switch ON and the 10k barometric switches closed (below 10,000 feet altitude).

### NOTE

The ELS AUTO switch must be in the AUTO position to allow the 14 second timer to expire before the MAIN CHUTE RELEASE switch will operate.

## ABORT CONTROLS

### TRANSLATIONAL CONTROLLER

Manual aborts will be commanded by counterclockwise (CCW) rotation of the translational controller T-handle. Clockwise (CW) rotation will transfer SC control from the command module computer (CMC) to the stability control system (SCS).

For LES aborts, the CCW position sends redundant engine cutoff commands (engine cutoff from the SC is inhibited for the first 30 seconds of flight) to the launch vehicle, initiates CM/SM separation, fires the LES motors, resets the sequencer, and initiates the post abort sequence.

For Service Propulsion System (SPS) aborts, the CCW rotation commands LV engine cutoff, resets the spacecraft sequencer and initiates the CSM/LV separation sequence.

The T-handle also provides CSM translation control along one or more axes. The control is mounted approximately parallel to the SC axis; therefore, T-handle movement will cause corresponding SC translation. Translation in the +X axis can also be accomplished by use of the direct ullage pushbutton; however, rate damping is not available when using this method.

## SEQUENCER EVENT MANUAL PUSHBUTTON

The LES MOTOR FIRE, CANARD DEPLOY, CSM/LV SEP, APEX COVER JETT, DROGUE DEPLOY, MAIN DEPLOY, and CM RCS He DUMP, pushbuttons (X-51, figure 3-1) provide backup of sequenced events for both abort and normal reentry situations.

The MAIN DEPLOY pushbutton is the primary method of deploying the main parachutes for pad/near pad aborts to assure deceleration to terminal velocity at touchdown (downrange tipover).

## SPS SWITCHES AND DISPLAYS

### DIRECT ULLAGE Pushbutton

When the DIRECT ULLAGE pushbutton (Y-38, figure 3-1) is depressed, a +X translation utilizing all four quads results. It is the backup method for ullage maneuvers prior to an SPS burn (the prime method for ullage is the translational controller). The DIRECT ULLAGE switch is momentary and must be held until the ullage maneuver is complete. It will not provide rate damping.

### THRUST ON Pushbutton

The THRUST ON pushbutton (AA-38, figure 3-1) can be used to start the SPS engine under the following conditions:

1. S/C control is in the SCS mode.

2. Ullage is provided.

3. Either of two ΔV THRUST switches (Z-41, figure 3-1) are in the NORMAL position

NOTE

Both switches must be OFF to shut off the engine.

4. When backup for guidance and navigation start command is required.

### SPS Engine Shutdown

The SPS engine can be shut off (when fired as described above) in the following manner:

1. Flight combustion stability monitor (FCSM) shuts it down automatically.

2. ΔV = 0 (SCS or MTVC).

3. ΔV THRUST switches (both) OFF.

NOTE

The SPS THRUST light (J-49, figure 3-1) will illuminate when the engine is firing.

### SPS THRUST DIRECT Switch

The SPS THRUST DIRECT Switch (Z-36, figure 3-1) is a two-position, lever lock, toggle switch. The ON position provides a ground for the solenoid valve power and all of the SCS logic. The engine must be turned off manually by placing the SPS THRUST DIRECT switch in the NORMAL position. The ΔV THRUST switches must be in the NORMAL position (at least one) to apply power to the solenoids for the SPS THRUST DIRECT switch to operate.

**WARNING**

The SPS THRUST DIRECT switch is a single point failure when the ΔV THRUST switches are in the NORMAL position.

### SPS Gimbal Motors/Indicators

Four gimbal motors control the SPS engine position in the pitch and yaw planes (two in each plane). These motors are activated by four switches (AC-43, figure 3-1). The motors should be activated one at a time due to excessive current drain during the start process.

The gimbal thumbwheels (Z-46, figure 3-1) can be used to position the gimbals to the desired attitude as shown on the indicators. The indicators are analog displays time-shared with the booster fuel and oxidizer pressure readings.

The other methods of controlling the gimbal movement is through the hand controller in the MTVC mode or by automatic SCS logic.

### ΔV THRUST (Prevalves and Logic)

The two guarded ΔV THRUST switches (Z-41, figure 3-1) apply power to the SPS solenoid prevalves and to the SCS logic for SPS ignition. These switches must be in the NORMAL position before the SPS engine can be started, even by the SPS THRUST DIRECT switch.

**WARNING**

Both switches must be in the OFF position to stop the engine.

### SCS TVC Switches

The SCS TVC switches (AC-39, figure 3-1) provide thrust vector control only in the SCS mode. The pitch and yaw channels can be used independently, i.e., pitch control in SCS AUTO position and yaw in either RATE CMD or ACCEL

CMD position. The three available modes are:

1. AUTO: The TVC is directed by the SCS electronics.

2. RATE CMD: MTV with rate damping included.

3. ACCEL CMD: MTVC without rate damping.

## EMS MODE and EMS FUNCTION Switches

In order for the ΔV counter to operate during an SPS burn, the EMS MODE switch (H-44, figure 3-1) and the EMS FUNCTION switch(F-45) must be in the following positions:

1. EMS MODE - AUTO.

2. EMS FUNCTION - ΔV.

To set the ΔV counter for a desired ΔV burn the switches would be as follows:

1. EMS MODE - AUTO

2. EMS FUNCTION - ΔV SET

The ΔV/EMS SET switch (I-52, figure 3-1) is then used to place the desired quantity on the ΔV display.

## SCS SYSTEM SWITCHES

### AUTO RCS SELECT Switches

Power to the RCS control box assembly is controlled by 16 switches (I-15, figure 3-1). Individual engines may be enabled or disabled as required. Power to the attitude control logic is also controlled in this manner, which thereby controls all attitude hold and/or maneuvering capability using SCS electronics (automatic coils). The direct solenoids are not affected as all SCS electronics are bypassed by activation of the DIRECT RCS switches.

NOTE

The automatic coils cannot be activated until the RCS ENABLE is activated either by the MESC or manually.

### DIRECT Switches

Two DIRECT switches (W-35, figure 3-1) provide for manual control of the SM RCS engines. Switch 1 controls power to the direct solenoid switches in rotational controller 1 and switch 2 controls power to the direct solenoid switches in rotational controller 2. In the down position switch 1 receives power from MNA and switch 2 receives power from MNB. In the up position both switches receive power from both MNA and MNB. Manual control is achieved by positioning the rotational control hardover to engage the direct solenoids for the desired axis change.

### ATT SET Switch

The ATT SET switch (Q-40, figure 3-1) selects the source of total attitude for the ATT SET resolvers as outlined below.

| Position | Function | Description |
|---|---|---|
| UP | IMU | Applies inertial measurement unit (IMU) gimbal resolver signal to ATT SET resolvers. FDAI error needles display differences. Needles are zeroed by maneuvering S/C or by moving the ATT SET dials. |
| DOWN | GDC | Applies GDC resolver signal to ATT SET resolvers. FDAI error needles display differences resolved into body coordinates. Needles zeroed by moving S/C or ATT SET dials. New attitude reference is established by depressing GDC ALIGN button. This will cause GDC to drive to null the error; hence, the GDC and ball go to ATT SET dial value. |

### MANUAL ATTITUDE Switches

The three MANUAL ATTITUDE switches (T-34, figure 3-1) are only operative when the S/C is in the SCS mode of operation.

| Position | Description |
|---|---|
| ACCEL CMD | Provides direct RCS firing as a result of moving the rotational controller out of detent (2.5degrees) to apply direct inputs to the solenoid driver amplifiers. |
| RATE CMD | Provides proportional rate command from rotational controller with inputs from the BMAG's in a rate configuration. |
| MIN IMP | Provides minimum impulse capability through the rotational controller. |

### LIMIT CYCLE Switch

The LIMIT CYCLE switch, when placed in the LIMIT CYCLE position, inserts a psuedo-rate function which provides the capability of maintaining low SC rates while holding the SC attitude within the selected deadband limits (limit cycling). This is accomplished by pulse-width modulation of the switching amplifier outputs. Instead of driving the SC from limit-to-limit with high rates by firing the RCS engines all the time, the engines are fired in spurts proportional in length and repetition rate to the switching amplifier outputs.

Extremely small attitude corrections could be commanded which would cause the pulse-width of the resulting output command to be of too short a duration to activate the RCS solenoids. A one-shot multivibrator is connected in parallel to ensure a long enough pulse to fire the engines.

## RATE and ATT DEADBAND Switches

The switching amplifier deadband can be interpreted as a rate or an attitude (minimum) deadband. The deadband limits are a function of the RATE switch (T-39, figure 3-11). An additional deadband can be enabled in the attitude control loop with the ATT DEADBAND switch (T-38, figure 3-1). See figure 3-3 for relative rates.

| ATTITUDE DEADBAND SWITCH POSITION | | | |
|---|---|---|---|
| RATE SWITCH POSITION | RATE DEADBAND °/SEC | ATT DEADBAND SWITCH POSITION | |
| | | MINIMUM | MAXIMUM |
| LOW | +0.2 | +0.2° | +4.2° |
| HIGH | +2.0 | +2.0° | +8.0° |

Figure 3-3

The rate commanded by a constant stick deflection (Proportional Rate Mode only) is a function of the RATE switch position. The rates commanded at maximum stick deflection (soft stop) are shown in figure 3-4.

| MAXIMUM PROPORTIONAL RATE COMMAND | | |
|---|---|---|
| RATE SWITCH POSITION | MAXIMUM PROPORTIONAL RATE COMMAND | |
| | PITCH AND YAW | ROLL |
| LOW | 0.65°/sec | 0.65°/sec |
| HIGH | 7.0°/sec | 20.0 °/sec |

Figure 3-4

## SC CONT Switch

The SC CONT switch (W-38, figure 3-1) selects the spacecraft control as listed below:

| Position | Description |
|---|---|
| CMC | Selects the G&N system computer controlled SC attitude and TVC through the digital auto-pilot. An auto-pilot control discrete is also applied to CMC. |
| SCS | The SCS system controls the SC attitude and TVC. |

## BMAG MODE Switches

The BMAG MODE switches (Y-33, figure 3-1) select displays for the FDAI using SCS inputs.

| Position | Description |
|---|---|
| RATE 2 | BMAG Set No. 2 provides the rate displays on the FDAI. There is no Body Mounted Attitude Gyro (BMAG) attitude reference available. |
| ATT 1 | BMAG Set No. 1 provides attitude reference on the FDAI, while |
| RATE 2 | Set No. 2 provides the rate display. |
| RATE 1 | BMAG Set No. 1 provides the rate displays on the FDAI. There is no BMAG attitude reference available. |

## ENTRY EMS ROLL Switch

The ENTRY EMS ROLL switch (AE-37, figure 3-1) enables the EMS roll display for the earth reentry phase of the flight.

## ENTRY, .05 G Switch

Illumination of the .05 G light located on the EMS panel is the cue for the crew to actuate the .05 G switch (AE-38, figure 3-1). During atmospheric reentry (after .05 G), the SC is maneuvered about the stability roll axis rather than the body roll axis. Consequently, the yaw rate gyro generates an undesirable signal. By coupling a component of the roll signal into the yaw channel, the undesirable signal is cancelled. The .05 G switch performs this coupling function.

## EMS DISPLAYS

### Threshold Indicator (.05 G Light)

The threshold indicator (.05G light) (J-48, figure 3-1) provides the first visual indication of total acceleration sensed at the reentry threshold (approximately 290,000 feet). Accelerometer output is fed to a comparison network and will illuminate the .05 G lamp when the acceleration reaches .05 G. The light will come on not less than 0.5 seconds or more than 1.5 seconds after the acceleration reaches .05 G and turns off when it falls below .02 G (skipout).

### Corridor Indicators

By sensing the total acceleration buildup over a given period of time, the reentry flight path angle can be evaluated. This data is essential to determine whether or not the entry angle is steep enough to prevent superorbital "skipout."

The two corridor indicator lights (J-45 and L-45, figure 3-1) are located on the face of the roll stability indicator (K-45).

If the acceleration level is greater than 0.2 G at the end of a ten second period after threshold (.05 G light ON), the upper light will be illuminated. It remains ON until the G-level reaches 2 G's and then goes OFF. The lower light illuminates if the acceleration is equal to or less than 0.2 G at the end of a ten second period after threshold. This indicates a shallow entry angle and that the lift vector should be down for controlled entry, i.e., skipout will occur.

### Roll Stability Indicator

The roll stability indicator (K-45, figure 3-1) provides a visual indication of the roll attitude of the CM about the stability axis. Each revolution of the indicator represents 360 degrees of vehicle rotation. The display is capable of continuous rotation in either direction. The pointer up position (0 degrees) indicates maximum lift-up vector (positive lift) and pointer down (180 degrees) indicates maximum lift-down vector (negative lift).

## G-V Plotter

The G-V plotter assembly (G-49, figure 3-1) consists of a scroll of mylar tape and a G-indicating stylus. The tape is driven from right to left by pulses which are proportional to the acceleration along the velocity vector. The stylus which scribes a coating on the back of the mylar scroll, is driven in the vertical direction in proportion to the total acceleration.

The front surface of the mylar scroll is imprinted with patterns consisting of "high-G-rays" and "exit rays." The

"high-G-rays" must be monitored from initial entry velocity down to 4000 feet per second. The "exit rays" are significant only between the entry velocity and circular orbit velocity and are, therefore, only displayed on that portion of the pattern.

The imprinted "high-G-rays" and "exit rays" enable detection of primary guidance failures of the type that would result in either atmospheric exits at supercircular speeds or excessive load factors at any speed. The slope of the G-V trace is visually compared with these rays. If the trace becomes tangent to any of these rays, it indicates a guidance malfunction and the need for manual takeover.

## EMS FUNCTION Switch

The EMS FUNCTION switch (F-45, figure 3-1) is a 12 position selector mode switch, used as outlined in figure 3-5.

| EMS FUNCTION SWITCH OPERATION | | | |
|---|---|---|---|
| Operational Mode | Switch Selection | Switch Position | Description |
| ΔV Mode | Start at ΔV and rotate clockwise | ΔV | Operational mode for monitoring ΔV maneuvers |
| | | ΔV Set | Establish circuitry for slewing ΔV counter for self test or as operational |
| | | ΔV Test | Operational mode for self test of the ΔV subsystem |
| Self Test and Entry Mode | Start at No. 1 and rotate counter-clockwise | No. 1 | Tests lower trip point of .05 G threshold comparator |
| | | No. 2 | Tests higher trip point of .05 G threshold comparator |
| | | No. 3 | Tests lower trip point of corridor verification comparator |
| | | No. 4 | Tests velocity integration circuitry, g-servo circuitry, G-V plotter, and the range-to-go subsystem |
| | | No. 5 | Tests higher trip point of corridor verification comparator |
| | | RNG Set | Establish circuitry for slewing range-to-go counter for operational and test modes |
| | | Vo Set | Establish circuitry for slewing G-V plotter scroll for operational mode |
| | | Entry | Operational mode for monitoring entry mode |
| | | OFF | Turns OFF all power except to the SPS thrust light and switch lighting. |

Figure 3-5

## EMS MODE Switch

The EMS MODE switch (H-44, figure 3-1) performs the following functions in the positions indicated:

### AUTO

1. Acts as backup display for G&N entry.

2. Initiates the function selected by the EMS FUNCTION switch.

### STBY

1. Resets circuits following tests.

2. Removes power if EMS FUNCTION switch is OFF.

### MAN

1. Position for manual entry and TVC modes, or auto entry backup display.

2. Does not permit negative acceleration spikes into countdown circuits.

## ABORT MODES AND LIMITS

The abort mode and limits listed in figures 3-6 and 3-7 are based on a nominal launch trajectory. More specific times can be obtained from current mission documentation.

## NOMINAL LAUNCH CALLOUTS

The nominal launch callouts are listed in figure 3-8 for the boost phase only.

## EMERGENCY MODES

Aborts performed during the ascent phase of the mission will be performed by either of the two following methods:

## LAUNCH ESCAPE SYSTEM

The Launch Escape System (LES) consists of a solid propellant launch escape (LE) motor used to propel the CM a safe distance from the launch vehicle, a tower jettison motor, and a canard subsystem. A complete description on use of the system can be found in the specific mission Abort Summary Document (ASD). A brief description is as follows:

### Mode IA Low Altitude Mode

In Mode IA, a pitch control (PC) motor is mounted normal to the LE motor to propel the vehicle downrange to ensure water landing and escape the "fireball." The CM RCS propellants are dumped through the aft heat shield during this mode to prevent a possible fire source at landing.

The automatic sequence of major events from abort initiation is as follows:

| Time | Event |
|---|---|
| 00:00 | Abort |
|  | Ox Rapid Dump |
|  | LE and PC Motor Fire |
| 00:05 | Fuel Rapid Dump |
| 00:11 | Canards Deploy |
| 00:14 | ELS Arm |
| 00:14.4 | Apex Cover Jett |
| 00:16 | Drogue Deploy |
| 00:18 | He Purge |
| 00:28 | Main Deploy |

| APOLLO ABORT MODES | | |
|---|---|---|
| Time Period | Mode | Description |
| Pad to 42 sec | Mode IA | LET<br>Low Alt |
| 42 Sec to 1 min 50 sec | Mode IB | LET<br>MED Alt |
| 1 min 50 sec to 3 min 07 sec | Mode IC | LET<br>High Alt |
| 3 min 07 sec to 10 min 30 sec | Mode II | Full Lift |
| 10 min 30 sec to insertion | Mode III<br>for CSM<br>No Go | SPS Retro<br>Half Lift |
| 10 min 7 sec to insertion | Mode IV<br>for CSM<br>Go | SPS to Orbit |

Figure 3-6

## ABORT LIMITS

RATES

    1.  Pitch and Yaw

        L/O to S-IC/S-II Staging        4° per second

        Staging to SECO             9.2° per second
        (Excluding staging)

    2.  Roll

        L/O to SECO                20° per second

MAX Q REGION

    NOTE:  The following limits represent single cues and are restricted
            to the time period from 50 seconds to 1 minute 40 seconds.
            Abort action should be taken only after both have reached
            threshold.

    1.  Angle of Attack = 100 percent

    2.  Roll Error = 6 degrees

AUTOMATIC ABORT LIMITS
L/O until deactivate (time to be determined)

    1.  Pitch and Yaw           4.0° ± 0.5° per second

        Roll                   20.0° ± 0.5° per second

    2.  Any Two Engines Out

    NOTE:  Between L/O and 2 + 0, switch TWO ENG OUT AUTO to OFF
            following confirmation of ONE ENG OUT.

    3.  CM/IU breakup

ENGINE FAILURE
(L/O to CECO)

    1.  One Engine Out           Continue Mission

    2.  Simultaneous Loss two
        or more engines         Abort

    3.  Second engine loss fol-
        lowing confirmation of
        one engine out         Continue Mission

S-IVB TANK PRESSURE LIMITS
(L/O to CSM/LV SEP)

    $\Delta P$  $LH_2 > LO_2$ = 26 PSID

        $LO_2 > LH_2$ = 36 PSID

Figure 3-7

| NOMINAL LAUNCH PHASE VOICE CALLOUTS (BOOST ONLY) | | | |
|------|---------|--------|-------|
| TIME | STATION | REPORT | EVENT |
| -0:09 | LCC | IGNITION | S-IC IGNITION |
| 0:00 | LCC | LIFT-OFF | UMBILICAL DISCONNECT |
| 0:01 | CDR | LIFT-OFF | CMD TO P11 DET START |
| 0:11 | CDR | ROLL COMMENCE | ROLL PROGRAM STARTS |
| 0:21 | CDR | PITCH TRACKING | PITCH RATE DETECTION |
| 0:29 | CDR | ROLL COMPLETE | ROLL COMPLETE |
| 0:42 | MCC | MARK, MODE IB | PRPLNT DUMP - RCS CMD |
| 1:50 | MCC | MARK, MODE IC | h = 100,000 FT, 16.5 NM |
| 2:00 | CDR | EDS MANUAL | EDS RATES - OFF |
|      |     |           | EDS ENG - OFF |
|      |     |           | EDS AOA - Pc |
| 2:00 | MCC | GO/NO GO FOR STAGING | STAGING STATUS-TWR JETT STATUS IF REQUIRED |
| 2:00 | CDR | GO/NO GO FOR STAGING | |
| 2:05 | CDR | INBOARD OFF | S-IC INBOARD ENG - OFF |
| 2:31 | CDR | OUTBOARD OFF | S-IC OUTBOARD ENG - OFF |
| 2:32 | CDR | STAGING | S-II LIGHTS OFF |
| 2:33 |     |           | S-II IGNITION COMMAND |
| 2:36 | CDR | S-II 65% | S-II 65% |
| 3:01 | CDR | S-II SEP LIGHT OUT | S-II SEP LIGHT OUT |
| 3:07 | CDR | TOWER JETT | TOWER JETTISONED |
|      |     | MARK, MODE II | MAN ATT (P) - RATE CMD |
| 3:12 | CDR | GUID INITIATE | IGM STARTS |
| 4:00 | CDR | S/C GO/NO GO | |
|      | MCC | GUIDANCE GO/NO GO | IGM LOOKS GOOD |
| 4:30 | MCC | TRAJECTORY GO/NO GO | TRAJECTORY STATUS |
| 5:00 | CDR | S/C GO/NO GO | |
| 5:53 | MCC | S-IVB TO ORBIT CAPABILITY | |
| 6:00 | CDR | S/C GO/NO GO | |
| 7:00 | CDR | S/C GO/NO GO | |
| 8:00 | CDR | S/C GO/NO GO | |
| 8:20 | MCC | GO/NO GO FOR STAGING | STAGING STATUS |
| 8:40 | CDR | S-II OFF | S-II LIGHTS - ON |
| 8:41 | CDR | STAGING | S-IVB LIGHT - OFF |
| 8:45 | CDR | S-IVB IGNITION | S-IVB IGNITION |
| 8:46 | CDR | S-IVB 65% | S-IVB 65% |
| 9:00 | CDR | S/C GO/NO GO | |
| 9:45 | MCC | MODE IV | |
| 10:00 | MCC CDR | S/C GO/NO GO FOR ORBIT | |
| 11:21 | CDR | SECO | S-IVB LIGHT ON |
| 11:31 | MCC | INSERTION | |

Figure 3-8

The automatic sequence can be prevented, interrupted, or replaced by crew action.

### Mode IB Medium Altitude

Mode IB is essentially the same as Mode IA with the exception of deleting the rapid propellant dump and PC motor features. The canard subsystem was designed specifically for this altitude region to initiate a tumble in the pitch plane. The CM/tower combination CG is located such that the vehicle will stabilize (oscillations of ± 30 degrees) in the blunt-end-forward (BEF) configuration. Upon closure of barometric switches, the tower would be jettisoned and the parachutes automatically deployed.

As in Mode IA, the crew intervention can alter the sequence of events if desired.

### Mode IC High Altitude

During Mode IC the LV is above the atmosphere. Therefore, the canard subsystem cannot be used to induce a pitch rate to the vehicle. The crew will, therefore, introduce a five degree per second pitch rate into the system. The CM/tower combination will then stabilize BEF as in Mode IB. The ELS would likewise deploy the parachutes at the proper altitudes.

An alternate method (if the LV is stable at abort) is to jettison the tower manually and orient the CM to the reentry attitude. This method provides a more stable entry, but requires a functioning attitude reference.

## SERVICE PROPULSION SYSTEM

The Service Propulsion System (SPS) aborts utilize the Service Module SPS engine to propel the CSM combination away from the LV, maneuver to a planned landing area, or boost into a contingency orbit. The SPS abort modes are:

### Mode II

The SM RCS engines are used to propel the CSM away from the LV unless the vehicle is in danger of exploding or excessive tumble rates are present at LV/CSM separation. In these two cases the SPS engine would be used due to greater $\Delta V$ and attitude control authority. When the CMS is a safe distance and stable, the CM is separated from the SM and maneuvered to a reentry attitude. A normal entry procedure is followed from there.

### Mode III

The SPS engine is used to slow the CSM combination (retrograde maneuver) so as to land at a predetermined point in the Atlantic Ocean. The length of the SPS burn is dependent upon the time of abort initiation. Upon completion of the retro maneuver, the CM will separate from the SM, assume the reentry attitude, and follow normal entry procedures.

### Mode IV

The SPS engine can be used to make up for a deficiency in insertion velocity up to approximately 3000 feet per second. This is accomplished by holding the CSM in an inertial attitude and applying the needed $\Delta V$ with the SPS to acquire the acceptable orbital velocity. If the inertial attitude hold mode is inoperative, the crew can take over manual control and maneuver the vehicle using onboard data.

# SECTION IV

# S-IC STAGE

## TABLE OF CONTENTS

## INTRODUCTION

The S-IC stage (figure 4-1) is a cylindrical booster, 138 feet long and 33 feet in diameter, powered by five liquid propellant F-1 rocket engines. These engines develop a nominal sea level thrust of 7,650,000 pounds total, and have a burn time of 150.7 seconds. The stage dry weight is approximately 305,100 pounds and total weight at ground ignition is approximately 4,792,200 pounds.

The S-IC stage provides first stage boost of the Saturn V launch vehicle to an altitude of about 200,000 feet (approximately 38 miles), and provides acceleration to increase the vehicle velocity to 7,700 feet per second (approximately 4,560 knots). It then separates from the S-II stage and falls to earth about 360 nautical miles downrange.

The stage interfaces structurally and electrically with the S-II stage. It also interfaces structurally, electrically, and pneumatically with two umbilical service arms, three tail service masts, and certain electronic systems by antennae. The stage consists of: the structural airframe (figure 4-1); five F-1 engines; 890 vehicle monitoring points; electrical, pneumatic control, and emergency flight termination equipment; and eight retrorockets. The major systems of the stage are: structures, propulsion, environmental control, fluid power, pneumatic control, propellants, electrical, instrumentation, and ordnance.

## STRUCTURE

The S-IC structure design reflects the requirements of F-1 engines, propellants, control, instrumentation and interfacing systems. The structure maintains an ultimate factor of safety of at least 1.40 applied to limit load and a yield factor of safety of 1.10 on limit load. Aluminum alloy is the primary structural material. The major components, shown in figure 4-1, are the forward skirt, oxidizer tank, intertank section, fuel tank, and thrust structure.

### FORWARD SKIRT

The aft end of the forward skirt (figure 4-1) is attached to the oxidizer (lox) tank and the forward end interfaces with the S-II stage. The forward skirt has accomodations for the forward umbilical plate, electrical and electronic canisters, and the venting of the lox tank and interstage cavity. The skin panels, fabricated from 7075-T6 aluminum, are stiffened and strengthened by ring frames and stringers.

### OXIDIZER TANK

The 345,000 gallon lox tank is the structural link between the forward skirt and the intertank section. The cylindrical tank skin is stiffened by "integrally machined" T stiffeners. Ring baffles (figure 4-1) attached to the skin stiffeners stabilize the tank wall and serve to reduce lox sloshing. A cruciform baffle at the base of the tank serves to reduce both slosh and vortex action. Support for four helium bottles is provided by the ring baffles. The tank is a 2219-T87 aluminum alloy cylinder with ellipsoidal upper and lower bulkheads. The skin thickness is decreased in eight steps from .254 inches at the aft section to .190 inches at the forward section.

### INTERTANK SECTION

The intertank structure provides structural continuity between the lox and fuel tanks. This structure provides a lox fill and drain interface to the intertank umbilical. One opening vents the fuel tank. The corrugated skin panels and circumferential ring frames are fabricated from 7075-T6 aluminum.

### FUEL TANK

The 216,000 gallon fuel tank (figure 4-1) provides the load carrying structural link between the thrust and intertank structures. It is cylindrical with ellipsoidal upper and lower bulkheads. Antislosh ring baffles are located on the inside wall of the tank and antivortex cruciform baffles are located in the lower bulkhead area (figure 4-1). Five lox ducts (figure 4-1) run from the lox tank, through the RP-1 tank, and terminate at the F-1 engines. The fuel tank has an exclusion riser, made of a lightweight foam material, bonded to the lower bulkhead of the tank to minimize unusable residual fuel. The 2219-T87 aluminum skin thickness is decreased in four steps from .193 inches at the aft section to .170 inches at the forward section.

### THRUST STRUCTURE

The thrust structure assembly (figure 4-1) redistributes locally applied loads of the five F-1 engines into uniform loading about the periphery of the fuel tank. It also provides support for the five F-1 engines, engine accessories, base heat shield, engine fairings and fins, propellant lines, retrorockets, and environmental control ducts. The lower thrust ring has four holddown points which support the fully loaded Saturn/Apollo (approximately 6,000,000 pounds) and also, as necessary, restrain the vehicle from lifting off at full F-1 engine thrust. The skin segments are fabricated from 7075-T6 aluminum alloy.

The base heat shield is located at the base of the S-IC stage, forward of the engine gimbal plane. The heat shield provides thermal shielding for critical engine components and base

# S-IC STAGE

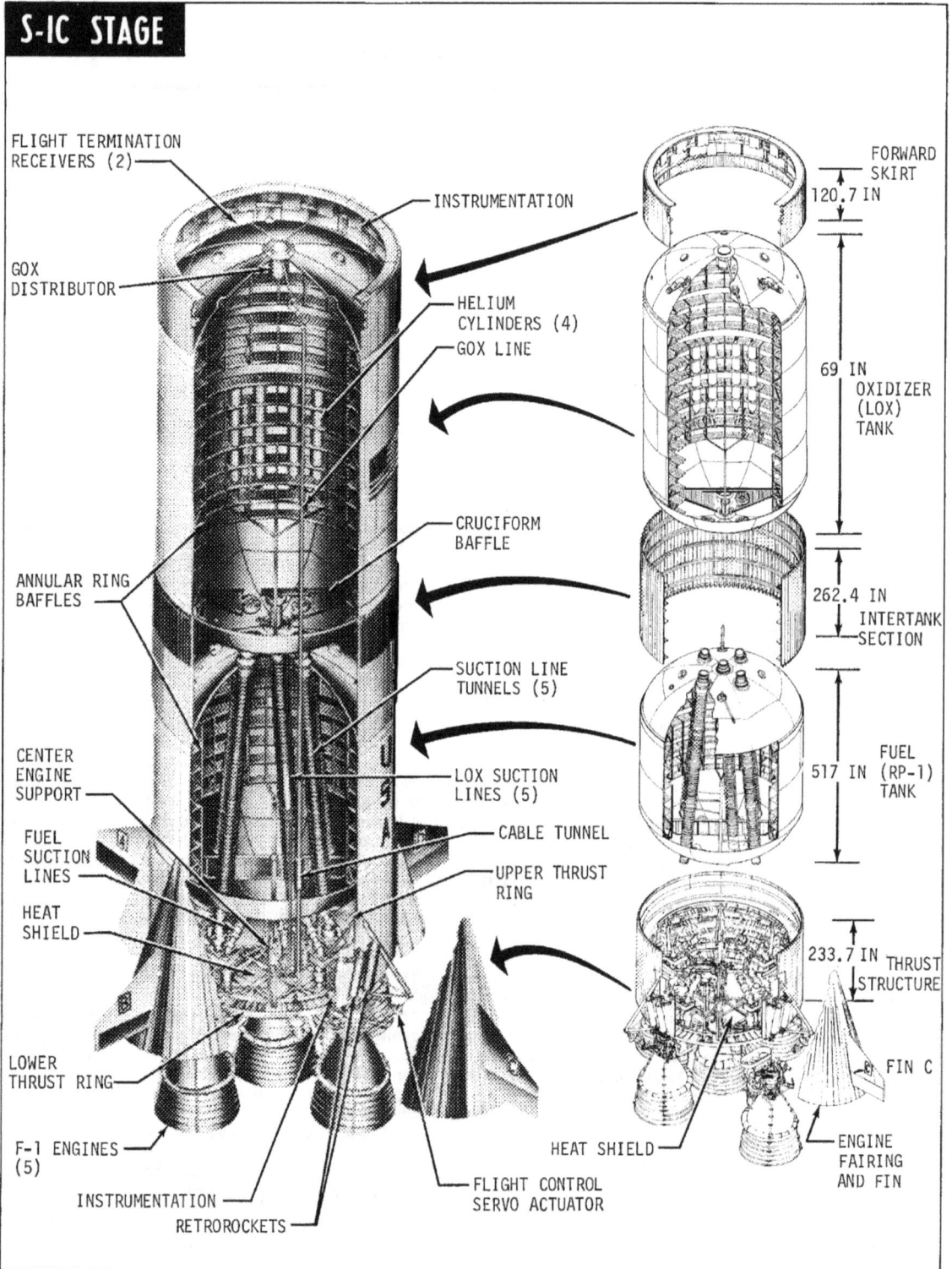

FLIGHT TERMINATION RECEIVERS (2)

GOX DISTRIBUTOR

INSTRUMENTATION

HELIUM CYLINDERS (4)

GOX LINE

CRUCIFORM BAFFLE

ANNULAR RING BAFFLES

SUCTION LINE TUNNELS (5)

CENTER ENGINE SUPPORT

LOX SUCTION LINES (5)

FUEL SUCTION LINES

CABLE TUNNEL

UPPER THRUST RING

HEAT SHIELD

LOWER THRUST RING

F-1 ENGINES (5)

INSTRUMENTATION

RETROROCKETS

FLIGHT CONTROL SERVO ACTUATOR

HEAT SHIELD

FORWARD SKIRT
120.7 IN

69 IN OXIDIZER (LOX) TANK

262.4 IN INTERTANK SECTION

FUEL (RP-1) TANK
517 IN

233.7 IN THRUST STRUCTURE

FIN C

ENGINE FAIRING AND FIN

Figure 4-1

region structural components for the duration of the flight. The heat shield panels are constructed of 15-7 PH stainless steel honeycomb, 1.00-inch thick, brazed to .010 inch steel face sheets.

Each outboard F-1 engine is protected from aerodynamic loading by a conically shaped engine fairing (figure 4-1). The fairings also house the retrorockets and the engine actuator supports. The fairing components are primarily titanium alloy below station 115.5 and aluminum alloy above this station. Four fixed, titanium covered, stabilizing fins augment the stability of the Saturn V vehicle.

## ENVIRONMENTAL CONTROL

During launch preparations the environmental control systems (ECS) protects the S-IC stage and stage equipment from temperature extremes, excessive humidity, and hazardous gases. Conditioned air, provided by the ground support equipment environmental control unit (GSE-ECU), is forced into the forward skirt and thrust structure where it is used as a temperature and humidity control medium. Approximately 20 minutes before the two upper stages are loaded with cryogenic fluids gaseous nitrogen (GN$_2$) replaces conditioned air and is introduced into the S-IC as the conditioning medium. The GN$_2$ flow terminates at umbilical disconnect since the system is not needed in flight.

### FORWARD SKIRT COMPARTMENT - ECS

The environmental control system distributes air or GN$_2$ to 18 electrical/electronic equipment module canisters located in the forward skirt. Onboard probes control the temperature of the flow medium to maintain canister temperature at 80 degrees (±20)F. Three phases of conditioning/purge flow are provided to compensate for the three environmental imbalances generated by ambient air changes, internal heat and lox load chill effects. Prior to launch the first phase supplies cool, conditioned air to the canisters when onboard electrical systems are energized before cryogenic loading. The second phase occurs when relatively warm GN$_2$ is substituted for the cool air to offset temperature differences caused by the cryogenic loading. The third phase uses a warmer GN$_2$ flow to offset temperature decreases caused by second stage J-2 engine thrust chamber chilldown. The air or GN$_2$ is vented from the canisters and overboard through vent openings in the forward skirt of the S-IC stage. A by-product of the use of the inert GN$_2$ purge is the reduction of gaseous hydrogen or oxygen concentrations.

### THRUST STRUCTURE COMPARTMENT - ECS

The environmental control system discharges air or GN$_2$ through 22 orificed duct outlets directly into the upper thrust structure compartment. The GSE-ECU supplies conditioned air at two umbilical couplings during launch preparations. At 20 minutes before cryogenic loading commences, the flow medium is switched to GN$_2$ and the temperature varied as necessary to maintain the compartment temperature at 80 degrees (±10)F. The controlled temperature compensates for temperature variations caused by ambient air change, chill effects from lox in the suction ducts, prevalves, and inter-connect ducts. The GN$_2$ prevents the oxygen concentration in the compartment from going above 6 percent.

## HAZARDOUS GAS DETECTION

The hazardous gas detection system monitors the atmosphere in the forward skirt and the thrust structure compartment of the S-IC (figures 4-2 and 4-3). This system is not redundant, however, large leaks may be detected by propellant pressure indications displayed in the Launch Control Center.

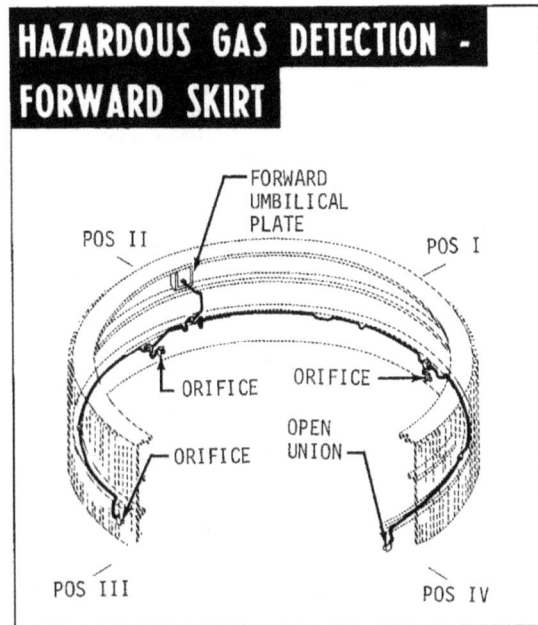

Figure 4-2

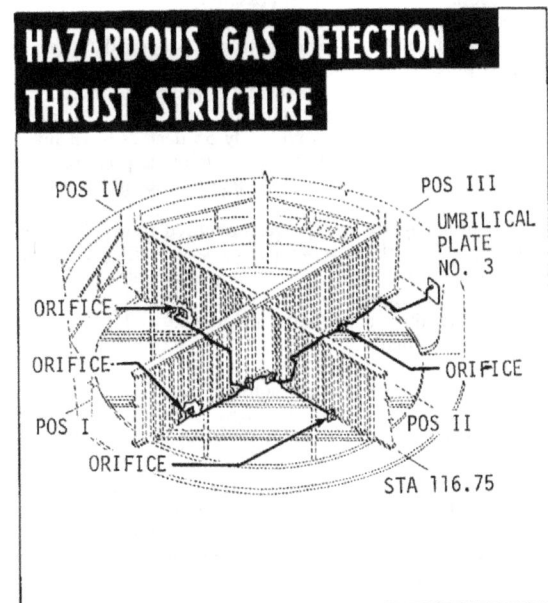

Figure 4-3

## PROPULSION

The F-1 engine is a single start, 1,530,000 pound fixed thrust, calibrated, bipropellant engine which uses liquid oxygen as the oxidizer and RP-1 as the fuel. Engine features include a bell shaped thrust chamber with a 10:1 expansion ratio, and detachable, conical nozzle extension which increases the thrust chamber expansion ratio to 16:1. The thrust chamber is cooled regeneratively by fuel, and the nozzle extension is cooled by gas generator exhaust gases. Liquid oxygen and RP-1 fuel are supplied to the thrust chamber by a single turbopump powered by a gas generator which uses the same propellant combination. RP-1 fuel is also used as the turbopump lubricant and as the working fluid for the engine fluid power system. The four outboard engines are capable of gimbaling and have provisions for supply and return of RP-1 fuel as the working fluid for a thrust vector control system. The engine contains a heat exchanger system to condition engine supplied liquid oxygen and externally supplied helium for stage propellant tank pressurization. An instrumentation system monitors engine performance and operation. External thermal insulation provides an allowable engine environment during flight operation.

### ENGINE OPERATING REQUIREMENTS

The engine requires a source of pneumatic pressure, electrical power, and propellants for sustained operation. A ground hydraulic pressure source, an inert thrust chamber prefill solution, gas generator igniters, gas generator exhaust igniters, and hypergolic fluid are required during the engine start sequence. The engine is started by ground support equipment (GSE) and is capable of only one start before reservicing.

### PURGE, PREFILL, AND THERMAL CONDITIONING

A gaseous nitrogen purge is applied for thermal conditioning and elimination of explosive hazard under each engine cocoon. Because of the possibility of low temperatures existing in the space between the engine and its cocoon of thermal insulation, heated nitrogen is applied to this area. This purge is manually operated, at the discretion of launch operations, whenever there is a prolonged hold of the countdown with lox onboard and with an ambient temperature below approximately 55 degrees F. In any case, the purge will be turned on five minutes prior to ignition command and continue until umbilical disconnect.

A continuous nitrogen purge is required to expel propellant leakage from the turbopump lox seal housing and the gas generator lox injector. The purge pressure also improves the sealing characteristic of the lox seal. The purge is required from the time propellants are loaded and is continuous throughout flight.

A nitrogen purge prevents contaminants from accumulating on the radiation calorimeter viewing surfaces. The purge is started at T-52 seconds and is continued during flight.

A gaseous nitrogen purge is required to prevent contaminants from entering the lox system through the engine lox injector or the gas generator lox injector. The purge system is activated prior to engine operation and is continued until umbilical disconnect.

At approximately T-13 hours, an ethylene glycol solution fills the thrust tubes and manifolds of all five engines. This inert solution serves to smooth out the combustion sequence at engine start. Flow is terminated by a signal from an observer at the engines. At approximately T-5 minutes, 50 gallons are supplied to top off the system to compensate for liquid loss that occurred during engine gimbaling.

### POGO SUPPRESSION SYSTEM

The POGO suppression system (figure 4-4) utilizes the lox prevalve cavities as surge chambers to suppress the POGO phenomenon. The lox prevalve cavities are pressurized with gaseous helium (GHe) at T-11 minutes from ground supply by opening the POGO suppression control valves. During the initial fill period (T-11 to T-9 minutes), the filling of the valves is closely monitored utilizing measurements supplied by the liquid level resistance thermometers $R_3$ (primary) and $R_2$ (backup). The GHe ground fill continues to maintain the cavity pressure until umbilical disconnect. Following umbilical disconnect the cavity pressure is maintained by the cold helium spheres located in the lox tank.

Status on system operation is monitored through two pressure transducers and four liquid level resistance thermometers. One pressure transducer (0-800 psia) monitors system input pressure. A second pressure transducer (0-150 psia) monitors the pressure inside the No. 1 engine lox prevalve cavity. These pressure readings are transmitted via telemetry to ground monitors. The liquid level within the prevalves is monitored by four liquid level resistance thermometers in each prevalve. These thermometers transmit a "wet" (colder than -165 degrees centigrade) and a "dry" (warmer than -165 degrees centigrade) reading to ground monitors.

### ENGINE SUBSYSTEMS

The subsystems of the F-1 engine shown in figure 4-5 are the turbopump, checkout valve, hypergol manifold, heat exchanger, main fuel valve and main lox valve. Subsystems not shown are the gas generator, 4-way control valve, and pyrotechnic igniters.

#### Hypergol Manifold

The hypergol manifold consists of a hypergol container, an ignition monitor valve (IMV), and an igniter fuel valve (IFV). The hypergol solution is forced into the thrust chamber by the fuel where combustion is initiated upon mixing with the lox. The IFV prevents thrust chamber ignition until the turbopump pressure has reached 375 psi. The IMV prevents opening of the main fuel valves prior to hypergolic ignition. Position sensors indicate if the hypergol cartridge has been installed. The loaded indication is a prerequisite to the firing command.

#### Control Valve - 4-Way

The 4-way control valve directs hydraulic fluid to open and close the fuel, lox, and gas generator valves. It consists of a filter manifold, a start and stop solenoid valve, and two check valves.

#### Turbopump

The turbopump is a combined lox and fuel pump driven through a common shaft by a single gas turbine.

## Gas Generator

The gas generator (GG) provides the gases for driving the turbopump. Its power output is controlled by orifices in its propellant feed lines. The gas generator system consists of a dual ball valve, an injector, and a combustor. Combustion is initiated by two pyrotechnic igniters. Total propellant flow rate is approximately 170 lb/sec at a lox/RP-1 mixture ratio of 0.42:1. The dual ball valve must be closed prior to fuel loading and must remain closed to meet an to meet an interlock requirement for engine start.

## Heat Exchanger

The heat exchanger expands lox and cold helium for propellant tank pressurization. The cold fluids, flowing through separate heating coils, are heated by the turbopump exhaust. The warm expanded gases are then routed from the heating coils to the propellant tanks.

## Main Fuel Valve

There are two main fuel valves per engine. They control flow of fuel to the thrust chamber. The main fuel valve is a fast acting, pressure balanced, poppet type, hydraulically operated valve. Movement of the poppet actuates a switch which furnishes valve position signals to the telemetry system. This valve is designed to remain open, at rated engine pressures and flowrates, if the opening control pressure is lost. Both valves must be in the closed position prior to fuel loading or engine start.

## Main Lox Valve

The two main lox valves on each engine control flow to the thrust chamber. These valves are fast acting, pressure balanced, poppet type, hydraulically operated valves. A sequence valve operated by the poppet allows opening pressure to be applied to the GG valve only after both main lox valve poppets have moved to a partially open position. This valve is designed to remain open, at rated engine pressures and flowrates, if the opening control pressure is lost. Both main lox valves must be in the closed position prior to lox loading or engine start.

## Checkout Valve

The checkout valve directs ground supplied control fluid

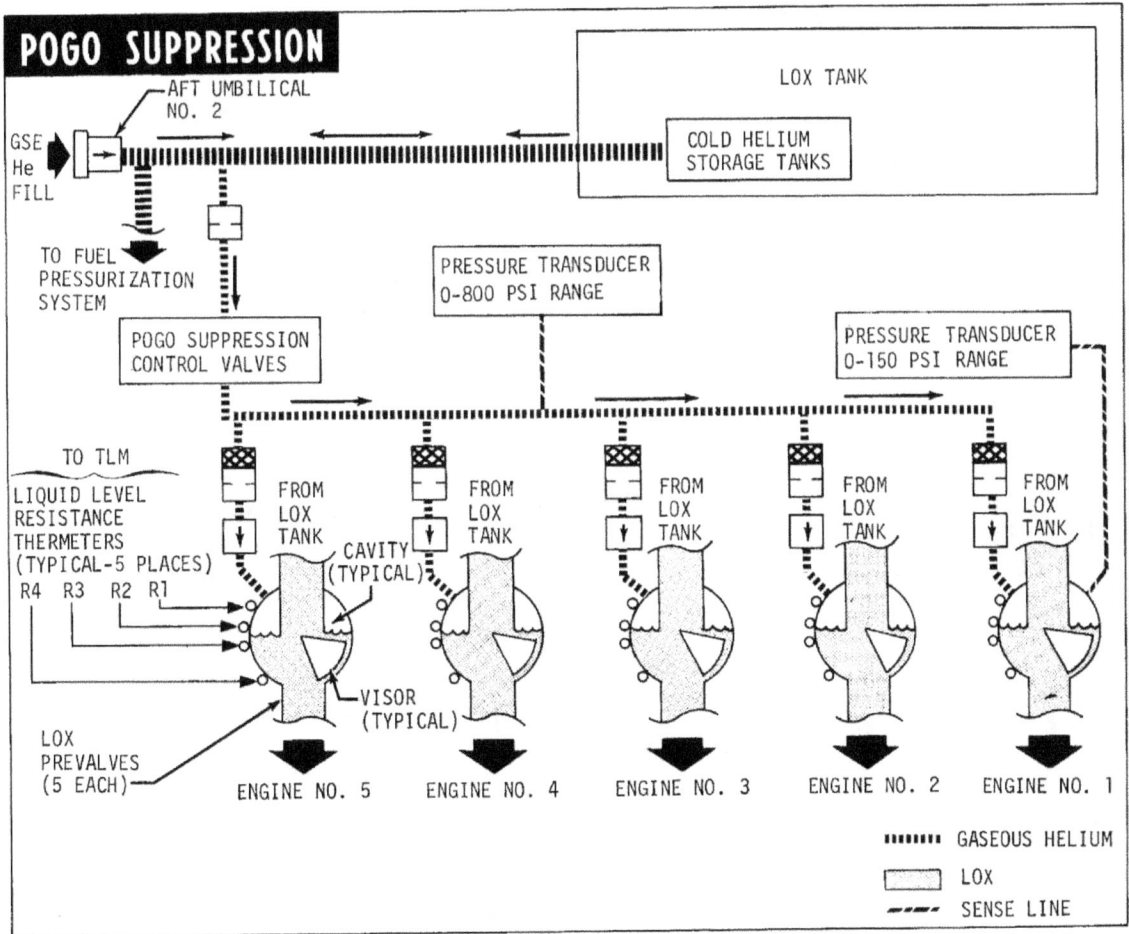

Figure 4-4

from the engine back to ground during engine checkout. Approximately 30 seconds prior to the firing command the valve is actuated to the engine position. In this position it directs control fluid to the No. 2 turbopump inlet. An ENGINE POSITION indication is required from this valve prior to, and is interlocked with, forward umbilical disconnect command.

### High Voltage Igniters

Four high voltage igniters, two in the gas generator (GG) body and two in the engine thrust chamber nozzle extension ignite the GG and the fuel rich turbopump exhaust gases. They are ignited during the F-1 engine start sequence by application of a nominal 500 volts to the igniter squibs.

### Engine Operation

Engine operation is illustrated in figure 4-6 in terms of engine start sequence and transition to mainstage for a typical single engine.

### ENGINE CUTOFF

The normal inflight cutoff sequence is center engine first,

followed by the four outboard engines. At 2 minutes, 5 seconds the center engine is programmed by the LVDC for cutoff. This command also initiates time base No. 2 ($T_2$ + 0.0). The LVDC provides a backup center engine shutoff signal. The outboard engine cutoff signal initiates time base No. 3 ($T_3$ +0.0). Outboard engine cutoff circuitry is enabled by a signal from the LVDC at $T_2$ + 19.5 seconds. Energizing two or more of the four lox level sensors starts a timer. Expiration of the timer energizes the 4-way control valve stop solenoid on each outboard engine. The remaining shutdown sequence of the outboard engines is the same as for the center engine as explained in Figure 4-7.

### EMERGENCY ENGINE CUTOFF

In an emergency, the engine will be cut off by any of the following methods: Ground Support Equipment (GSE) Command Cutoff, Range Safety Command Cutoff, Thrust Not OK Cutoff, Emergency Detection System, Outboard Cutoff System.

GSE has the capability of initiating engine cutoff anytime until umbilical disconnect. Separate command lines are supplied through the aft umbilicals to the engine cutoff relays and prevalve close relays.

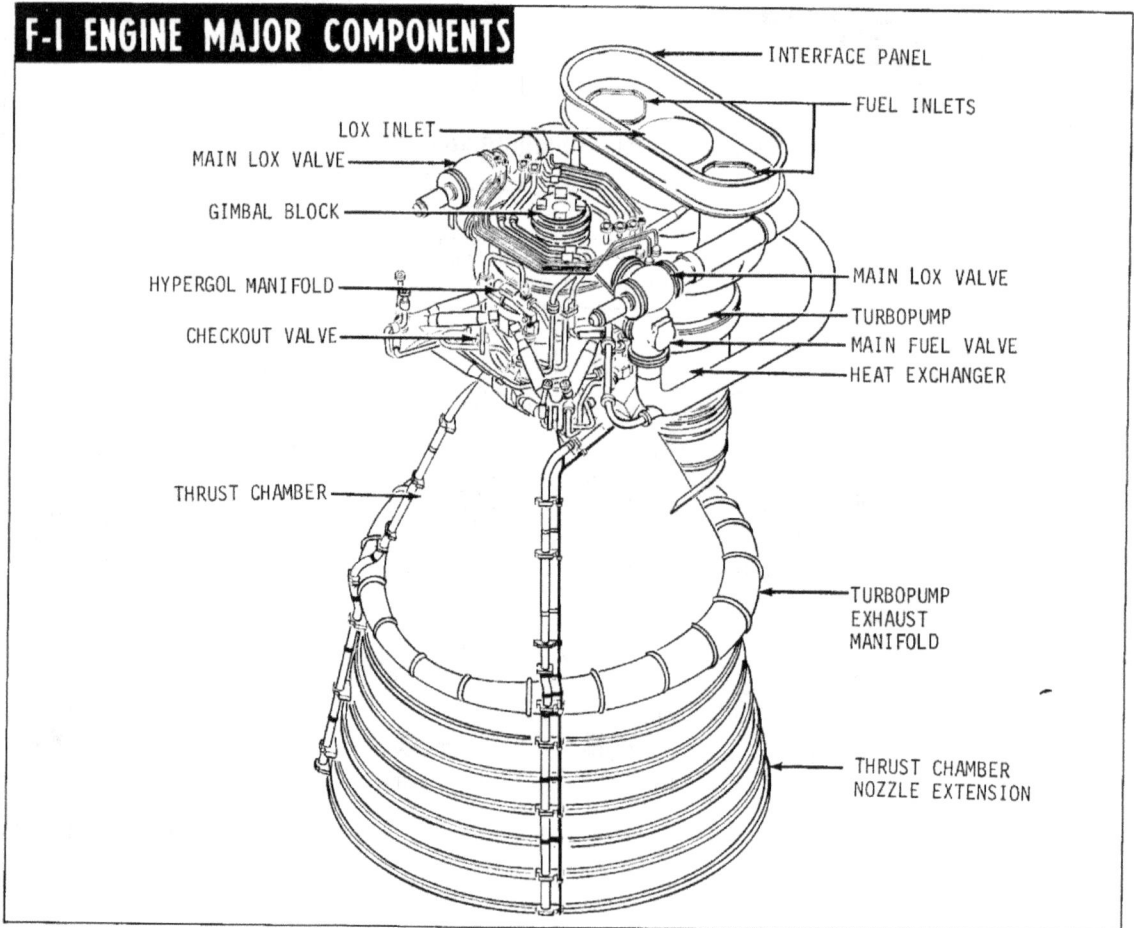

Figure 4-5

Range Safety Cutoff has the capability of engine cutoff anytime after liftoff. If it is determined during flight that the vehicle has gone outside the established corridor, the Range Safety Officer will send commands to effect engine cutoff and propellant dispersion.

Three thrust OK pressure switches are located on each F-1 engine thrust chamber fuel manifold and sense main fuel injection pressure. If the pressure level drops below the deactivation level of two of the three pressure switches, an engine cutoff signal is initiated. The circuitry is enabled at $T_1$ +14.0 seconds to allow the vehicle to clear the launch pad.

Prior to its own deactivation, the Emergency Detection System (EDS) initiates engine cutoff when it is determined that two or more engines have shutdown prematurely. When the IU receives signals from the thrust OK logic relays that two or more engines have shutdown, the IU initiates a signal to relays in the S-IC stage to shutdown the remaining engines. S-IC engine EDS cutoff is enabled at $T_1$ + 30.0 seconds and continues until 0.8 second before center engine cutoff or until deactivated.

Following EDS deactivation, the outboard engine cutoff system is activated 0.1 second before center engine cutoff, and is similar in function to the EDS. Whereas the EDS initiates emergency engine cutoff when any two engines are shutdown, the outboard cutoff system monitors only the outboard engines and provides outboard engine cutoff if the thrust OK pressure switches cause shutdown of two adjacent outboard engines.

NOTE

Loss of two adjacent outboard engines, after center engine cutoff, could cause stage breakup.

## FLIGHT CONTROL SYSTEM

The S-IC flight control system gimbals the four outboard engines to provide attitude control during the S-IC burn phase. See Section VII for a detailed discussion of the Saturn V flight control.

### FLUID POWER

There are five fluid power systems on the S-IC stage, one for each engine (Figure 4-8 shows a typical outboard engine fluid power system.) The hydraulic pressure is supplied from a GSE pressure source during test, prelaunch checkout and engine start. When the engine starts, hydraulic pressure is generated by the engine turbopump. Pressure from either source is made available to the engine valves such as the main fuel and lox valves and igniter fuel valve. These valves are sequenced and controlled by the terminal countdown sequencer, stage switch selector and by mechanical or fluid pressure means. A discussion of these components may be found under the titles "PROPULSION" or "ELECTRICAL". Fluid under pressure also flows through a filter (figure 4-8) and to the two flight control servoactuators on each outboard engine.

The fluid power system uses both RJ-1 ramjet fuel and RP-1 rocket propellant as hydraulic fluid. The RJ-1 is used by the Hydraulic Supply and Checkout Unit (GSE pressure source). RP-1 of course, is the fuel used in the S-IC stage. It is pressurized by the engine turbopump. The two pressure sources are separated by check valves while return flow is directed to GSE or stage by the ground checkout valve. Drilled passages in the hydraulic components (valves and servoactuators) permit a flow of fluid to thermally condition the units and to bleed gases from the fluid power system.

## HYDRAULIC SERVOACTUATOR

The servoactuator (figure 4-9) is the power control unit for converting electrical command signals and hydraulic power into mechanical outputs to gimbal the engines on the S-IC stage. The flight control computer (IU) receives inputs from the guidance system in the IU and sends signals to the servoactuators to gimbal the outboard engines in the direction and magnitude required. An integral mechanical feedback varied by piston position modifies the effect of the IU control signal. A built-in potentiometer senses the servoactuator position and transmits this information to the IU for further transmission via telemetry to the ground.

The servoactuators are mounted 90 degrees apart on each outboard engine and provide for engine gimbaling at a rate of 5 degrees per second and a maximum angle of +5.0 degrees square pattern.

## PNEUMATIC CONTROLS

The pneumatic control system (figure 4-10) provides a pressurized nitrogen supply for command operation of various pneumatic valves. Pneumatic control of the fuel and lox fill and drain valves and the No. 2 lox interconnect valve is provided directly from GSE. Lox interconnect valves No. 1,3 and 4 are controlled by the onboard pneumatic system.

The pneumatic control system for those valves which must be controlled during flight (fuel and lox prevalves, lox and fuel vent valves) is supplied by a GSE nitrogen source at 3200 psi. The system is charged through onboard control valves and filters, a storage bottle, and a pressure regulator which reduces the supplied pressure to 750 psi. There are direct lines from the GSE to the prevalve solenoid valves which provide pressure for emergency engine shutdown. Orifices in the fuel and lox prevalve lines control closing time of the valves.

## PROPELLANTS

Propellants for the S-IC stage are RP-1 (fuel) and liquid oxygen (lox). The propellant system includes hardware for fuel fill and drain operations, tank pressurization prior to and during flight and delivery of propellants to the engines. The system is divided into two systems, the fuel system and the lox system. Figures 4-11 and 4-13 show the components of the systems.

### FUEL LOADING AND DELIVERY

Fuel loading starts at approximately 126 hours before lift-off (figure 4-12) and continues at a rate of 2000 gpm until 99% full and then uses a 200 gpm rate until the total mass load reaches 102% of the desired load. At T-20 minutes the propellant management GSE gives the command to begin fuel level adjust to the prescribed flight load level. This initiates a limited drain. The fuel loading probe (figure 4-11), senses the mass level. The fuel vent and relief valve is opened during gravity drain but must be closed for pressurized drain.

# ENGINE START

STOP SOLENOID

RJ-1 FROM GSE

4-WAY CONTROL VALVE

START SOLENOID

CHECKOUT VALVE

RETURN FLUID TO ENGINE

IGNITION SEQUENCER

GAS GENERATOR

FROM LOX TANK

FROM FUEL TANK

FROM FUEL TANK

IGNITION MONITOR VALVE

MAIN LOX VALVE

HYPERGOL MANIFOLD

IGNITER

MAIN LOX VALVE

MAIN FUEL VALVE

MAIN LOX VALVE

MAIN FUEL VALVE

TURBOPUMP

THRUST O.K. PRESSURE SWITCH (3)

IGNITER

IGNITER

| | |
|---|---|
| ———— | ELECTRICAL |
| —·—·— | COMBUSTION GASES |
| ╱╱╱ | SENSE LINE |
| ·········· | STAGE FUEL |
| ▨▨▨ | LOX |
| ········· | HYPERGOLIC |

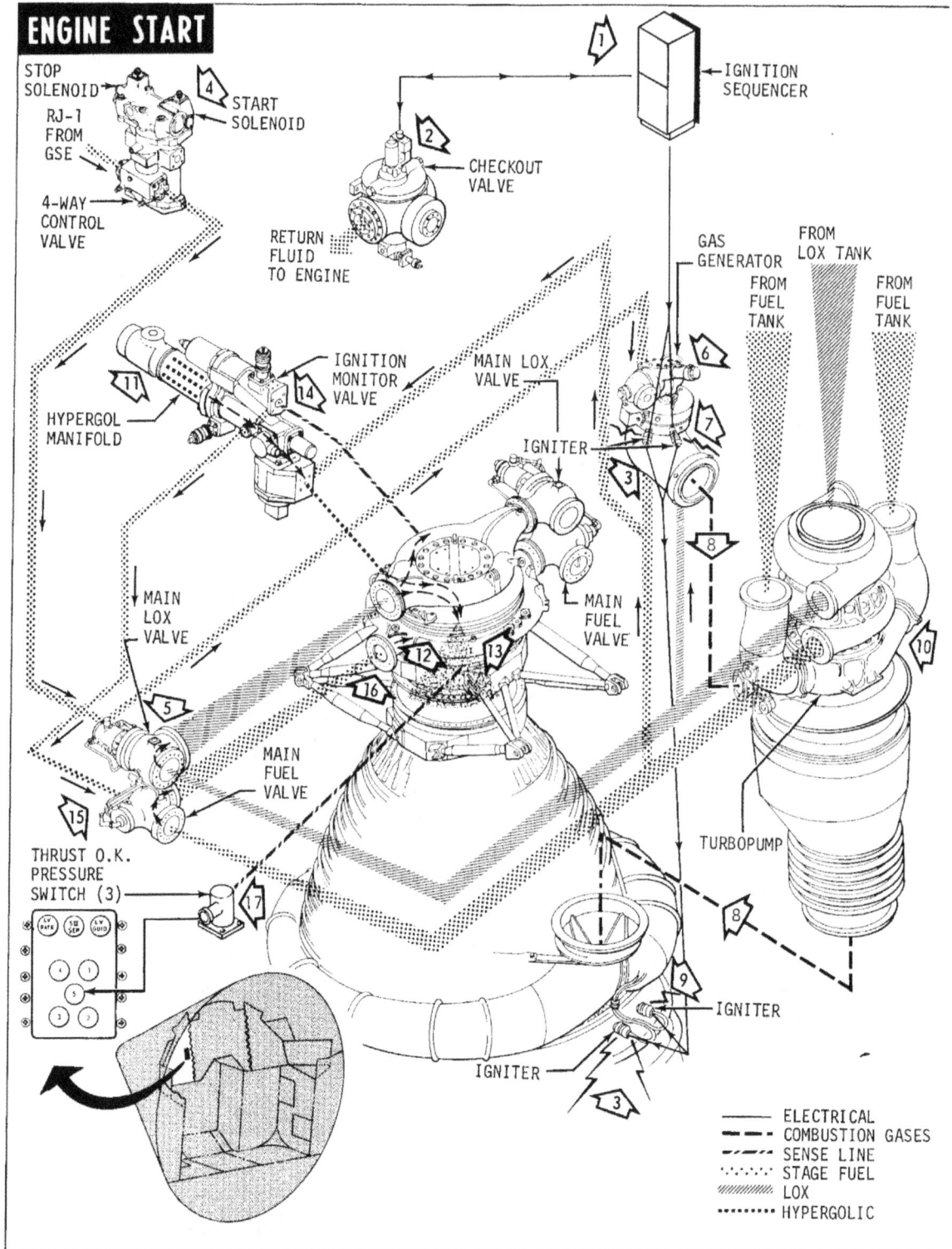

Figure 4-6 (Sheet 1 of 2)

1 ▷ Engine start is part of the terminal count-down sequence. When this point in the count-down is reached, the ignition sequencer con-trols starting of all five engines.

2 ▷ Checkout valve moves to engine return position.

3 ▷ Electrical signal fires igniters (4 each engine).

    a) Gas generator combustor and turbine exhaust igniters burn igniter links to trigger electrical signal to start solenoid of 4-way control valve.

    b) Igniters burn approximately six seconds.

4 ▷ Start solenoid of 4-way control valve directs GSE hydraulic pressure to main lox valves.

5 ▷ Main lox valves allow lox to flow to thrust chamber and GSE hydraulic pressure to flow through sequence valve to open gas generator ball valve.

6 ▷ Propellants, under tank pressure, flow into gas generator combustor.

7 ▷ Propellants are ignited by flame of igniters.

8 ▷ Combustion gas passes through turbopump, heat exchanger, exhaust manifold and nozzle extension.

9 ▷ Fuel rich turbine combustion gas is ignited by flame from igniters.

    a) Ignition of this gas prevents back-firing and burping.

    b) This relatively cool gas (approxi-mately 1,000°F) is the coolant for the nozzle extension.

10 ▷ Combustion gas accelerates the turbopump, causing the pump discharge pressure to increase.

11 ▷ As fuel pressure increases to approximately 375 psig, it ruptures the hypergol cart-ridge.

12 ▷ The hypergolic fluid and fuel are forced into the thrust chamber where they mix with the lox to cause ignition.

TRANSITION TO MAINSTAGE

13 ▷ Ignition causes the combustion zone pres-sure to increase.

14 ▷ As pressure reaches 20 psig, the ignition monitor valve directs fluid pressure to the main fuel valves.

15 ▷ Fluid pressure opens main fuel valves.

16 ▷ Fuel enters thrust chamber. As pressure increases the transition to mainstage is accomplished.

17 ▷ The thrust OK pressure switch (which senses fuel injection pressure) picks up at ap-proximately 1060 psi and provides a THRUST OK signal to the IU.

Figure 4-6 (Sheet 2 of 2)

# ENGINE CUTOFF

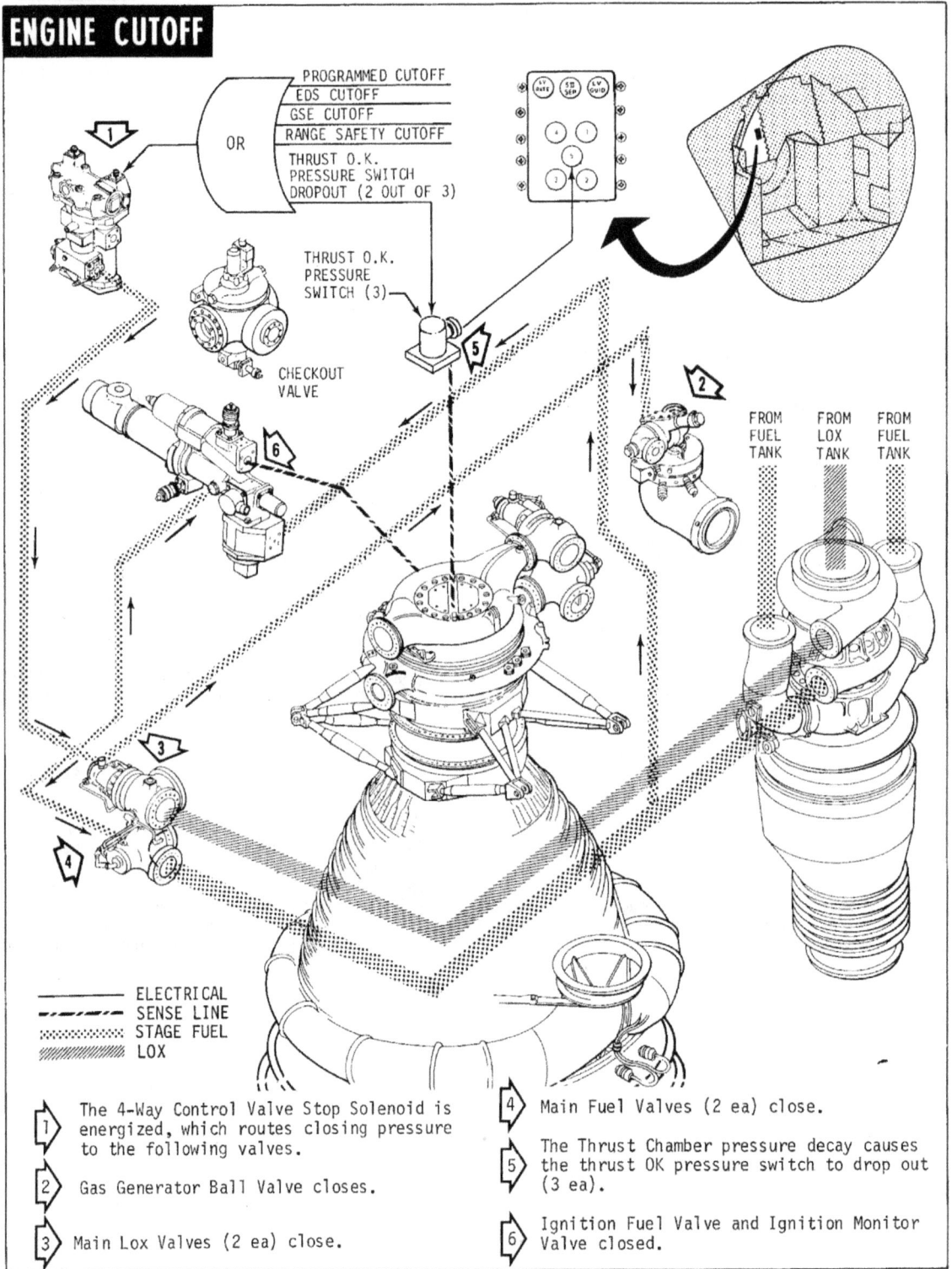

PROGRAMMED CUTOFF
EDS CUTOFF
GSE CUTOFF
RANGE SAFETY CUTOFF

OR

THRUST O.K.
PRESSURE SWITCH
DROPOUT (2 OUT OF 3)

THRUST O.K.
PRESSURE
SWITCH (3)

CHECKOUT
VALVE

FROM    FROM    FROM
FUEL    LOX    FUEL
TANK    TANK    TANK

———— ELECTRICAL
–––––– SENSE LINE
:::::::::: STAGE FUEL
////////// LOX

1. The 4-Way Control Valve Stop Solenoid is energized, which routes closing pressure to the following valves.

2. Gas Generator Ball Valve closes.

3. Main Lox Valves (2 ea) close.

4. Main Fuel Valves (2 ea) close.

5. The Thrust Chamber pressure decay causes the thrust OK pressure switch to drop out (3 ea).

6. Ignition Fuel Valve and Ignition Monitor Valve closed.

Figure 4-7

# FLIGHT CONTROL SYSTEM

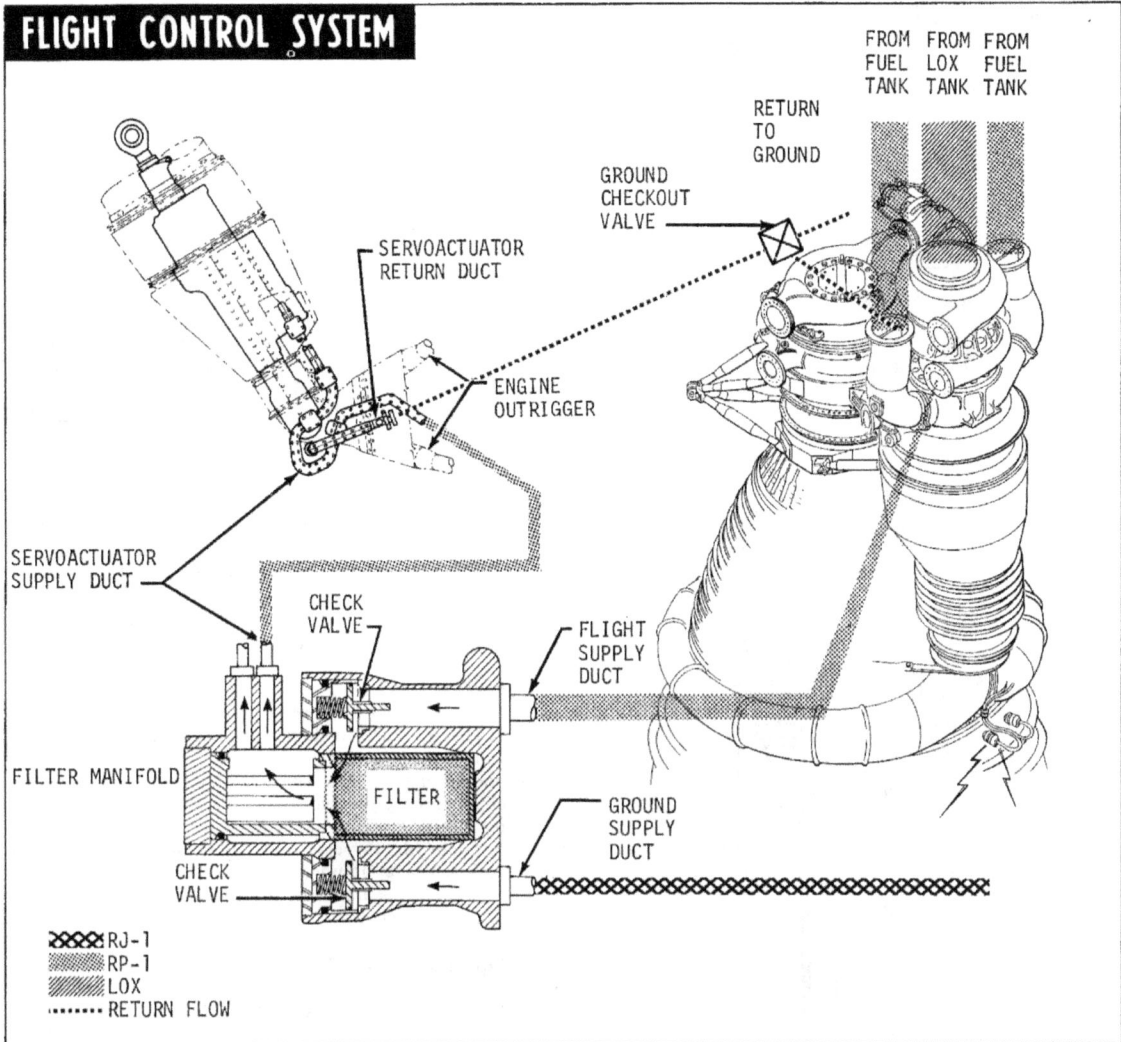

Figure 4-8

## RP-1 Pressurization

Fuel tank pressurization (figure 4-11) is required from engine start through stage flight to establish and maintain a net positive suction head at the fuel inlet to the engine turbopumps. Ground supplied helium for prepressurization is introduced into the cold helium line downstream from the flow controller resulting in helium flow through the engine heat exchanger and the hot helium line to the fuel tank distributor. During flight, the source of fuel tank pressurization is helium from storage bottles mounted inside the lox tank.

Fuel tank pressure switches control the fuel vent and relief valve, the GSE pressure supply during filling operations, prepressurization before engine ignition, and pressurization during flight. The flight pressurization pressure switch actuates one of the five control valves in the flow controller to ensure a minimum pressure of 24.2 psia during flight. The

other four valves are sequenced by the IU to establish an adequate helium flow rate with decreasing storage bottle pressure.

The onboard helium storage bottles are filled through a filtered fill and drain line upstream from the flow controller. The storage bottles are filled to a pressure of 1400 psi prior to lox loading. Fill is completed to 3150 psi after lox loading when the bottles are cold.

## RP-1 Delivery

Fuel feed (figure 4-11) is accomplished through two 12-inch ducts which connect the fuel tank to each F-1 engine. The ducts are equipped with gimbaling and sliding joints to compensate for motions from engine gimbaling and stage stresses. Prevalves, one in each fuel line, serve as an emergency backup to the main engine fuel shutoff valves. The prevalves also house flowmeters which provide flowrate

# HYDRAULIC SERVOACTUATOR

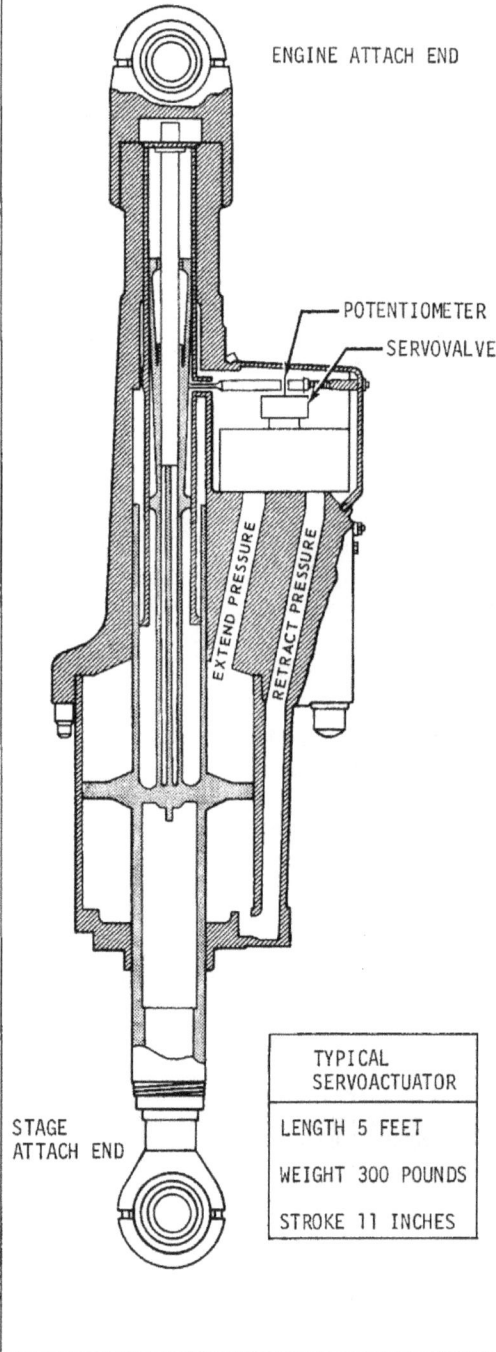

ENGINE ATTACH END

— POTENTIOMETER
— SERVOVALVE

EXTEND PRESSURE

RETRACT PRESSURE

STAGE
ATTACH END

TYPICAL
SERVOACTUATOR

LENGTH 5 FEET

WEIGHT 300 POUNDS

STROKE 11 INCHES

Figure 4-9

data via telemetry to GSE. A fuel level engine cutoff sensor in the bottom of the fuel tank, initiates engine shutdown when fuel is depleted if the lox sensors have failed to cut off the engines.

## LOX LOADING AND DELIVERY

As the oxidizer in the bi-propellant propulsion system, lox is contained and delivered through a separate tank and delivery system (figure 4-13). The 345,000 gallon tank is filled through two 6-inch fill and drain lines. Shortly after T-6 hours lox loading begins, three fill rates are used sequentially; a 300 gpm for tank chilldown, a 1500 gpm slow fill rate to stabilize the liquid level and thus prevent structural damage, and a fast fill rate of 10,000 gpm. At approximately 95% full the rate is reduced to 1500 gpm and ceases when the lox loading level sensor automatically stops the fill mode. Lox boiloff is replenished at 500 gpm until prepressurization occurs.

### Lox Drain

The lox is drained through the two fill and drain lines (figure 4-13) and a lox suction duct drain line in the thrust structure. The total drain capability is 7,500 gpm. During lox drain, positive ullage pressure is maintained by a GSE pressure source and two vent valves which are kept closed except when overpressure occurs.

Prior to launch, boil off in the lox tank may be harmlessly vented overboard; however, excessive geysering from boiling in the lox suction ducts can cause structural damage, and high lox temperatures near the engine inlets may prevent normal engine start. The lox bubbling system (figure 4-13) eliminates geysering and maintains low pump inlet temperatures. The helium induced convection currents circulate lox through the suction ducts and back into the tank. Once established, thermal pumping is self sustaining and continues until the interconnect valves are closed just prior to launch.

### Lox Pressurization System

Lox tank pressurization (figure 4-13) is required to ensure proper engine turbopump pressure during engine start, thrust buildup, and fuel burn. The pressurization gas, prior to flight (prepressurization), is helium. It is supplied at 240 lbs/min for forty seconds to the distributor in the lox tank. Lox tank pressure is monitored by three pressure switches. They control the GSE pressure source, the lox vent valve, and the lox vent and relief valve to maintain a maximum of 18 psig ullage pressure. Prepressurization is maintained until T-3 seconds and gox is used for pressurizing the lox tank during flight. A portion of the lox supplied to each engine is diverted from the lox dome into the engine heat exchanger where the hot turbine exhaust transforms lox into gox. The heated gox is delivered through the gox pressurization line and a flow control valve to the distributor in the lox tank. A sensing line provides pressure feedback to the flow control valve to regulate the gox flow rate and maintain ullage pressure between 18 and 20 psia.

### Lox Delivery

Lox is delivered to the engine through five suction lines (figure 4-13). The ducts are equipped with gimbals and sliding joints to compensate for motions from engine

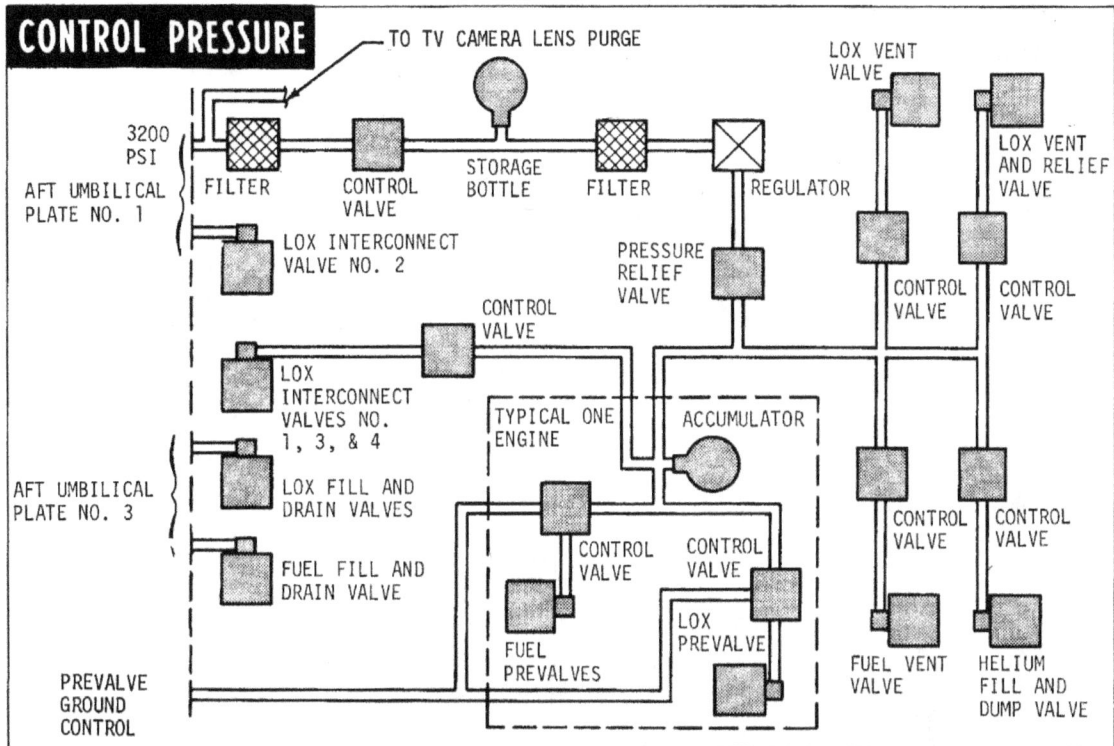

**CONTROL PRESSURE**

Figure 4-10

gimbaling and stage stresses. Pressure volume compensating ducts ensure constant lox flowrate regardless of the gimbaled position of the engine. Each suction line has a lox prevalve which is a backup to the engine lox valve. The prevalve cavity is charged with helium and functions as an accumulator which absorbs engine induced pulses to prevent POGO.

## ELECTRICAL

The electrical power system of the S-IC stage is divided into three basic subsystems: an operational power subsystem a measurement power subsystem and a visual instrumentation power subsystem. On board power is supplied by five 28-volt batteries, one each for the operation and measurement power systems and three for the visual instrumentation power system as shown in figure 4-14. The battery characteristics are shown in figure 4-15.

In figure 4-16, power distribution diagram, battery number 1 is identified as the operational power system battery. It supplies power to operational loads such as valve controls, purge and venting systems, pressurization systems, and sequencing and flight control. Battery number 2 is identified as the measurement power system battery. It supplies power to measurements loads such as telemetry systems, transducers, multiplexers, and transmitters. Both batteries supply power to their loads through a common main power distributor but each system is completely isolated from the other.

In the visual instrumentation system, two batteries provide power for the lox tank strobe lights. The third battery

energizes the control circuits, camera motors, and thrusters of the film camera portion of the visual instrumentation system.

During the prelaunch checkout period power for all electrical loads, except range safety receivers, is supplied from GSE. The range safety receivers are hardwired to batteries 1 and 2 in order to enhance the safety and reliability of the range safety system. At T-50 seconds a ground command causes the power transfer switch to transfer the S-IC electrical loads to onboard battery power. However, power for engine ignition and for equipment heaters (turbopump and lox valves) continues to come from the GSE until terminated at umbilical disconnect.

### DISTRIBUTORS

There are six power distributors on the S-IC stage. They facilitate the routing and distribution of power and also serve as junction boxes and housing for relays, diodes, switches and other electrical equipment.

There are no provisions for switching or transferring power between the operational power distribution system and the measurement power system. Because of this isolation, no failure of any kind in one system can cause equipment failure in the other system.

Main Power Distributor

The main power distributor contains a 26-pole power transfer switch, relays, and the electrical distribution busses. It serves

# FUEL SUBSYSTEMS

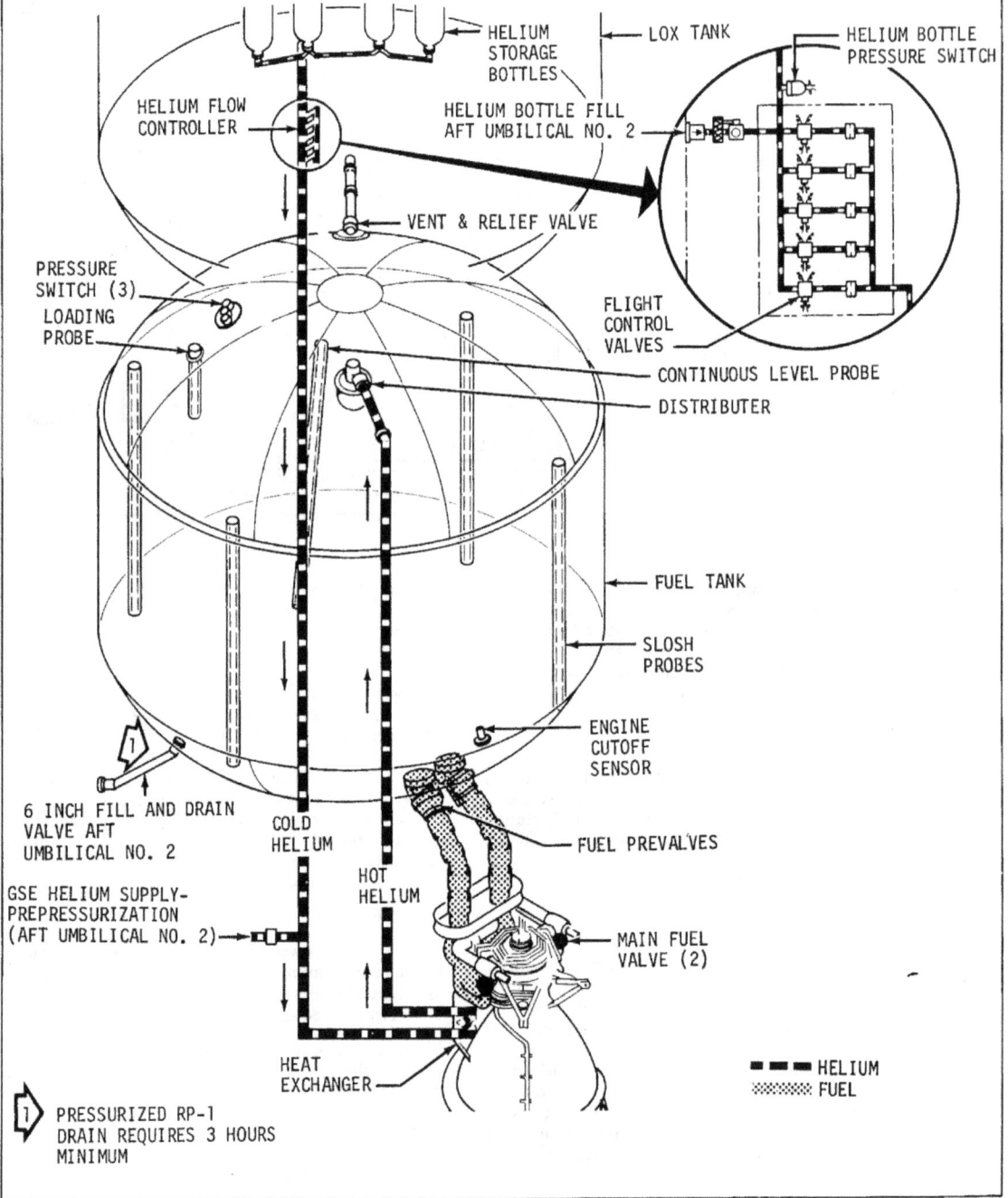

HELIUM
STORAGE
BOTTLES

LOX TANK

HELIUM BOTTLE
PRESSURE SWITCH

HELIUM FLOW
CONTROLLER

HELIUM BOTTLE FILL
AFT UMBILICAL NO. 2

VENT & RELIEF VALVE

PRESSURE
SWITCH (3)
LOADING
PROBE

FLIGHT
CONTROL
VALVES

CONTINUOUS LEVEL PROBE

DISTRIBUTER

FUEL TANK

SLOSH
PROBES

ENGINE
CUTOFF
SENSOR

6 INCH FILL AND DRAIN
VALVE AFT
UMBILICAL NO. 2

COLD
HELIUM

FUEL PREVALVES

HOT
HELIUM

GSE HELIUM SUPPLY-
PREPRESSURIZATION
(AFT UMBILICAL NO. 2)

MAIN FUEL
VALVE (2)

HEAT
EXCHANGER

▪▪▪ HELIUM
▒▒▒ FUEL

1  PRESSURIZED RP-1
DRAIN REQUIRES 3 HOURS
MINIMUM

Figure 4-11

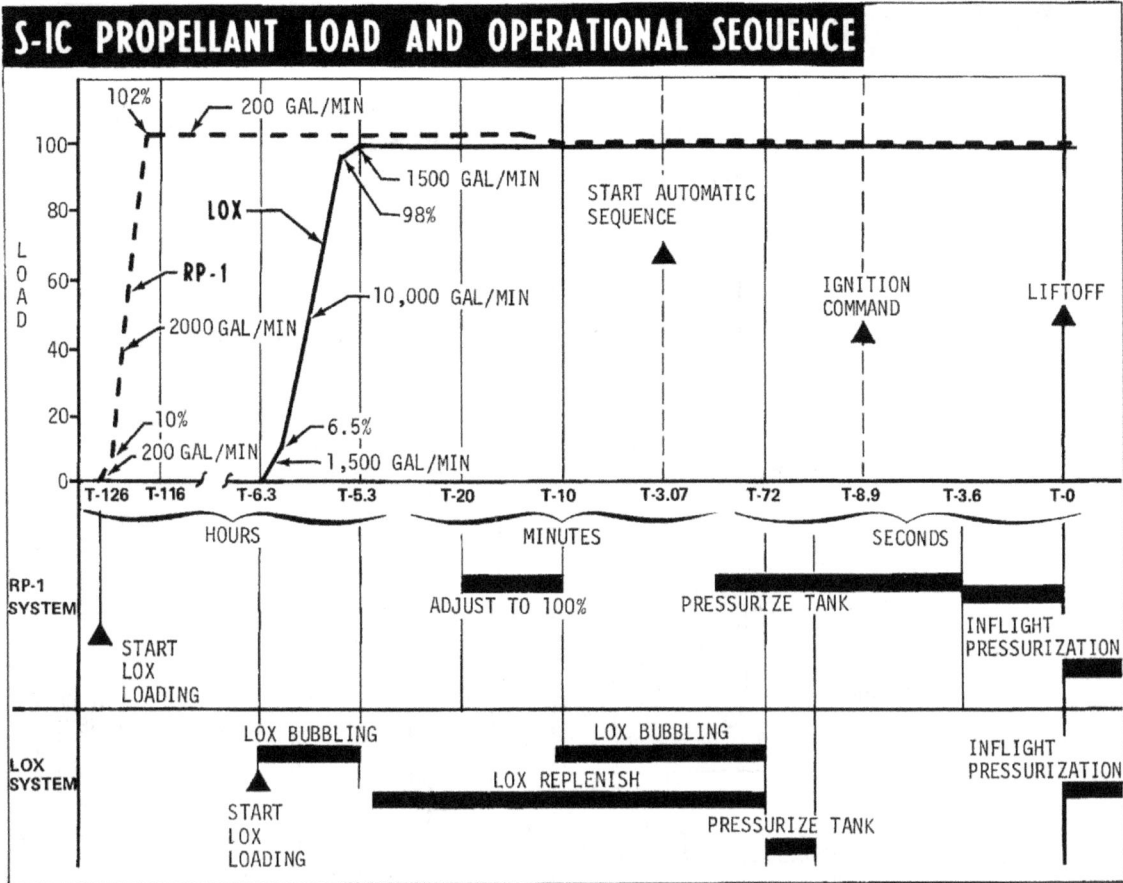

# S-IC PROPELLANT LOAD AND OPERATIONAL SEQUENCE

Figure 4-12

as a common distributor for both operational and measurement power subsystems. However, each of these systems is completely independent of the other. The power load is transferred from the ground source to the flight batteries at T-50 seconds. Inflight operation of the multicontact make-before-break power transfer switch is prevented by a brake, by mechanical construction, and by electrical circuitry. Operation of the switch several times during countdown verifies performance of the brake, motor, contacts, and mechanical components.

### Sequence and Control Distributor

The sequence and control distributor accepts command signals from the switch selector and through a series of magnetically latching relays provides a 28-volt dc command to initiate or terminate the appropriate stage function. The input from the switch selector latches a relay corresponding to the particular command. A 28-volt dc signal is routed through the closed contacts of the relay to the stage components being commanded. The relays, one for each command function, may be unlatched by a signal from GSE. The normally closed contacts of the relays are connected in series. A 28-volt dc signal is routed through the series connected relay contacts to indicate to GSE when all

sequence and control relays are in the reset state.

### Propulsion Distributor

The propulsion distributor contains relays, diodes, and printed circuit boards for switching and distributing propulsion signals during launch preparation and flight.

### THRUST OK Distributor

The THRUST OK distributor contains relays and printed circuit assemblies which make up the THRUST OK logic networks and timers required to monitor engine THRUST OK pressure switches and initiate engine shutdown. Signals from two of the three THRUST OK pressure switches on a particular engine will result in an output from a two-out-of-three voting network. This output activates a 0.044 second timer. If the THRUST OK condition is missing longer than 0.044 seconds the timer output sends a signal to initiate engine shutdown.

### Timer Distributor

Circuits to time the operation of relays, valves, and other electromechanical devices are mounted in the timer distributor.

4-15

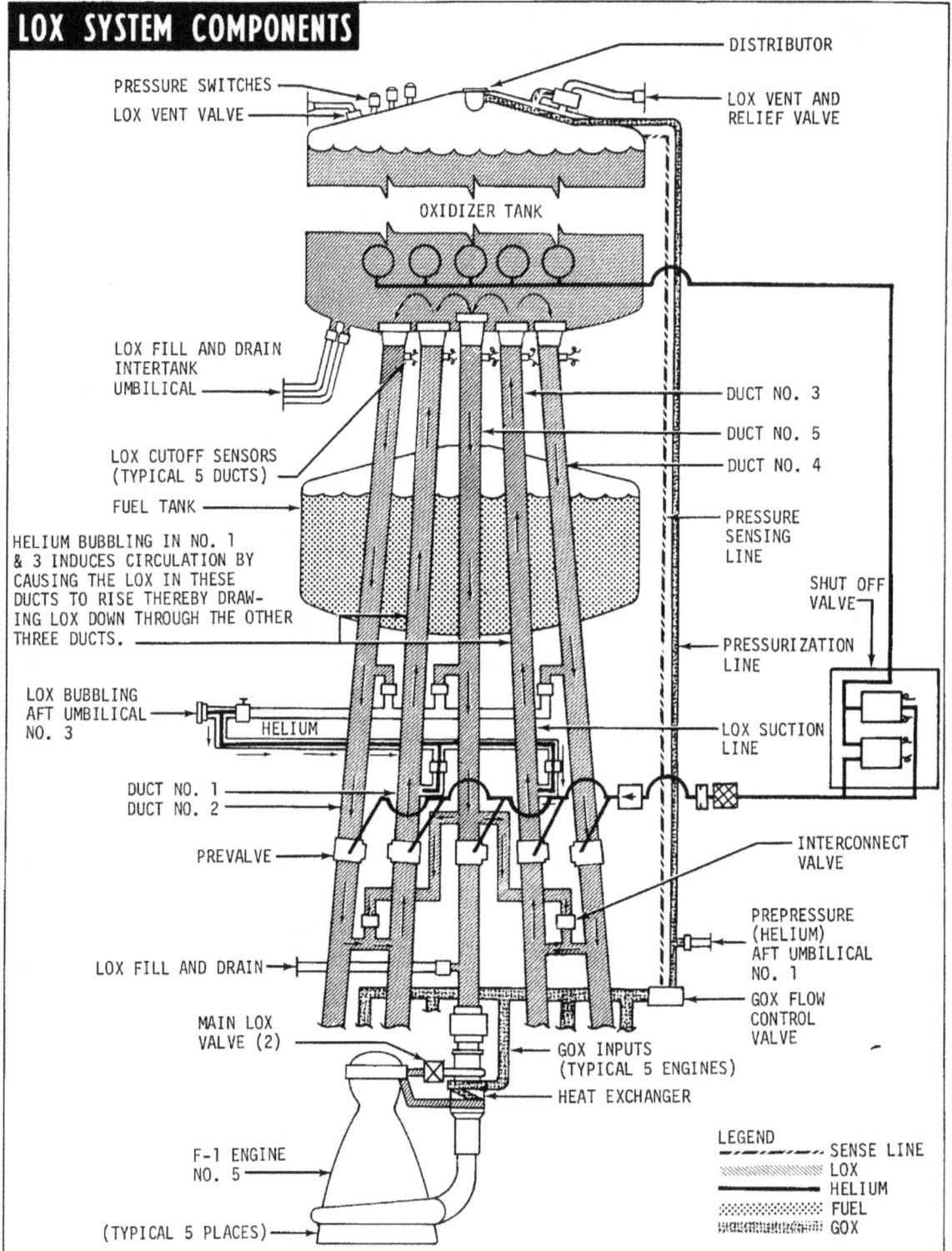

# LOX SYSTEM COMPONENTS

DISTRIBUTOR

PRESSURE SWITCHES

LOX VENT VALVE

LOX VENT AND RELIEF VALVE

OXIDIZER TANK

LOX FILL AND DRAIN INTERTANK UMBILICAL

DUCT NO. 3

DUCT NO. 5

DUCT NO. 4

LOX CUTOFF SENSORS (TYPICAL 5 DUCTS)

FUEL TANK

PRESSURE SENSING LINE

HELIUM BUBBLING IN NO. 1 & 3 INDUCES CIRCULATION BY CAUSING THE LOX IN THESE DUCTS TO RISE THEREBY DRAWING LOX DOWN THROUGH THE OTHER THREE DUCTS.

SHUT OFF VALVE

PRESSURIZATION LINE

LOX BUBBLING AFT UMBILICAL NO. 3

HELIUM

LOX SUCTION LINE

DUCT NO. 1
DUCT NO. 2

PREVALVE

INTERCONNECT VALVE

PREPRESSURE (HELIUM) AFT UMBILICAL NO. 1

LOX FILL AND DRAIN

GOX FLOW CONTROL VALVE

MAIN LOX VALVE (2)

GOX INPUTS (TYPICAL 5 ENGINES)

HEAT EXCHANGER

F-1 ENGINE NO. 5

LEGEND

SENSE LINE
LOX
HELIUM
FUEL
GOX

(TYPICAL 5 PLACES)

Figure 4-13

# S-IC ELECTRICAL POWER EQUIPMENT LOCATIONS

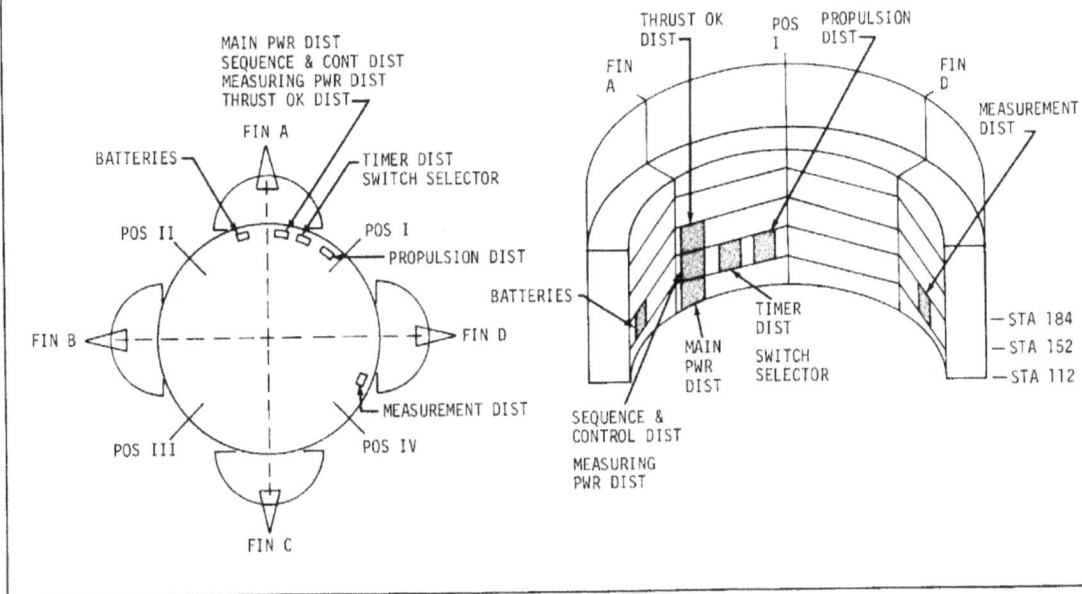

Figure 4-14

S-IC BATTERY CHARACTERISTICS

| TYPE | DRY CHARGE |
|---|---|
| MATERIAL | ZINC-SILVER OXIDE |
| ELECTROLYTE | POTASSIUM HYDROXIDE (KOH) IN PURE WATER |
| CELLS | 20 WITH TAPS FOR SELECTING 18 OR 19 TO REDUCE OUTPUT VOLTAGE AS REQUIRED |
| NOMINAL VOLTAGE | 1.5 VDC PER CELL: 28 + 2 VDC PER 18 TO 20 CELL GROUP |
| CURRENT RATINGS | BATTERY NO. 1 - OPERATIONAL LOADS = 640 AMPERE/MINUTE BATTERY NO. 2 - MEASUREMENT LOADS = 1250 AMPERE/MINUTE |
| GROSS WEIGHT | BATTERY NO. 1 = 22 LBS. BATTERY NO. 2 = 55 LBS. |

Figure 4-15

## Measuring Power Distributor

Each regulated 5-volt dc output from the seven measuring power supplies is brought to an individual bus in the measuring power distributor and then routed to the measuring and telemetry systems.

## SWITCH SELECTOR

The S-IC stage switch selector is the interface between the LVDC in the IU and the S-IC stage electrical circuits. Its function is to sequence and control various flight activities such as TM calibration, retrorocket initiation, and pressurization as shown in figure 4-17.

A switch selector is basically a series of low power transistor switches individually selected and controlled by an eight-bit binary coded signal from the LVDC in the IU. A coded word, when addressed to the S-IC switch selector, is accepted and stored in a register by means of magnetically latching relays. The coded transmission is verified by sending the complement of the stored word back to the LVDC in the IU. At the proper time an output signal is initiated via the selected switch selector channel to the appropriate stage operational circuit. The switch selector can control 112 circuits.

LVDC commands activate, enable, or switch stage electrical circuits as a function of elapsed flight time. Computer commands include:

1.  Telemetry calibration.

2.  Remove telemetry calibration.

3.  Open helium flow control valve No. 2.

4.  Open helium flow control valve No. 3.

5.  Open helium flow control valve No. 4.

6.  Enable center engine cutoff.

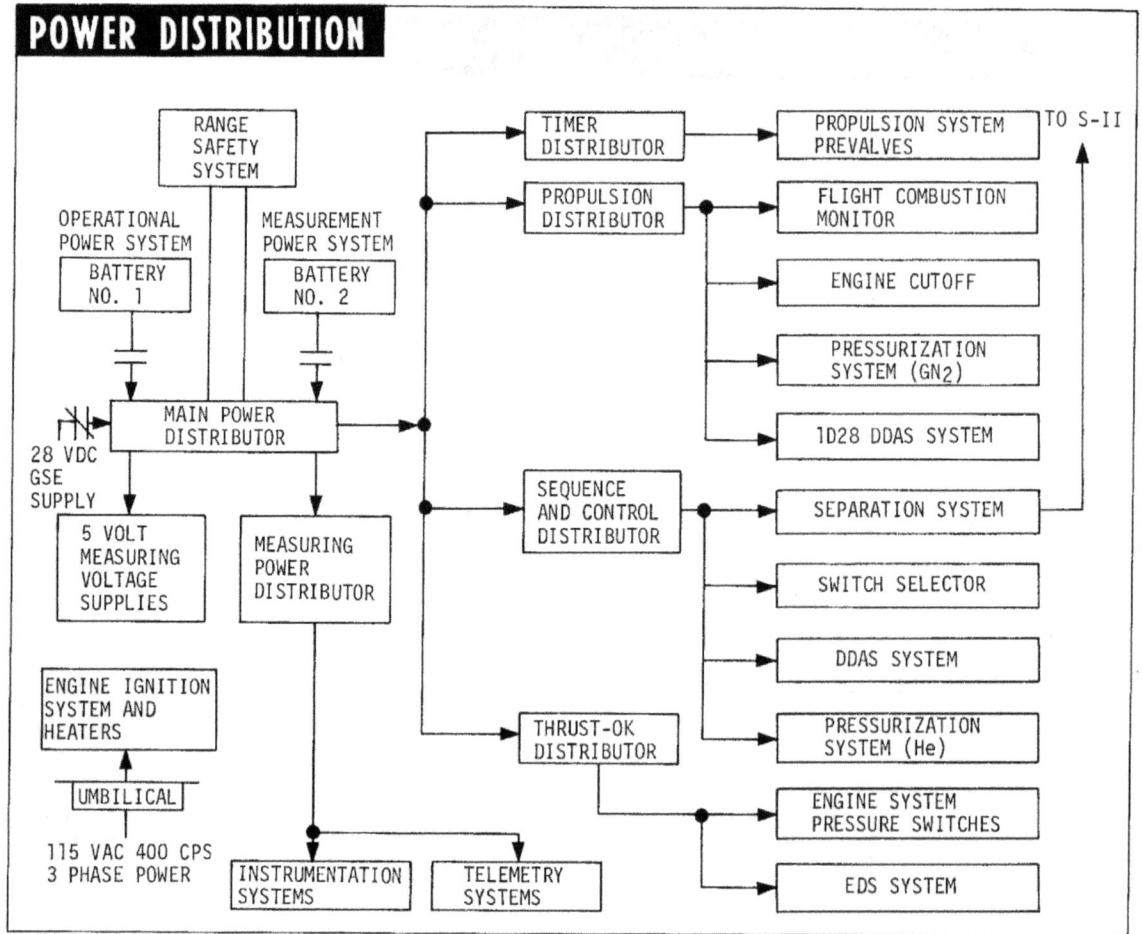

## POWER DISTRIBUTION

Figure 4-16

7. Enable outboard engine cutoff.

8. Arm EBW firing unit, retrorockets, and separation system.

9. Fire EBW firing unit, retrorockets, and separation system.

10. Measurement switchover.

In addition, a command from the emergency detection system in the IU can shut down all S-IC stage engines.

## INSTRUMENTATION

The S-IC stage instrumentation system (figure 4-18) monitors functional operations of stage systems and provides signals for vehicle tracking during the S-IC burn. Prior to liftoff, measurements are telemetered by coaxial cable to ground support equipment. During flight, data is transmitted to ground stations over RF links. The ODOP system uses the doppler principle to provide vehicle position and acceleration data during flight. Section VII provides a detailed discussion of the telemetry system.

## TELEMETRY SYSTEM

The telemetry system accepts the 890 signals produced by the measuring portion of the instrumentation system and transmits them to ground stations. The telemetry equipment includes multiplexers, subcarrier oscillators, amplifiers, modulators, transmitters, and an omnidirectional system of four antennae. The telemetry system uses multiplex techniques (time sharing) to transmit large quantities of measurement data over a relatively small number of basic RF links. This equipment also includes tape recorders for recording critical data which would otherwise be lost due to telemetry blackout during S-II ullage and S-IC retrorocket firing. The recorders, which have a 3 minute record capability, will playback the critical data during stage free fall.

There are three basic types of telemetry systems in the S-IC stage (figure 4-19). The high frequency data such as vibration and acoustics measurements are transmitted via two independent single sideband (SSB) FM telemetry links (3). Three pulse amplitude modulated/frequency modulated/frequency modulated (PAM/FM/FM) links (1) are used for telemetering low-to-medium frequency data such as

## SEQUENCE AND CONTROL

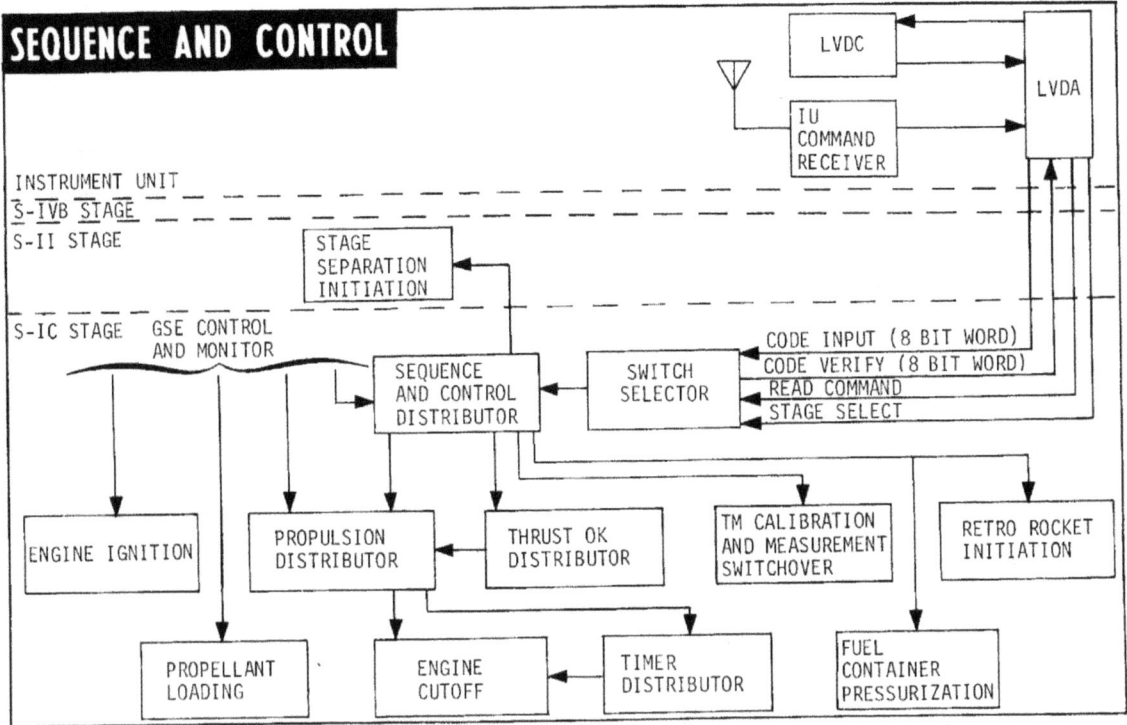

Figure 4-17

## S-IC INSTRUMENTATION

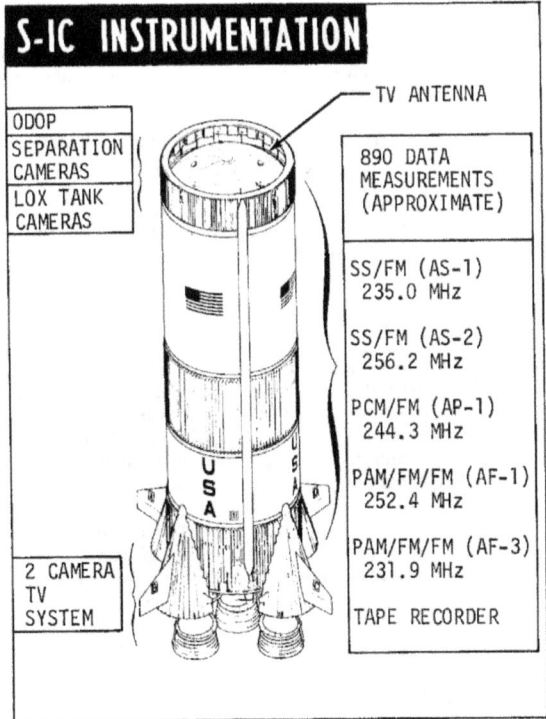

Figure 4-18

pressure, temperature or strain indications. Time multiplexed data from the PAM links are also routed through the PCM links at one third sampling rate for DDAS transmission during preflight testing and for redundant RF transmission during flight. A pulse code modulated/digital data acquisition system (PCM/DDAS) link (2) provides for acquisition of analog and digital flight data, provides a hardwire link for obtaining PCM data and PAM time multiplexed data during test and checkout and permits the redundant monitoring of PAM data during flight.

The PCM/DDAS system assembles and formats PCM/FM time shared data so it can be sent over coaxial cables for automatic ground checkout or over an RF link during flight.

### MEASUREMENT SYSTEM

The measurement system senses performance parameters and feeds signals to the telemetry system. It includes transducers, signal conditioning, and distribution equipment necessary to provide the required measurement ranges and suitably scaled voltage signals to the inputs of the telemetry system.

The S-IC measuring system performs three main functions:

1. Detection of the physical phenomena to be measured and transformation of these phenomena into electrical signals.

2. Process and condition the measured signals into the proper form for telemetering.

3. Distribution of the data to the proper channel of the telemetry system.

Measurements fall into a number of basic categories depending upon the type of measured variable, the variable rate of change with time, and other considerations. Since the vehicle is subjected to maximum buffeting forces during the S-IC boost phase, a number of strain and vibration measurements are required. Figure 4-20 summarizes the measurements to be taken on the S-IC stage.

### Remote Automatic Calibration System (RACS)

The RACS is used to verify measurement circuit operation and continuity by stimulating the transducer directly, or by inserting a simulated transducer signal in the signal conditioner circuit. Measurement operation is verified at 80 percent of the maximum transducer range (high level), at 20 percent of the maximum range (low level), and at the normal run level.

### Antennae

The S-IC stage telemetry system utilizes a total of four shunt fed stub antennae operating in pairs as two independent antenna systems shown as system 1 and system 2 in figure 4-19. Information from telemetry links operating at 240.2, 252.4 and 231.9 MHz is transmitted through system 1, and the information from telemetry links operating at 235.0, 244.3, and 256.2 MHz is transmitted through system 2.

### ODOP

An offset doppler (ODOP) frequency measurement system (figure 4-21) is an elliptical tracking system which measures the total doppler shift in an ultra high frequency (UHF) continuous wave (CW) signal transmitted to the S-IC stage. The ODOP system uses a fixed station transmitter, a vehicle borne transponder, and three or more fixed station receivers to determine the vehicle position. In this system the transmitter, transponder, and one receiver describe an ellipsoid whose intersection with the first ellipsoid is a line. The addition of the third receiver produces a third ellipsoid whose intersection with the line of intersection of the first and second ellipsoids is a point. This point is the transponder.

### FILM CAMERA

The film camera system (figure 4-22) consists of four individual film camera subsystems. Two pulse type cameras record lox behavior immediately preceding and during flight, and two movie (continuous running) type cameras film the separation of the Saturn V first and second stages. Two strobe lights illuminate the interior of the lox tank for the pulse cameras. Each of the four cameras is contained in a recoverable capsule which is ejected following stage separation.

When the ejection signal is given, $GN_2$ pressure, supplied to the ejection cylinder, causes camera ejection. Each camera has a flashing light beacon and a radio transmitter to assist in recovery.

$GN_2$ purges all camera capsules and optical system interfaces.

### LOX CAMERAS

Two pulse type cameras record the lox behavior through quartz windows in the forward dome of the lox tank. Visual information is transferred from each window to a camera by a fiber optic bundle with a coupling lens assembly on each end. One narrow angle lens (65 degrees) views the lower half of the lox tank and one wide angle lens (160 degrees) views the upper half of the lox tank. Each camera takes 5 frames-per-second on Ektachrome type MS (ASA 64) 16 mm color film for high speed cameras. A ground signal commands the camera timer and both lox cameras to begin operation at about 55 seconds before first ignition. The film is stationary during exposure and shortly after each flash. Film exposure is controlled by the strobe light flash duration and intensity. A narrow-angle pulsed strobe light operating from a 2,250-volt supply illuminates the lower half of the lox tank. A wide-angle strobe light operates similarly to illuminate the upper half of the lox tank. The strobe lights are turned off prior to engine cutoff (to prevent the light-sensitive lox cutoff sensors from initiating a premature engine cutoff signal) and are turned on again after engine cutoff.

### Separation Cameras

Two wide angle (160 degrees) movie cameras record separation of the Saturn V first and second stages on Ektachrome type MS (ASA 64) 16 mm color film at a 100 frame-per-second rate through windows in the camera capsules. Separation camera recording begins about four seconds before separation and continues until capsule ejection.

### AIRBORNE TELEVISION

The airborne television system (figure 4-23) provides inflight real-time visual performance information about the S-IC engines, as well as permanent storage of the pictures televised. The combined images from all the cameras are recorded on video tape at the TV checkout system within the RF checkout station. A separate monitor provides a real-time image from each of the cameras. A kinescope record is made up of the received video images as a permanent record backup to the videotape recording.

The transmitter and RF power amplifier are installed in an environmentally controlled equipment container. The remaining hardware, consisting of two TV cameras, associated optics, camera control units, and the video register, are located on the thrust structure and in the engine compartment. Figures 4-18 and 4-22 show the location of the TV system.

### TV Optical System

Four twelve-element, fixed-focus TV camera objective lens, with a sixty degree viewing angle, are mounted in four locations (figure 4-23). A removable aperture disc (waterhouse stop) can be changed from f/2 to f/22 to vary the image intensity. A quartz window, rotated by a dc motor through a friction drive protects each objective lens. Two fixed metallic mesh scrapers removes soot from each window. A nitrogen purge removes loose material from the windows and cools the protective "bird house" enclosure around the lens. Thermal insulation reduces temperature extremes.

Optical images are conducted from the engine area to the television camera through flexible, fiber optic bundles. Image intensity is reduced approximately seventy percent by the fiber optic bundles. Images conducted by fiber optic bundles from two objective lens are combined into a dual image in a larger bundle by a "T" fitting. A 14-element coupling lens adapts the large dual image bundle to the TV camera.

# TELEMETRY SYSTEMS

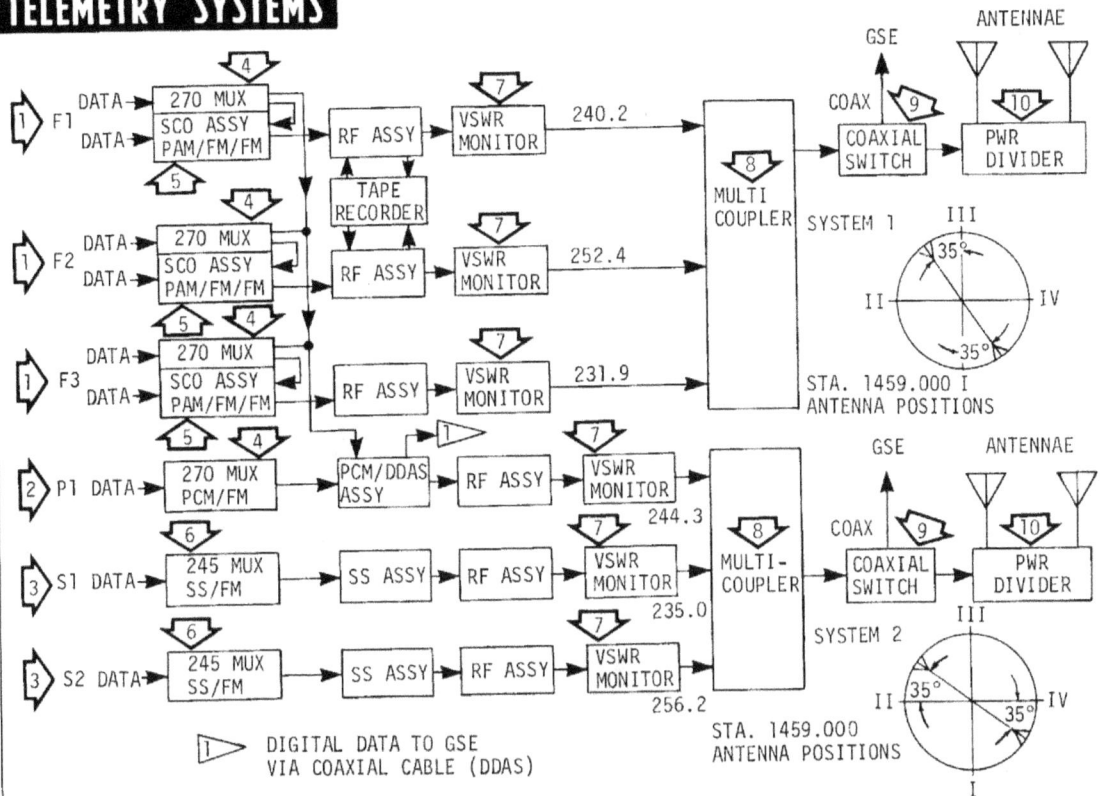

$\triangleright 1$ DIGITAL DATA TO GSE
VIA COAXIAL CABLE (DDAS)

$\triangleright 1$ The PAM/FM/FM system is used to transmit data in the frequency range below 1000 hertz. Each PAM/FM/FM link provides 14 inter-range instrumentation group (IRIG) continuous data channels with a maximum frequency response of 450 hertz and 27 time-sharing channels capable of transmitting a maximum of 230 measurements (10 per channel) with a sampling rate of 12 samples per second and 4 channels at 120 samples per second.

$\triangleright 2$ The PCM/FM system provides a primary data acquisition link for analog and digital data and a redundant means of monitoring data transmitted by the three PAM links. This system assemblies time-shared data into a serial PCM format which is transmitted by a RF assembly and a 600 kc voltage controlled oscillator (VCO) in the PCM/DDAS assembly.

$\triangleright 3$ SS/FM system links S1 and S2 transmit acoustical and vibration data in the frequency range of 50 to 3000 hertz. Total bandwidth of the system is 76 kHz. Each multiplexer can handle up to 80 measurements: each single sideband assembly handles 15 continuous channels. The SS/FM system provides frequency division multiplexing of fifteen AM data channels on a FM carrier.

$\triangleright 4$ The model 270 time division multiplexer accepts voltage inputs from 0 to 5 volts. It provides outputs to the PCM/DDAS assembly and the PAM Links. Of 30 channels, 3 channels (28, 39, and 30) are used for voltage reference and synchronization, 4 channels (24, 25, 26, and 27) are sampled at 120 samples per second the remaining 23 channels are sub-multiplexed and sampled at 12 samples per second to accomodate 10 measurements each.

Figure 4-19 (Sheet 1 of 2)

# TELEMETRY SYSTEMS

5 A SUB-CARRIER OSCILLATOR ASSEMBLY (SCO) provides the frequency division multiplexing for the PAM/FM/FM links. Channels 2 through 15 provide 14 continuous data channels and band X handles the time-division multiplexed data from the model 270 multiplexer. All oscillator outputs are mixed in an amplifier and routed to the FM transmitter in the RF assembly.

6 A model 245 multiplexer increases the number of measurements which may be transmitted over one RF link. The unit permits 80 data measurements to be time division multiplexed into 16 output channels. An auxiliary output is also provided for channel identification.

7 The voltage standing wave ratio (VSWR) monitor is used to monitor the performance of the telemetry antenna system and the output of the telemetry transmitter.

8 The multicoupler couples three simultaneous radio frequency (RF) signals into a common output without mutal interference and with minimum insertion loss (0.15 db or less).

9 The Coaxial Switch provides a means of connecting to the GSE coaxial transmission line or to the vehicle antennae. Switching between the two lines is accomplished by application or removal of 28-volts DC to the switch.

10 The Power Divider splits RF power between the two antennae while handling an average power level of 100 watts over a range of 215 to 260 MHz.

Figure 4-19 (Sheet 2 of 2)

| STAGE MEASUREMENTS | |
|---|---|
| TYPE | QTY |
| ACCELERATION | 3 |
| ACOUSTIC | 4 |
| TEMPERATURE | 252 |
| PRESSURE | 164 |
| VIBRATION | 150 |
| FLOWRATE | 35 |
| POSITION | 1 |
| DISCRETE SIGNALS | 147 |
| LIQUID LEVEL | 22 |
| VOLTAGE, CURRENT, FREQUENCY | 11 |
| TELEMETRY POWER | 19 |
| ANGULAR VELOCITY | 6 |
| STRAIN | 71 |
| RPM | 5 |
| TOTAL | 890 |

Figure 4-20

Video System

A video system (figure 4-23) converts the optical image to an electronic signal. The video system consists of a camera control unit, a transmitter and an antenna. Each camera houses a 28-volt vidicon tube, preamplifiers, and vertical sweep circuits for 30 frames-per-second scanning. A camera control unit for each camera houses the amplifiers, fly back, horizontal sweep, and synchronization equipment. A video register amplifies and samples every other frame from the two dual image video cameras. A 2.5 watt, 1710 mc, FM transmitter feeds the radome covered, seven-element yagi antenna array.

## ORDNANCE

The S-IC ordnance systems include the propellant dispersion (flight termination) system (figure 4-24) and the retrorocket system (figure 4-25).

## PROPELLANT DISPERSION SYSTEM

The S-IC propellant dispersion system (PDS) provides the means of terminating the flight of the Saturn V if it varies beyond the prescribed limits of its flight path or if it becomes a safety hazard during the S-IC boost phase. The system is installed on the stage in compliance with Air Force Eastern Test Range (AFETR) Regulation 127.9 and AFETR Safety Manual 127.1.

The PDS is a dual channel, parallel redundant system composed of two segments. The radio frequency segment receives, decodes, and controls the propellant dispersion commands. The ordnance train segment consists of two exploding bridgewire (EBW) firing units, two EBW detonators, one safety and arming (S&A) device (shared by both channels), six confined detonating fuse (CDF) assemblies, two CDF tees, two CDF/flexible linear shaped charge (FLSC) connectors, and two FLSC assemblies.

## ODOP

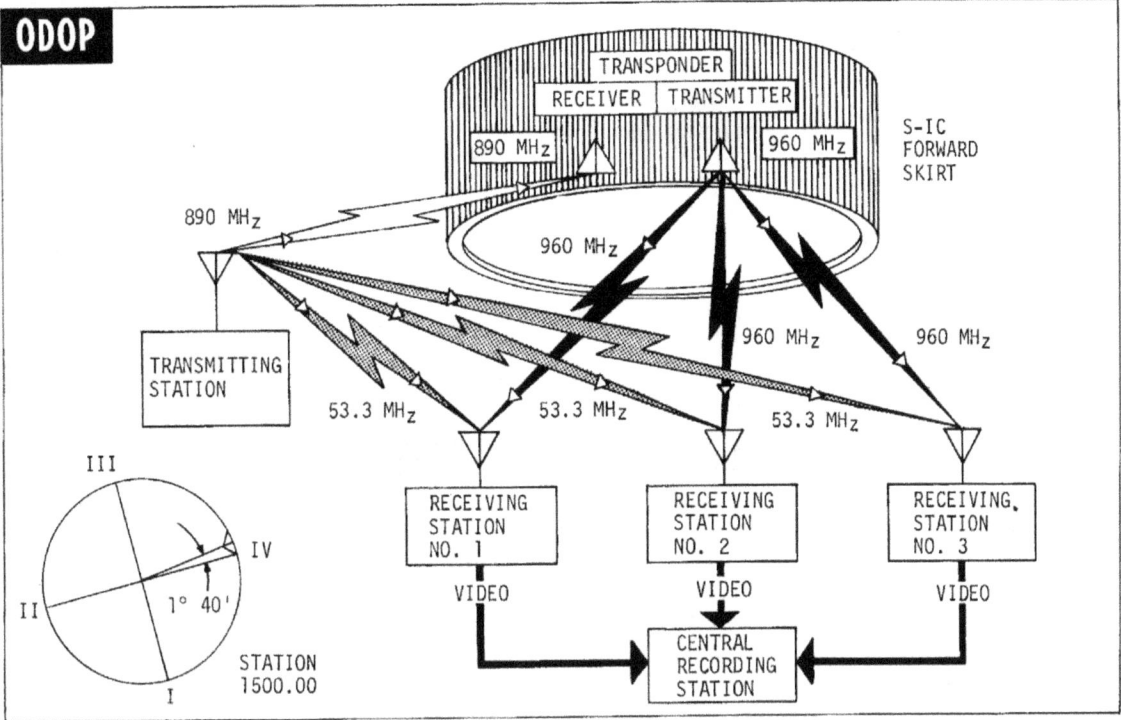

Figure 4-21

## FILM CAMERAS

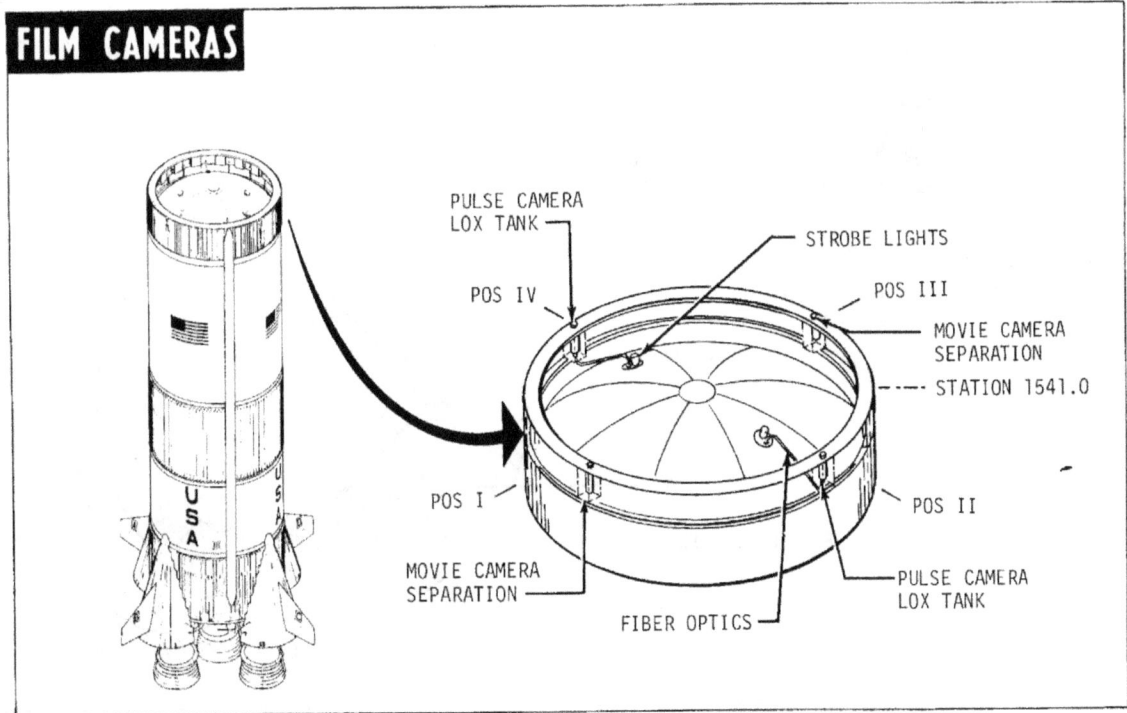

Figure 4-22

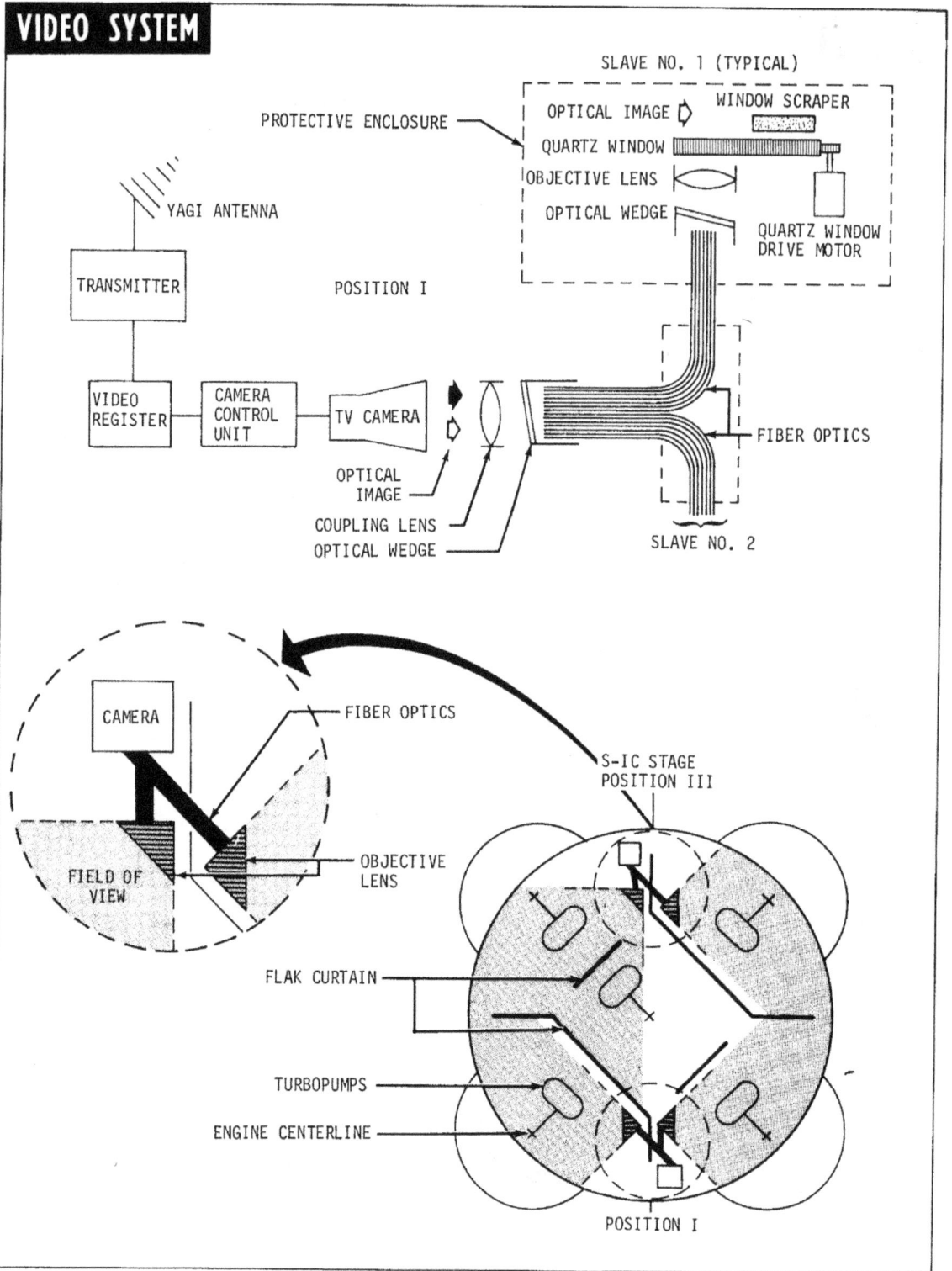

# VIDEO SYSTEM

SLAVE NO. 1 (TYPICAL)

PROTECTIVE ENCLOSURE

OPTICAL IMAGE
WINDOW SCRAPER
QUARTZ WINDOW
OBJECTIVE LENS
OPTICAL WEDGE
QUARTZ WINDOW
DRIVE MOTOR

YAGI ANTENNA

POSITION I

TRANSMITTER

VIDEO
REGISTER

CAMERA
CONTROL
UNIT

TV CAMERA

FIBER OPTICS

OPTICAL
IMAGE

COUPLING LENS

OPTICAL WEDGE

SLAVE NO. 2

CAMERA

FIBER OPTICS

S-IC STAGE
POSITION III

OBJECTIVE
LENS

FIELD OF
VIEW

FLAK CURTAIN

TURBOPUMPS

ENGINE CENTERLINE

POSITION I

Figure 4-23

# PROPELLANT DISPERSION

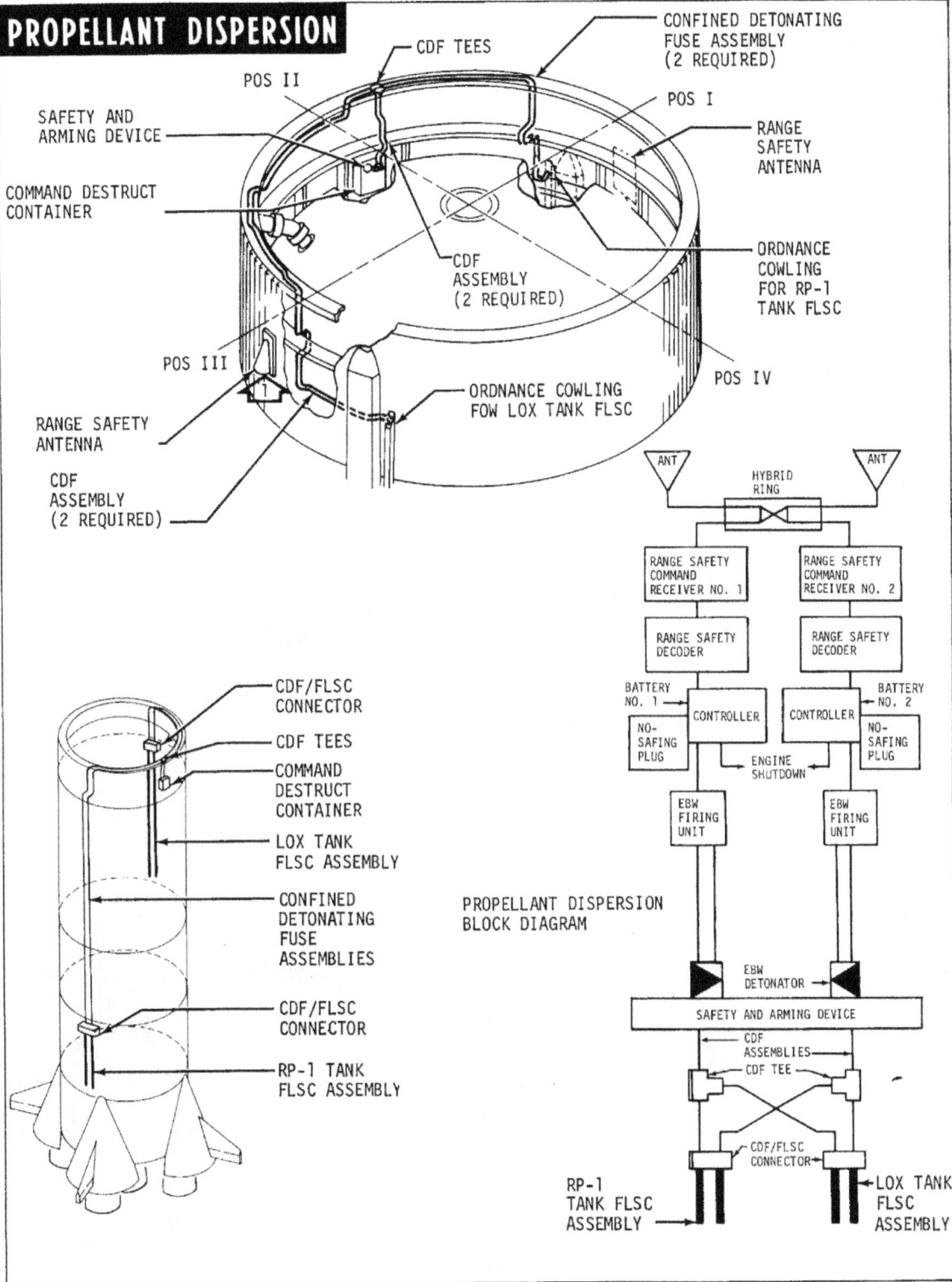

CDF TEES

CONFINED DETONATING
FUSE ASSEMBLY
(2 REQUIRED)

POS II

POS I

SAFETY AND
ARMING DEVICE

RANGE
SAFETY
ANTENNA

COMMAND DESTRUCT
CONTAINER

CDF
ASSEMBLY
(2 REQUIRED)

ORDNANCE
COWLING
FOR RP-1
TANK FLSC

POS III

POS IV

ORDNANCE COWLING
FOW LOX TANK FLSC

RANGE SAFETY
ANTENNA

CDF
ASSEMBLY
(2 REQUIRED)

CDF/FLSC
CONNECTOR

CDF TEES

COMMAND
DESTRUCT
CONTAINER

LOX TANK
FLSC ASSEMBLY

CONFINED
DETONATING
FUSE
ASSEMBLIES

CDF/FLSC
CONNECTOR

RP-1 TANK
FLSC ASSEMBLY

ANT

HYBRID
RING

ANT

RANGE SAFETY
COMMAND
RECEIVER NO. 1

RANGE SAFETY
COMMAND
RECEIVER NO. 2

RANGE SAFETY
DECODER

RANGE SAFETY
DECODER

BATTERY
NO. 1

CONTROLLER

CONTROLLER

BATTERY
NO. 2

NO-
SAFING
PLUG

ENGINE
SHUTDOWN

NO-
SAFING
PLUG

EBW
FIRING
UNIT

EBW
FIRING
UNIT

PROPELLANT DISPERSION
BLOCK DIAGRAM

EBW
DETONATOR

SAFETY AND ARMING DEVICE

CDF
ASSEMBLIES

CDF TEE

CDF/FLSC
CONNECTOR

RP-1
TANK FLSC
ASSEMBLY

LOX TANK
FLSC
ASSEMBLY

Figure 4-24

# RETROROCKETS

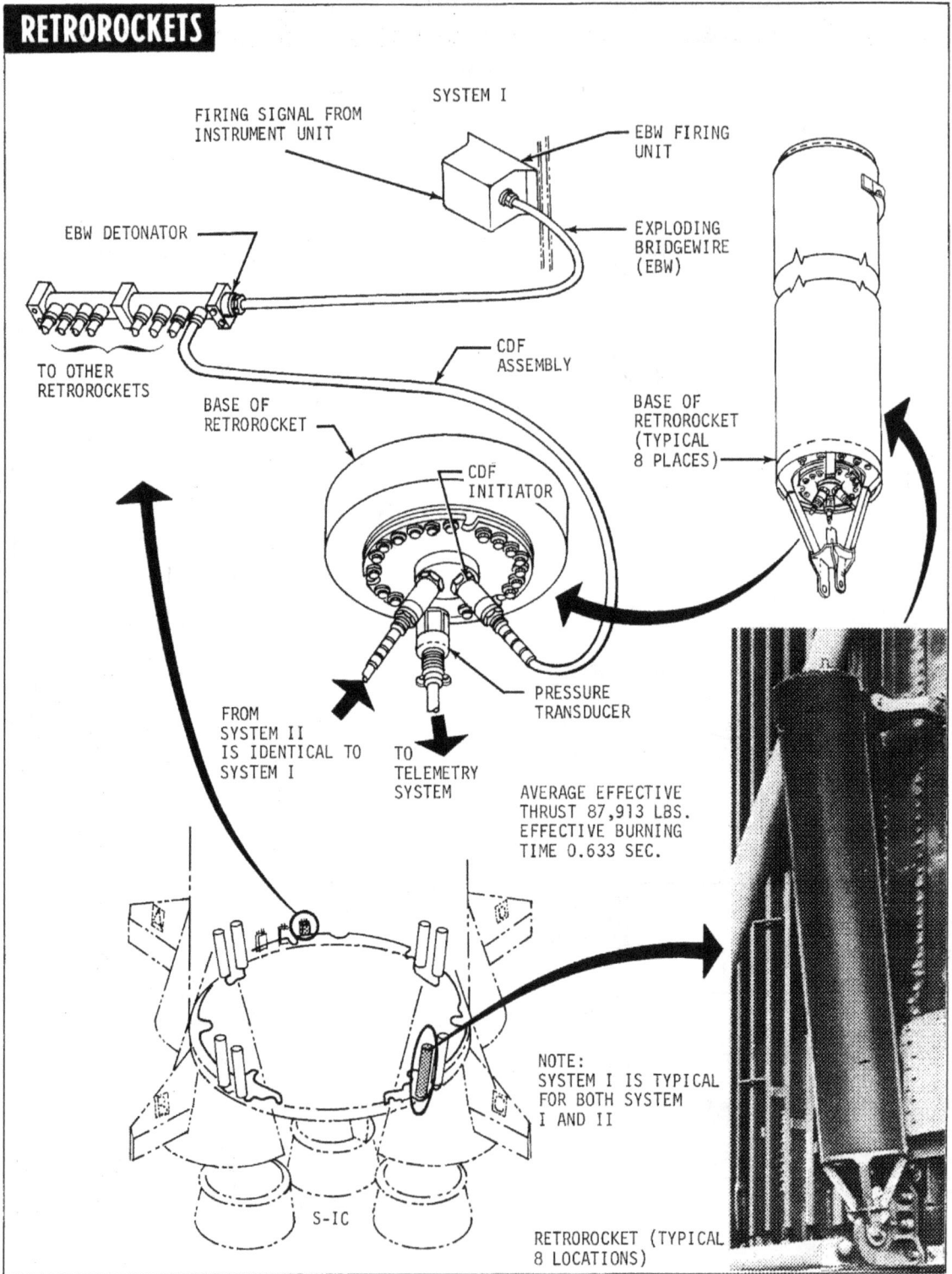

SYSTEM I

FIRING SIGNAL FROM
INSTRUMENT UNIT

EBW FIRING
UNIT

EBW DETONATOR

EXPLODING
BRIDGEWIRE
(EBW)

TO OTHER
RETROROCKETS

CDF
ASSEMBLY

BASE OF
RETROROCKET

BASE OF
RETROROCKET
(TYPICAL
8 PLACES)

CDF
INITIATOR

FROM
SYSTEM II
IS IDENTICAL TO
SYSTEM I

TO
TELEMETRY
SYSTEM

PRESSURE
TRANSDUCER

AVERAGE EFFECTIVE
THRUST 87,913 LBS.
EFFECTIVE BURNING
TIME 0.633 SEC.

NOTE:
SYSTEM I IS TYPICAL
FOR BOTH SYSTEM
I AND II

S-IC

RETROROCKET (TYPICAL
8 LOCATIONS)

Figure 4-25

The S&A device (figure 4-26) is a remotely controlled electro-mechanical ordnance device that is used to make safe and to arm the S-IC, S-II, and S-IVB stage PDS's. The device can complete and interrupt the explosive train by remote control, provide position indications to remote monitoring equipment, and provide a visual position indication. It also has a manual operation capability. The S&A device consists of a rotary solenoid assembly, a metal rotor shaft with two explosive inserts, and position sensing and command switches that operate from a rotor shaft cam. In the safe mode, the longitudinal axis of the explosive inserts are perpendicular to the detonating wave path, thus forming a barrier to the explosive train. To arm the device, the shaft is rotated 90 degrees to align the inserts between the EBW detonators and the CDF adapters to form the initial part of the explosive train.

Should emergency flight termination become necessary, two coded radio frequency commands are transmitted to the launch vehicle by the range safety officer. The first command arms the EBW firing units and initiates S-IC engine cutoff. (See figure 4-24 for block diagram of the PDS and location of PDS components.) The second command, which is delayed to permit charging of the EBW firing units, discharges the EBW firing units across the exploding bridgewire in the EBW detonators mounted on the S&A device (see figure 4-26). The resulting explosive wave propagates through the S&A device inserts to the CDF assemblies and to the CDF tees. The CDF tees propagate the wave through insulated CDF assemblies to the FLSC assemblies mounted on the lox and RP-1 tanks. The FLSC's provide the explosive force to longitudinally sever the propellant tanks and disperse the propellants. There are six 88-inch FLSC sections mounted on the lox tank and three 88-inch sections on the fuel tank. These sections are positioned on the propellant tanks to minimize mixing of the propellants after the tanks are severed.

## RETROROCKETS

The S-IC retrorockets are mounted, in pairs, (figure 4-25) in the fairings of the F-1 engine. At retrorocket ignition the forward end of the fairing is burned and blown through by the exhausting gases. Each retrorocket is pinned securely to the vehicle support and pivot support fittings at an angle of 7.5 degrees from center line. The thrust level developed by seven retrorockets (one retrorocket out) is adequate to separate the S-IC stage a minimum of six feet from the vehicle in less than one second.

The eight retrorockets (figure 4-25), provide separation thrust after S-IC burnout. The firing command originates in the Instrument Unit and activates redundant firing systems.

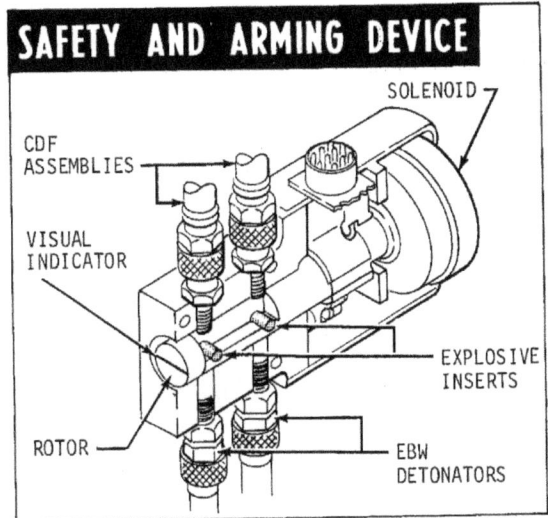

Figure 4-26

Additional redundancy is provided by interconnection of the two confined detonating fuse (CDF) manifolds with CDF assemblies. The exploding bridgewire (EBW) firing unit circuits are grounded by a normally closed relay until the firing command is initiated by the Instrument Unit. High voltage electrical signals are released from the two EBW firing units to the EBW detonators upon vehicle deceleration to 0.5g. The signals cause the detonator bridgewires to explode, thereby detonating the surrounding explosives. The explosion then propagates through the CDF manifold explosive and CDF assemblies into the igniter assembly. The igniter assembly located within the base of each retrorocket is then ignited, causing a buildup and release of the gases into the main grain of the retrorocket. Each retrorocket is ignited by either of two CDF initiators mounted on its aft structure. Operational ground check of the system through the firing unit is accomplished through use of pulse sensors which absorb the high voltage impulse from the firing unit and transmit a signal through the telemetry system. The pulse sensors are removed prior to launch.

Each retrorocket is a solid propellant rocket with a case bonded, twelve-point star, internal burning, composite propellant cast directly into the case and cured. The propellant is basically ammonium perchlorate oxidizer in a polysulfide fuel binder. The motor is 86 inches long by 15-1/4 inches diameter and weights 504 pounds, nominal, of which 278 pounds is propellant.

# SECTION V

# S-II STAGE

## TABLE OF CONTENTS

## INTRODUCTION

The S-II stage provides second stage boost for the Saturn V launch vehicle. The stage (figure 5-1) is 81.5 feet long, 33 feet in diameter, and is powered by five liquid propellant J-2 rocket engines which develop a nominal vacuum thrust of 228,000 pounds each for a total of 1,140,000 pounds. The four outer J-2 engines are equally spaced on a 17.5 foot diameter circle and are capable of being gimbaled through a plus or minus 7.0 degree pattern for thrust vector control. The fifth engine is mounted on the stage centerline and is fixed. Dry weight of the stage is approximately 88,400 pounds (98,659 pounds including the S-IC/S-II interstage). The stage approximate gross weight is 1,034,900 pounds.

At engine cutoff the S-II stage separates from the S-IVB and, following a suborbital path, reenters the atmosphere where it disintegrates due to reentry loads.

The stage consists of the structural airframe, the J-2 engines, piping, valves, wiring, instrumentation, electrical and electronic equipment, ordnance devices, and four solid propellant ullage rockets. These are collected into the following major systems: structural, environmental control, propulsion, flight control, pneumatic control, propellant, electrical, instrumentation, and ordnance. The stage has structural and electrical interfaces with the S-IC and S-IVB stages; and electrical, pneumatic, and fluid interfaces with ground support equipment through its umbilicals and antennae.

## STRUCTURE

The S-II airframe (figure 5-1) consists of a body shell structure (forward and aft skirts and interstage), a propellant tank structure (liquid hydrogen and liquid oxygen tanks), and a thrust structure. The body shell structure transmits first and second stage boost loads (axial, shear, and bending moment) and stage body bending and longitudinal forces between the adjacent stages, the propellant tank structure, and the thrust structure. The propellant tank structure holds the propellants, liquid hydrogen ($LH_2$) and liquid oxygen (lox), and provides structural support between the aft and forward skirts. The thrust structure transmits the thrust of the five J-2 engines to the body shell structure; compression loads from engine thrust; tension loads from idle engine

weight; and cantilever loads from engine weight during S-II boost.

## BODY SHELL STRUCTURE

The body shell structure units, the forward skirt, aft skirt, and interstage are of the same basic design except that the aft skirt and interstage are of generally heavier construction because of higher structural loads.

Each unit is a cylindrical shell of semimonocoque construction, built of 7075 aluminum alloy material, stiffened by external hat-section stringers and stabilized internally by circumferential ring frames. The forward skirt has a basic skin thickness of 0.040 inch and the aft skirt and interstage both have basic skin thicknesses of 0.071 inch.

## THRUST STRUCTURE

The thrust structure, like the body shell structure, is of semimonocoque construction but in the form of a truncated cone increasing in size from approximately 18 feet in diameter to the 33 foot outside diameter of the airframe. It is stiffened by circumferential ring frames and hat-section stringers. Four pairs of thrust longerons (two at each outboard engine location) and a center engine support beam cruciform assembly accept and distribute the thrust loads of the J-2 engines. The shell structure is of 7075 aluminum alloy. A fiberglass honeycomb heat shield, supported from the lower portion of the thrust structure, protects the stage base area from excessive temperatures during S-II boost.

The conical shell also serves to support the major portion of systems components carried on the S-II, either mounted in environmentally controlled equipment containers or directly to the airframe structure (See figure 5-2).

## PROPELLANT TANK STRUCTURE

The $LH_2$ tank consists of a long cylinder with a concave modified ellipsoidal bulkhead forward and a convex modified ellipsoidal bulkhead aft. The aft bulkhead is common to the lox tank. The $LH_2$ tank wall is composed of six cylindrical sections which incorporate stiffning members in both the longitudinal and circumferential directions. Wall sections and bulkheads are all fabricated from 2014 aluminum alloy and are joined together by fusion welding. The forward bulkhead has a 36 inch diameter access manhole built into its center.

The common bulkhead is an adhesive-bonded sandwich assembly employing facing sheets of 2014 aluminum alloy and fiberglass/phenolic honeycomb core to prevent heat transfer and retain the cryogenic properties of the two fluids to which it is exposed. Fiberglass core insulation thickness varies from approximately 5 inches at the apex to 0.080 inch at the outer extremity. No connections or lines pass through the common bulkhead. The forward skin has a "J" section return at the outer edge to permit peripheral attachment to the $LH_2$ tank while the lower facing is carried through to provide structural continuity with the lox tank aft bulkhead.

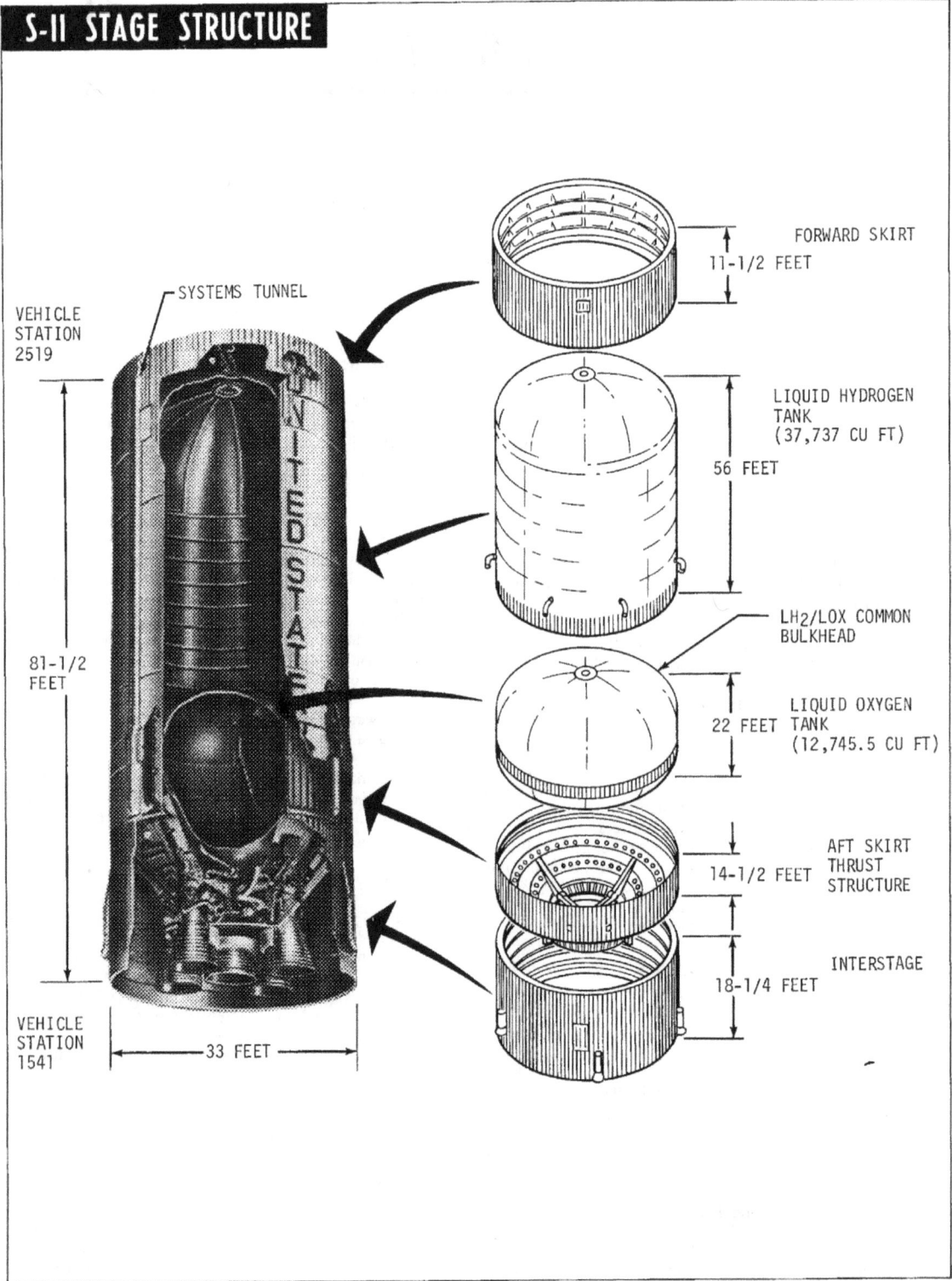

# S-II STAGE STRUCTURE

FORWARD SKIRT
11-1/2 FEET

SYSTEMS TUNNEL

VEHICLE
STATION
2519

LIQUID HYDROGEN
TANK
(37,737 CU FT)

56 FEET

LH2/LOX COMMON
BULKHEAD

81-1/2
FEET

LIQUID OXYGEN
TANK
(12,745.5 CU FT)

22 FEET

AFT SKIRT
THRUST
STRUCTURE

14-1/2 FEET

INTERSTAGE

18-1/4 FEET

VEHICLE
STATION
1541

33 FEET

Figure 5-1

# S-II STAGE EQUIPMENT LOCATIONS

1.  RANGE SAFETY COMMAND CONTAINER
2.  THERMAL CONTROL MANIFOLD
3.  LIQUID HYDROGEN TANK BULKHEAD
4.  INSTRUMENTATION SIGNAL CONDITIONER NO. 2
5.  INSTRUMENTATION SIGNAL CONDITIONER NO. 1
6.  FORWARD SKIRT
7.  MANHOLE COVER
8.  RADIO COMMAND ANTENNA - 4 PLACES
9.  TELEMETRY ANTENNA - 4 PLACES
10. UMBILICAL AND SERVICE CONNECTIONS
11. THRUST STRUCTURE
12. INSTRUMENTATION SIGNAL CONDITIONER NO. 2
13. INSTRUMENTATION SIGNAL CONDITIONER NO. 3
14. INSTRUMENTATION SIGNAL CONDITIONER NO. 1
15. PROPELLANT MANAGEMENT PACKAGE
16. ELECTRICAL ASSY (TIMING AND STAGING CENTER)
17. SERVOACTUATOR (TYPICAL)
18. ULLAGE ROCKET-4 PLACES
19. J-2 ENGINE (NO. 2)
20. HEATSHIELD
21. ACCESS DOOR
22. ELECTRICAL POWER AND CONTROL SYSTEM
23. INTERSTAGE
24. LOX FEEDLINE
25. FLIGHT CONTROL CONTAINER
26. AFT SKIRT
27. INSTRUMENTATION CONTAINER NO. 2
28. INSTRUMENTATION CONTAINER NO. 1
29. SYSTEMS TUNNEL
30. PROPELLANT DISPERSION SYSTEM
31. LH$_2$ VENTLINE AND VALVES
32. PROPELLANT MANAGEMENT ELECTRONIC PACKAGE
33. TELEMETERING CONTAINER NO. 2
34. INSTRUMENTATION SIGNAL CONDITIONER NO. 3

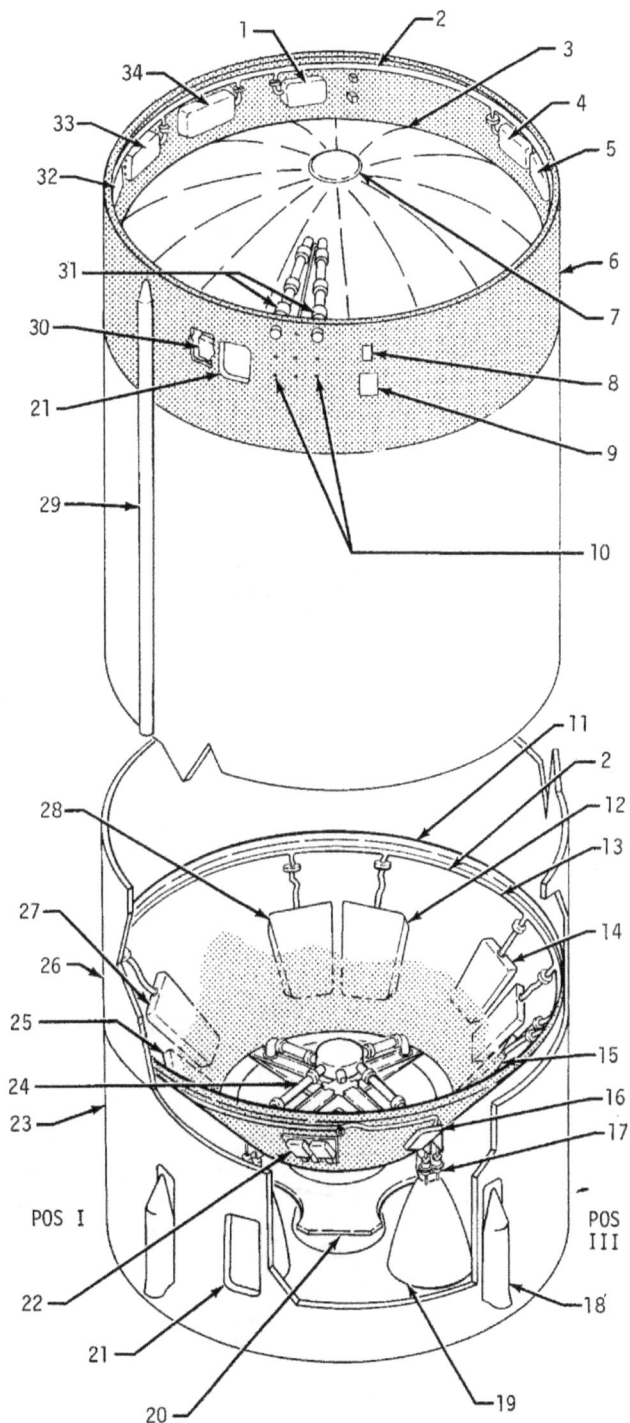

POS I

POS III

Figure 5-2

The liquid oxygen tank (figure 5-3) consists of ellipsoidal fore and aft halves with waffle-stiffened gore segments. The tank is fitted with three ring-type slosh baffles to control propellant sloshing and minimize surface disturbances and cruciform baffles to prevent the generation of vortices at the tank outlet ducts and to minimize residuals. A six-port sump assembly located at the lowest point of the lox tank provides a fill and drain opening and openings for five engine feed lines.

### SYSTEMS TUNNEL

A systems tunnel, housing electrical cables, pressurization lines, and the tank propellant dispersion ordnance, is attached externally from the S-II stage aft skirt area to the forward skirt. It has a semicircular shape 22 inches wide and about 60 feet long. Cabling which connects the S-IC stage to the instrument unit also runs through this tunnel (see figures 5-1 and 5-2).

### ENVIRONMENTAL CONTROL

The environmental control system is supplied dehumidified, thermally-conditioned air and nitrogen from a ground source for temperature control and purging of the compartments during prelaunch operations only.

### THERMAL CONTROL

The thermal control system (figure 5-2) provides temperature control to forward and aft skirt mounted equipment containers. The system is put into operation shortly after the vehicle is mated to pad facilities. Air is used as the conditioning medium until approximately twenty minutes prior to liquid hydrogen ($LH_2$) loading. At this time gaseous nitrogen ($GN_2$) is used until umbilical disconnect to preclude the possibility of an explosion in the event of $LH_2$ leakage. The change to $GN_2$ is made before propellant loading to ensure that all oxygen is expelled and dissipated before a hazard can arise. The nitrogen flow is terminated at liftoff, and no flow is provided during boost, since the equipment container insulation is capable of maintaining equipment temperatures throughout the S-II flight trajectory.

### ENGINE COMPARTMENT CONDITIONING

The engine compartment conditioning system purges the engine and interstage areas of explosive mixtures and maintains a proper temperature. Purging the compartment is accomplished prior to propellant tanking and whenever propellants are on board. A 98% $GN_2$ atmosphere circulating through the compartment maintains desired temperature while the danger of fire or explosion resulting from propellant leakage is minimized.

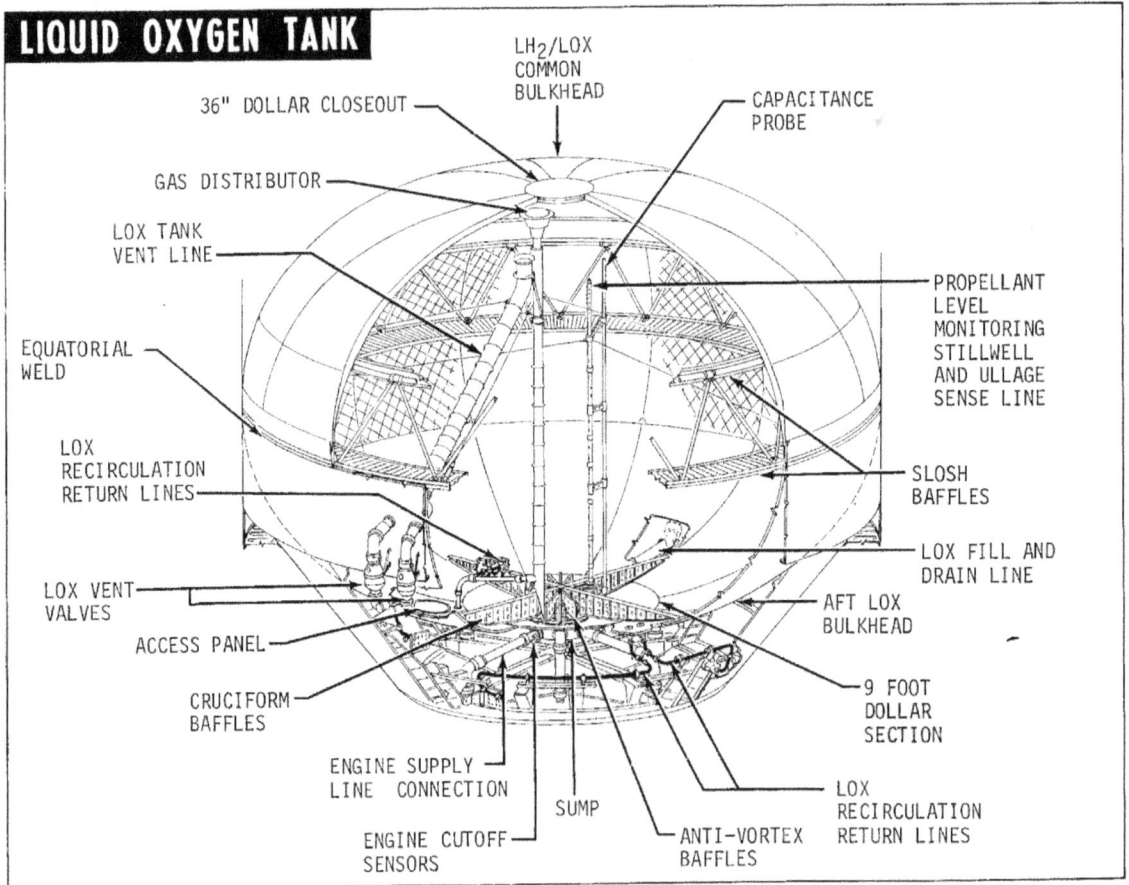

Figure 5-3

## INSULATION PURGE LEAK DETECTION

All exposed surfaces of the LH$_2$ tank require insulation to prevent condensation and to reduce temperature rise during cryogenic operations.

### Insulation/Purge

The insulation material is a foam-filled honeycomb, (figure 5-4) approximately 1.6 inches thick on the LH$_2$ tank sidewalls, and 0.5 inch thick on the forward bulkhead. The insulation has a network of passages through which helium gas is drawn. The areas to be purged are divided into several circuits: the LH$_2$ tank sidewalls, forward bulkhead, the common bulkhead, the LH$_2$ tank/forward skirt junction, and the lower LH$_2$ tank/bolting ring areas.

### Leak Detection

The purge system is used in conjunction with the leak detection system (figure 5-5) in the LH$_2$ tank sidewall, forward bulkhead, and common bulkhead areas to provide a means of detecting any hydrogen, oxygen, or air leaks while diluting or removing the leaking gases. From initiation of propellant loading until launch, the insulation is continuously purged of hazardous gases and a GSE gas analyzer determines any leakage in the purged gas.

## PROPULSION

The S-II stage engine system consists of five single-start J-2 rocket engines utilizing liquid oxygen and liquid hydrogen for propellants. The four outer J-2 engines are mounted parallel to the stage centerline. These engines are suspended by gimbal bearings to allow thrust vector control. The fifth engine is fixed and is mounted on the centerline of the stage.

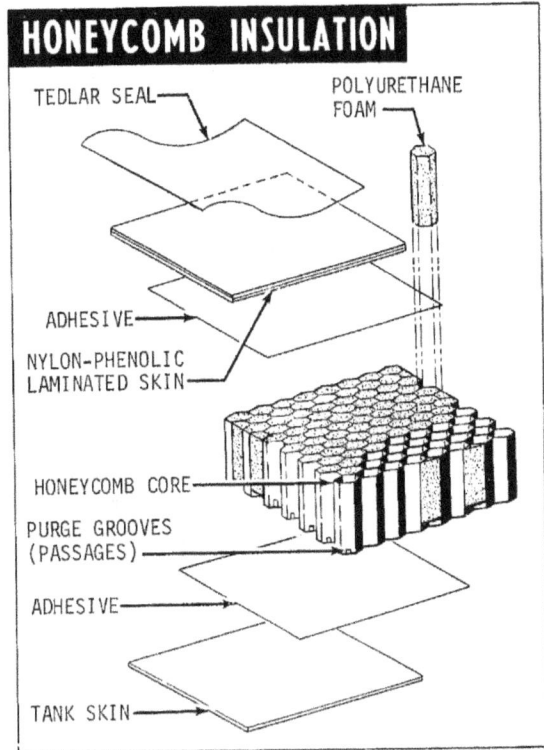

Figure 5-4

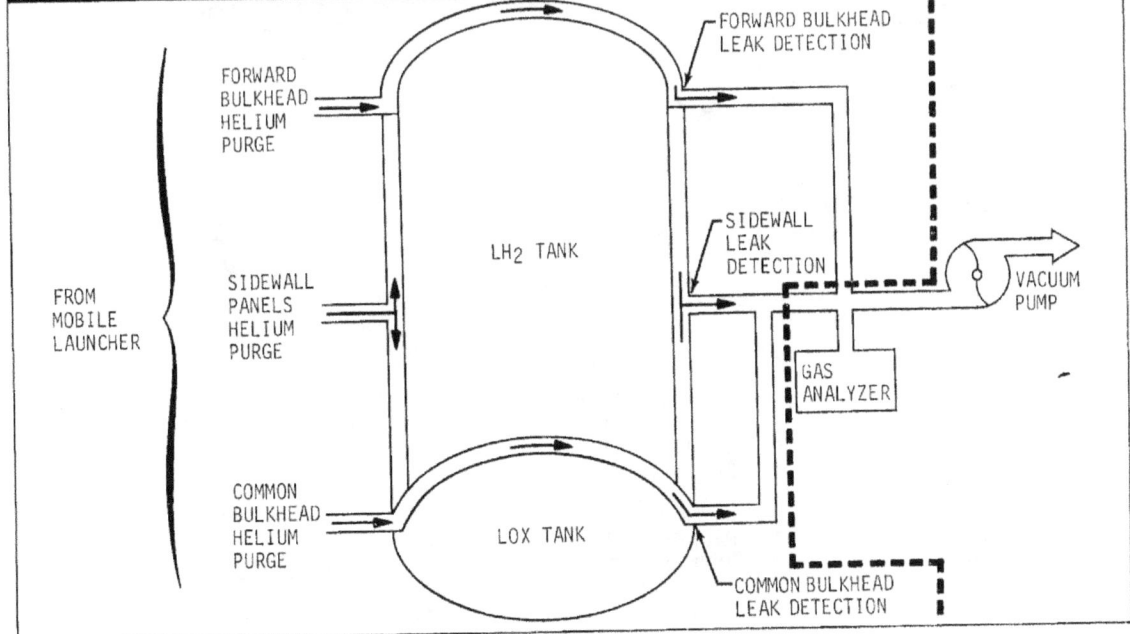

Figure 5-5

## J-2 ROCKET ENGINE

The J-2 rocket engine (figure 5-6) is a high performance, high altitude, engine utilizing liquid oxygen and liquid hydrogen as propellants. The only substances used in the engine are the propellants and helium gas. The extremely low operating temperature of the engine prohibits the use of lubricants or other fluids. The engine features a single tubular-walled, bell-shaped thrust chamber and two independently driven, direct drive, turbopumps for liquid oxygen and liquid hydrogen. Both turbopumps are powered in series by a single gas generator, which utilizes the same propellants as the thrust chamber. The main hydraulic pump is driven by the oxidizer turbopump turbine. The ratio of fuel to oxidizer is controlled by bypassing liquid oxygen from the discharge side of the oxidizer turbopump to the inlet side through a servovalve.

The engine valves are controlled by a pneumatic system powered by gaseous helium which is stored in a sphere inside the start tank. An electrical control system, which uses solid

state logic elements, is used to sequence the start and shutdown operations of the engine. Electrical power is stage supplied.

During the burn periods, the lox tank is pressurized by flowing lox through the heat exchanger in the oxidizer turbine exhaust duct. The heat exchanger heats the lox causing it to expand. The LH$_2$ tank is pressurized during burn periods by GH$_2$ from the thrust chamber fuel manifold.

Thrust vector control is achieved by gimbaling each engine with hydraulic actuators powered by engine mounted hydraulic pumps.

### Start Preparations

Preparations for an engine start include ascertaining the positions and status of various engine and stage systems and components. The J-2 engine electrical control system controls engine operation by means of electrical signals. The heart of the engine electrical control system is the electrical control package (17, figure 5-6) which sequences and times the engine start or cutoff functions.

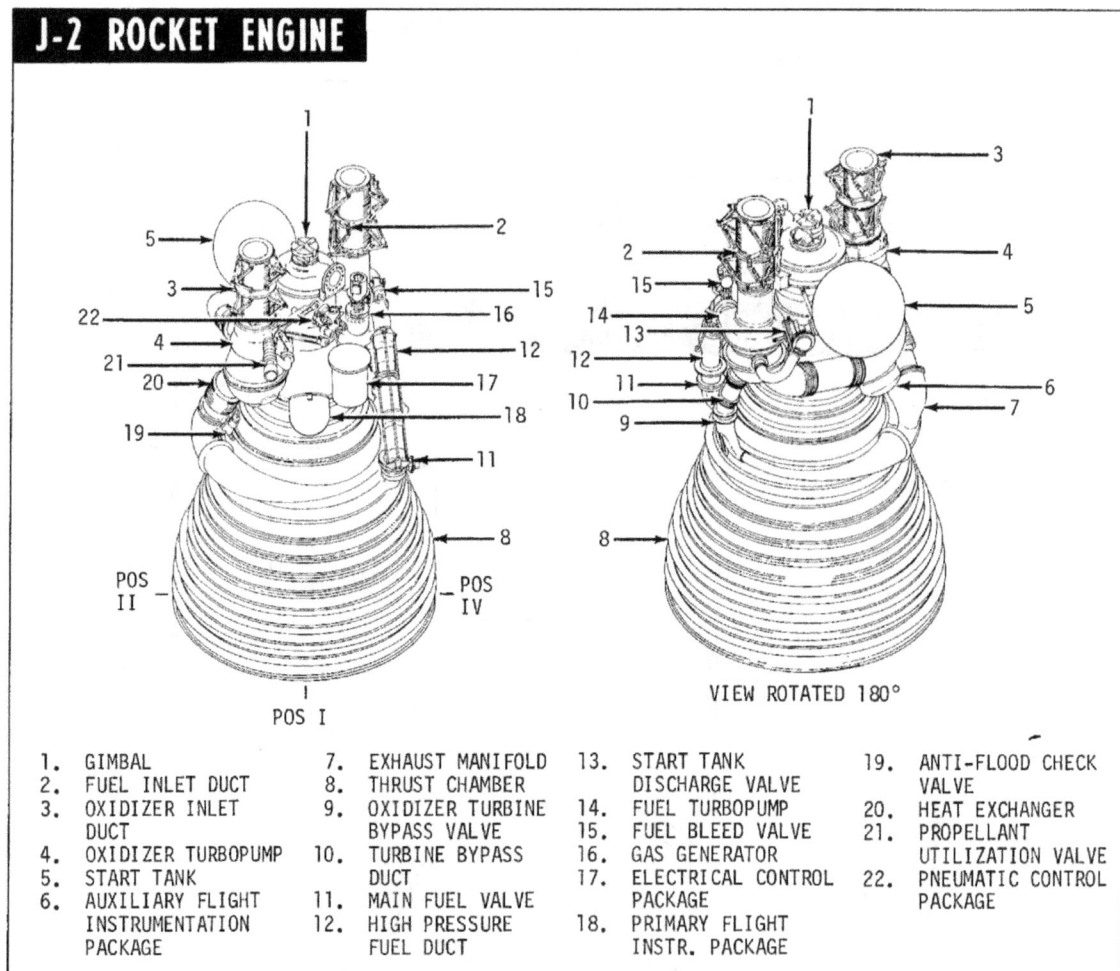

# J-2 ROCKET ENGINE

| | | | |
|---|---|---|---|
| 1. GIMBAL | 7. EXHAUST MANIFOLD | 13. START TANK | 19. ANTI-FLOOD CHECK |
| 2. FUEL INLET DUCT | 8. THRUST CHAMBER | DISCHARGE VALVE | VALVE |
| 3. OXIDIZER INLET | 9. OXIDIZER TURBINE | 14. FUEL TURBOPUMP | 20. HEAT EXCHANGER |
| DUCT | BYPASS VALVE | 15. FUEL BLEED VALVE | 21. PROPELLANT |
| 4. OXIDIZER TURBOPUMP | 10. TURBINE BYPASS | 16. GAS GENERATOR | UTILIZATION VALVE |
| 5. START TANK | DUCT | 17. ELECTRICAL CONTROL | 22. PNEUMATIC CONTROL |
| 6. AUXILIARY FLIGHT | 11. MAIN FUEL VALVE | PACKAGE | PACKAGE |
| INSTRUMENTATION | 12. HIGH PRESSURE | 18. PRIMARY FLIGHT | |
| PACKAGE | FUEL DUCT | INSTR. PACKAGE | |

Figure 5-6

Engine cutoff automatically causes the electrical control package circuitry to reset itself ready for start providing all reset conditions are met. The LVDC issues an engine ready bypass signal just prior to an engine start attempt. This bypass signal acts in the same manner as a cutoff would act. The reset signals engine ready and this allows the LVDC to send its start command. Receipt of the start command initiates the engine start sequence.

## ENGINE START SEQUENCE

When engine start is initiated (3, figure 5-7) the spark exciters in the electrical control package provide energy for the gas generator (GG) and augmented spark igniter (ASI) spark plugs (4). The helium control and ignition phase control valves, in the pneumatic control package (1), are simultaneously energized allowing helium from the helium tank (2) to flow through the pneumatic regulator to the pneumatic control system. The helium is routed through the internal check valve in the pneumatic control package (1) to ensure continued pressure to the engine valves in the event of helium supply failure. The regulated helium fills a pneumatic accumulator, closes the propellant bleed valves, (5) and purges (6) the oxidizer dome and gas generator oxidizer injector manifold. The oxidizer turbopump (12) intermediate seal cavity is continuously purged. The mainstage control valve holds the main oxidizer valve closed and opens the purge control valve which allows the oxidizer dome and gas generator oxidizer injector to be purged (6). The mainstage control valve also supplies opening control pressure to the oxidizer turbine bypass valve (13). An ignition phase control valve, when actuated, opens the main fuel valve (7) and the ASI oxidizer valve (8) and supplies pressure to the sequence valve located within the main oxidizer valve (14). Fuel is tapped from downstream of the main fuel valve for use in the ASI (4). Both propellants, under tank pressure, flow through the stationary turbopumps (12).

The sequence valve, in the main fuel valve (12), opens when the fuel valve reaches approximately 90% open and routes helium to the start tank discharge valve (STDV) (11) control valve. Simultaneously with engine start, the STDV delay timer is energized. Upon expiration of the STDV timer, and the receipt of a stage supplied mainstage enable signal, the STDV control valve and ignition phase timer are energized. As the STDV control valve energizes, the discharge valve opens, allowing pressurized $GH_2$ to flow through the series turbine drive system. This accelerates both turbopumps (12) to the proper operating levels to allow subsequent ignition and power build up of the gas generator (16). The relationship of fuel to lox turbopump speed buildup is controlled by an orifice in the oxidizer turbine bypass valve (13). During the start sequence the normally open oxidizer bypass valve (13) permits a percentage of the gas to bypass the oxidizer turbine.

During this period, ASI combustion is detected by the ASI ignition monitor. (Absence of the ignition detection signal or a start tank depressurized signal will cause cutoff at the expiration of the ignition phase timer.) With both signals present at ignition phase timer expiration, the mainstage control valve energizes. Simultaneously, the sparks deenergize timer is energized and the STDV control valve is deenergized, causing the STDV to close. Helium pressure is vented from the main oxidizer valve (14) and from the purge control valve through the mainstage control valve. The purge control valve closes, terminating the oxidizer dome and gas generator oxidizer injector manifold purges (6). Pressure from the mainstage control valve is routed to open the main oxidizer valve (14).

A sequence valve operated by the main oxidizer valve (14) permits GHe to open the gas generator control valve (4) and to close the oxidizer turbine bypass valve (13). (Flow to close the oxidizer turbine bypass valve (13) is restricted as it passes through an orifice. The orifice controls the closing speed of this valve.)

Propellants flowing into the gas generator,(16) are ignited by the sparkplugs (4). Combustion of the propellants cause the hot gases to drive the turbopump (12). The turbopump's rotation causes propellant pressure to build up. The propellant flow increases and is ignited in the thrust chamber by the torch from the ASI.

Transition into mainstage occurs as the turbopumps (12) accelerate to steadystate speeds. As oxidizer pump output pressure increases a thrust OK signal is generated by either of the two thrust OK pressure switches (17). (Cutoff occurs if no signal is received before expiration of the sparks deenergized timer.) The ASI and GG sparks exciters are deenergized at expiration of the sparks deenergized timer. Cutoff occurs if both pressure switch actuated signals (thrust OK) are lost during mainstage operation.

Steadystate operation is maintained until a cutoff signal is initiated. During this period, $GH_2$ is tapped from the fuel injection manifold to pressurize the $LH_2$ tank. The lox tank is pressurized by lox heated by the heat exchanger in the turbine exhaust duct.

Propellant utilization is discussed under a subsequent paragraph heading of PROPELLANTS.

Nominal engine thrust and specific impulse as a function of mixture ratio for the engines are shown in figure 5-8.

## ENGINE CUTOFF

The S-II J-2 engine may receive cutoff signals from several different sources. These sources include engine interlock deviations, EDS automatic and manual abort cutoffs and propellant depletion cutoff. Each of these sources signal the LVDC in the IU. The LVDC sends the engine cutoff signal to the S-II switch selector. The switch selector, in turn, signals the electrical control package. The electrical control package controls all the local signals for the cutoff sequence.

### Cutoff Sequence

The engine cutoff sequence is shown graphically in figure 5-9. The electrical control package receives the cutoff signal (1), and deenergizes the mainstage and ignition phase control

# S-II ENGINE START

START COMMAND FROM IU

FROM
LOX TANK

FROM
LH$_2$ TANK

ELECTRICAL
CONTROL
PACKAGE

STAGE SUPPLIED
MAINSTAGE ENABLE SIGNAL

TO
LH$_2$ TANK

TO
LOX TANK

TO LH$_2$
TANK

FROM GSE

LEGEND

SENSE LINE

HELIUM

FUEL

COMBUSTION GASES

LOX

GASEOUS HYDROGEN

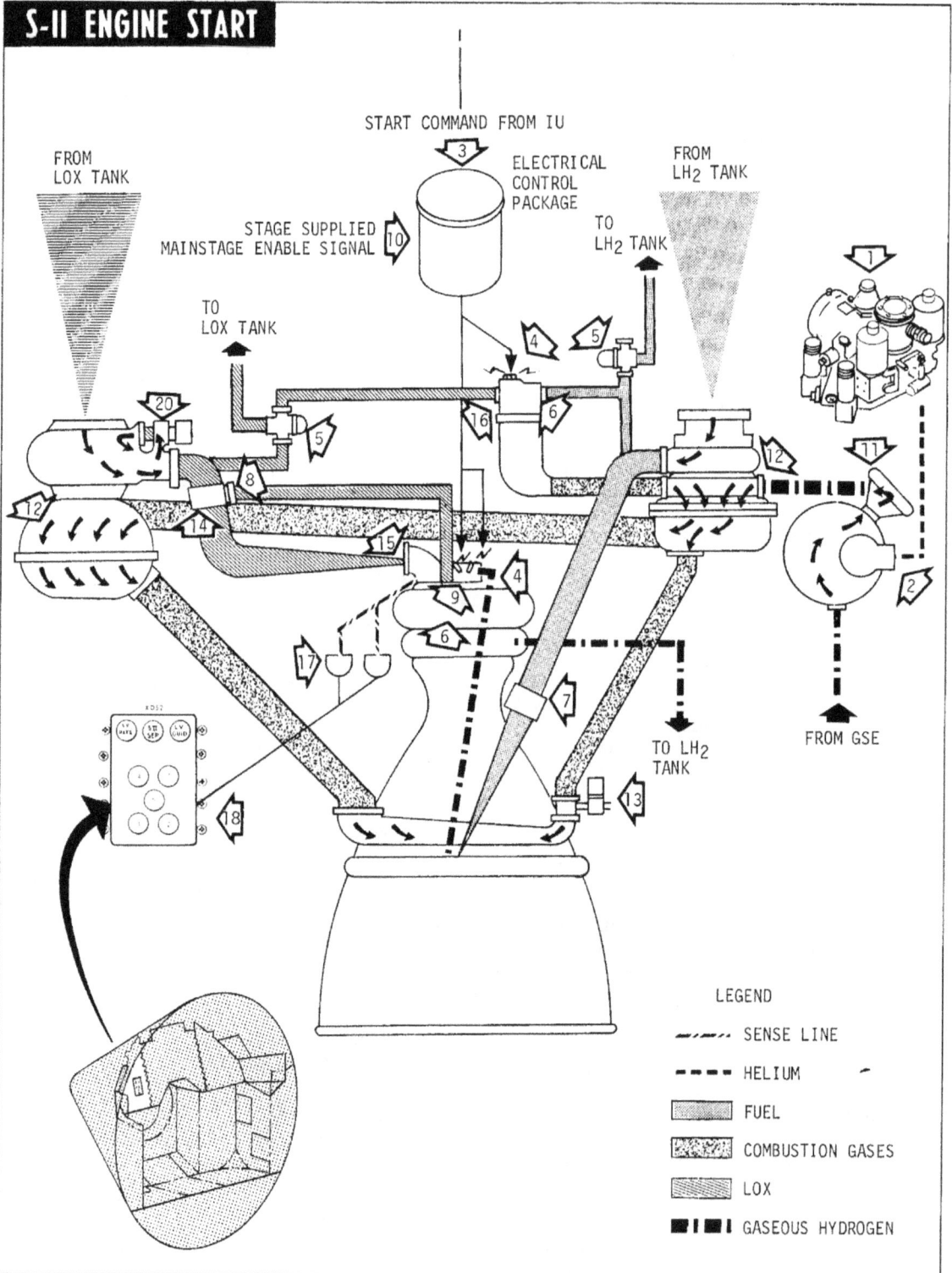

Figure 5-7 (Sheet 1 of 3)

# S-II ENGINE START

| SEQUENCE | EVENT | TIME IN SECONDS |
|----------|-------|-----------------|

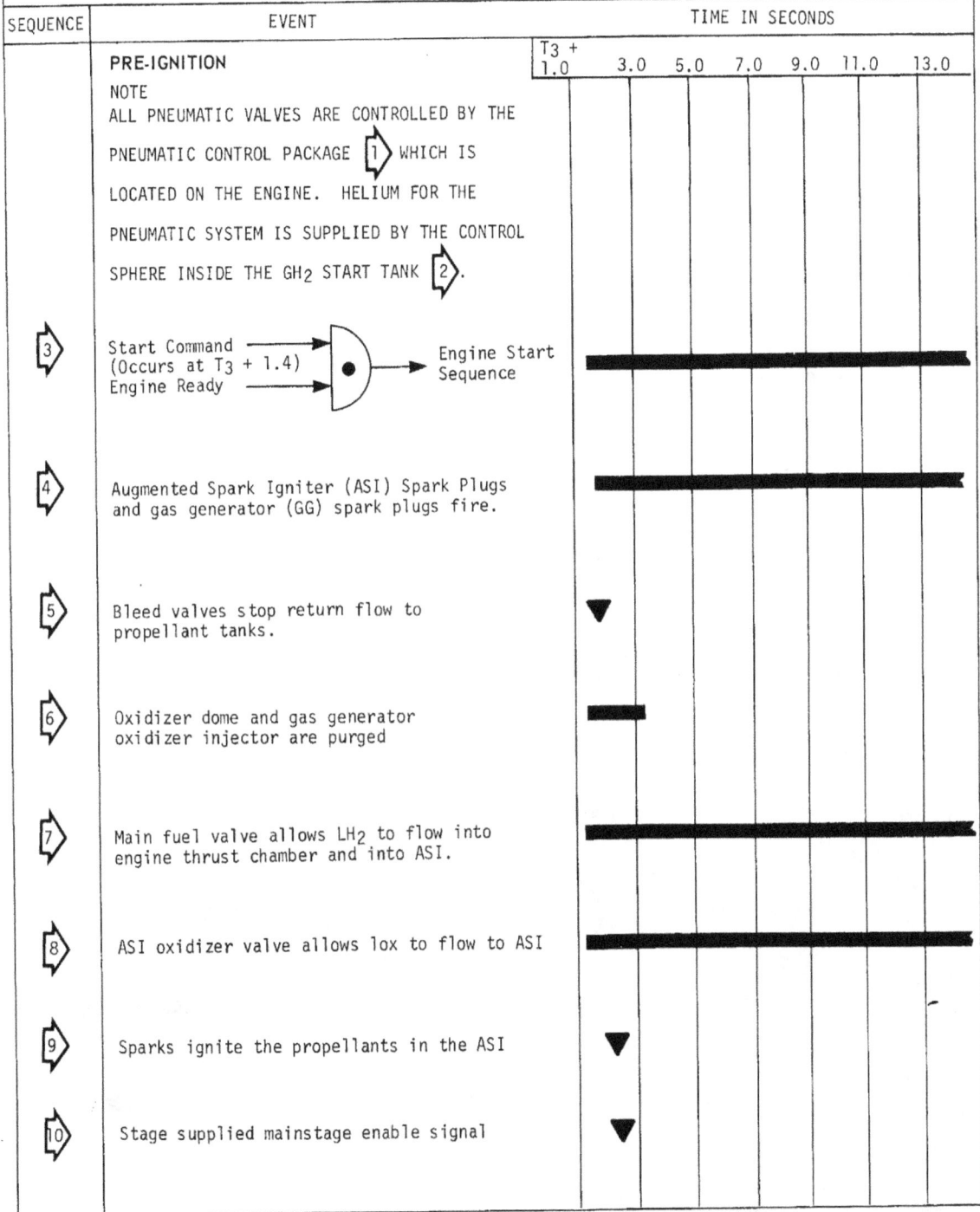

Figure 5-7 (Sheet 2 of 3)

# S-II ENGINE START

| SEQUENCE | EVENT | TIME IN SECONDS |
|---|---|---|
| 11 | Start tank discharges $GH_2$ causing | |
| 12 | the $LH_2$ and lox turbopumps turbines buildup propellant pressure | |
| 13 | Lox turbopump bypass valve opens to control lox pump speed. | |
| 14 | Main oxidizer valve opens allowing | |
| 15 | lox to be injected into thrust chamber | |
| 16 | G G valves admit propellants. (Spark ignites propellants causing pressure build up.) | |
| | **MAIN STAGE** | |
| 17 | OK pressure switches send mainstage OK signal to CM. | |
| 18 | Engine out lights go out. | |
| 19 | Engine reaches and maintains 90% thrust or more. | |
| 20 | P.U. valve controls mass ratio by returning lox from pump discharge to pump inlet. | |

Time scale: T3+ 1.0   3.0   5.0   7.0   9.0   11.0   13.0

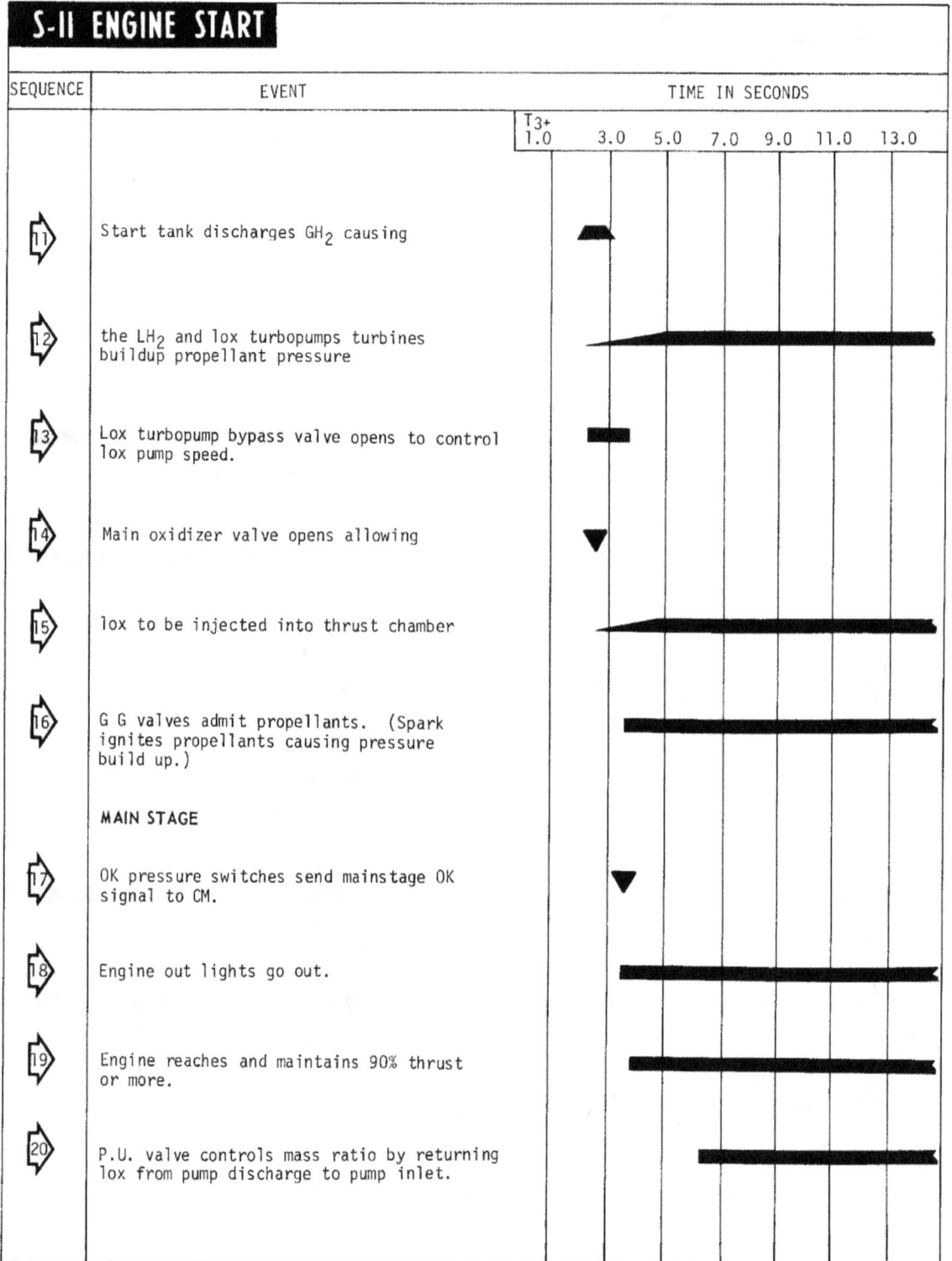

Figure 5-7 (Sheet 3 of 3)

# J-2 ENGINE P.U. EXCURSION EFFECT

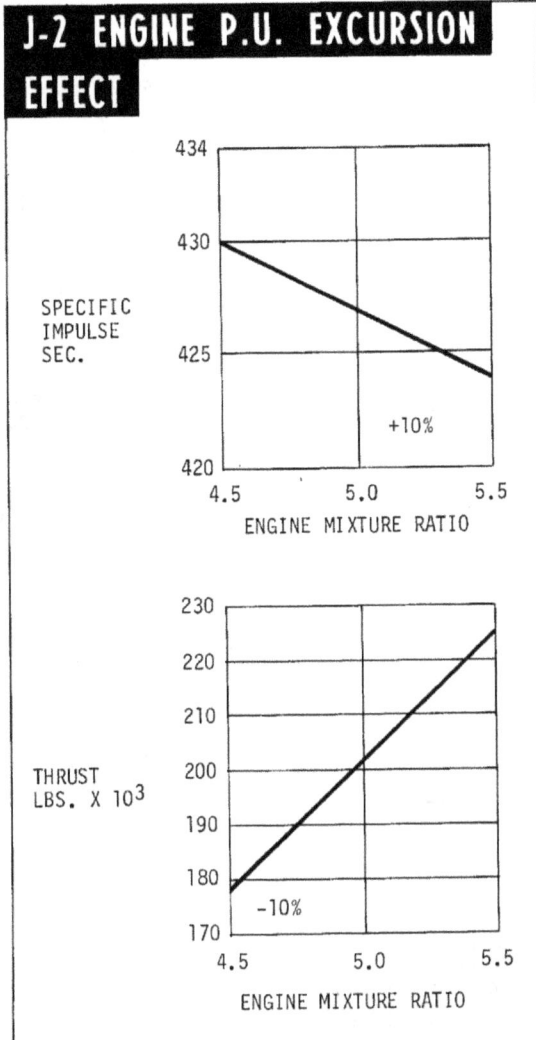

Figure 5-8

valves in the pneumatic control package (2), while energizing the helium control deenergize timer. The mainstage control valve closes the main oxidizer valve (3) and opens the purge control valve and the oxidizer turbine bypass valve (5). The purge control valve directs a helium purge (11) to the oxidizer dome and GG oxidizer injector. The ignition phase control valve closes the ASI oxidizer valve (4) and the main fuel valve (5) while opening the fast shutdown valve. The fast shutdown valve now rapidly vents the return flow from the GG control valve. All valves except the ASI oxidizer valve (4) and oxidizer turbine bypass valve (8), are spring loaded closed. This causes the valves to start moving closed as soon as the pressure to open them is released. GG combustion aids closing of the GG control valve.

Expiration of the helium control deenergize timer causes the helium control valve to close. When the helium control valve closes, it causes the oxidizer dome and GG oxidizer injector purges (11) to stop. An orifice in the locked up lines bleeds

off pressure from the propellant bleed valves (13). This loss of pressure allows springs to open the propellant bleed valves, allowing propellants to flow back to the propellant tanks.

## MALFUNCTION DETECTION

Each engine is provided with a system to detect malfunctions and to affect a safe shutdown. If neither mainstage OK pressure switch has indicated sufficient thrust for mainstage operation of the ignition phase timer, a shutdown of the particular engine is initiated. Once an engine attains mainstage operation, it is shut down if both mainstage OK pressure switches deactuate due to low level thrust.

## FLIGHT CONTROL

The center engine is fixed in place while the four outer engines are gimbaled in accordance with electrical signals from the flight control computer in the IU for thrust vector control. Each outboard engine is equipped with a separate, independent, closed-loop, hydraulic control system (figure 5-10). The system includes two servoactuators, mounted perpendicular to each other, that provide control over the vehicle pitch, roll and yaw axes. The servoactuators are capable of deflecting the engine $\pm$ 7 degrees in the pitch and yaw planes, at the rate of 8 degrees per second.

The primary components of the hydraulic control system are an auxiliary pump, a main pump, an accumulator/reservoir manifold assembly, and two servoactuators (figures 5-10 and 5-11). The auxiliary pump is used prior to launch to maintain hydraulic fluid temperature between 65 and 105 degrees F. The pump delivers two gallons per minute at 3650 psig, and is driven by a 400-cycle motor on GSE power.

The main pump is mounted to and driven by the engine lox turbopump. It is used during stage powered flight and delivers hydraulic fluid at 8 gallons per minute at 3500 psig. Prior to launch, the accumulator is pressurized with $GN_2$ and filled with hydraulic fluid from the pressurized auxiliary pump flow. The reservoir is, in turn, pressurized by the accumulator through a piston-type linkage. The accumulator/reservoir manifold assembly consists of a high pressure (3500 psig) accumulator which receives high pressure fluid from the pumps and a low pressure (88 psig) reservoir which receives return fluid from the servoactuators. During engine firing, hydraulic fluid is routed under pressure from the main pump to the pressure manifold of the accumulator/reservoir.

Hydraulic fluid under pressure in the accumulator furnishes high pressure fluid for sudden demands and smooths out pump pulsations. This pressurized hydraulic fluid is directed to the two identical controlled, hydraulically powered, servoactuators. The servoactuators have a nominal operating pressure of 3500 psig and provide the necessary forces and support to accurately position the engine in response to flight control system signals. The servoactuator is a power control unit that converts electrical command signals and hydraulic power into mechanical outputs that gimbal the engine. The developed force, direction, and velocity are determined by an electro-hydraulic servovalve.

Command signals received from the guidance system are interpreted as linear variations from a known piston position. Hydraulic fluid is then directed to either side of the actuator piston as required to satisfy the guidance command.

Figure 5-9 (Sheet 1 of 2)

# S-II ENGINE CUTOFF

| SEQUENCE | EVENTS | TIME IN SECONDS FROM CUTOFF SIGNAL |
|----------|--------|-------------------------------------|

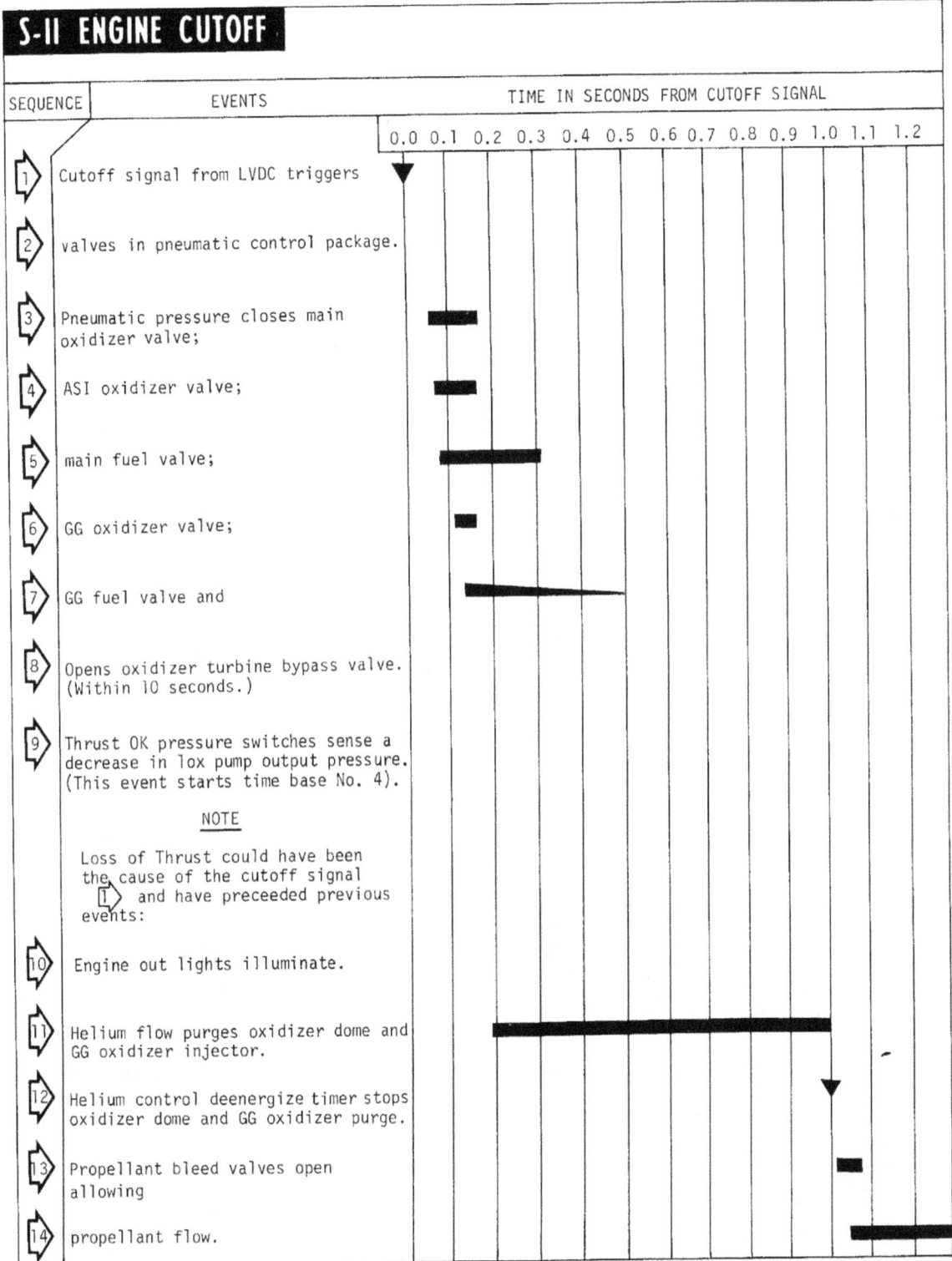

Figure 5-9 (Sheet 2 of 2)

# FLIGHT CONTROL SYSTEM

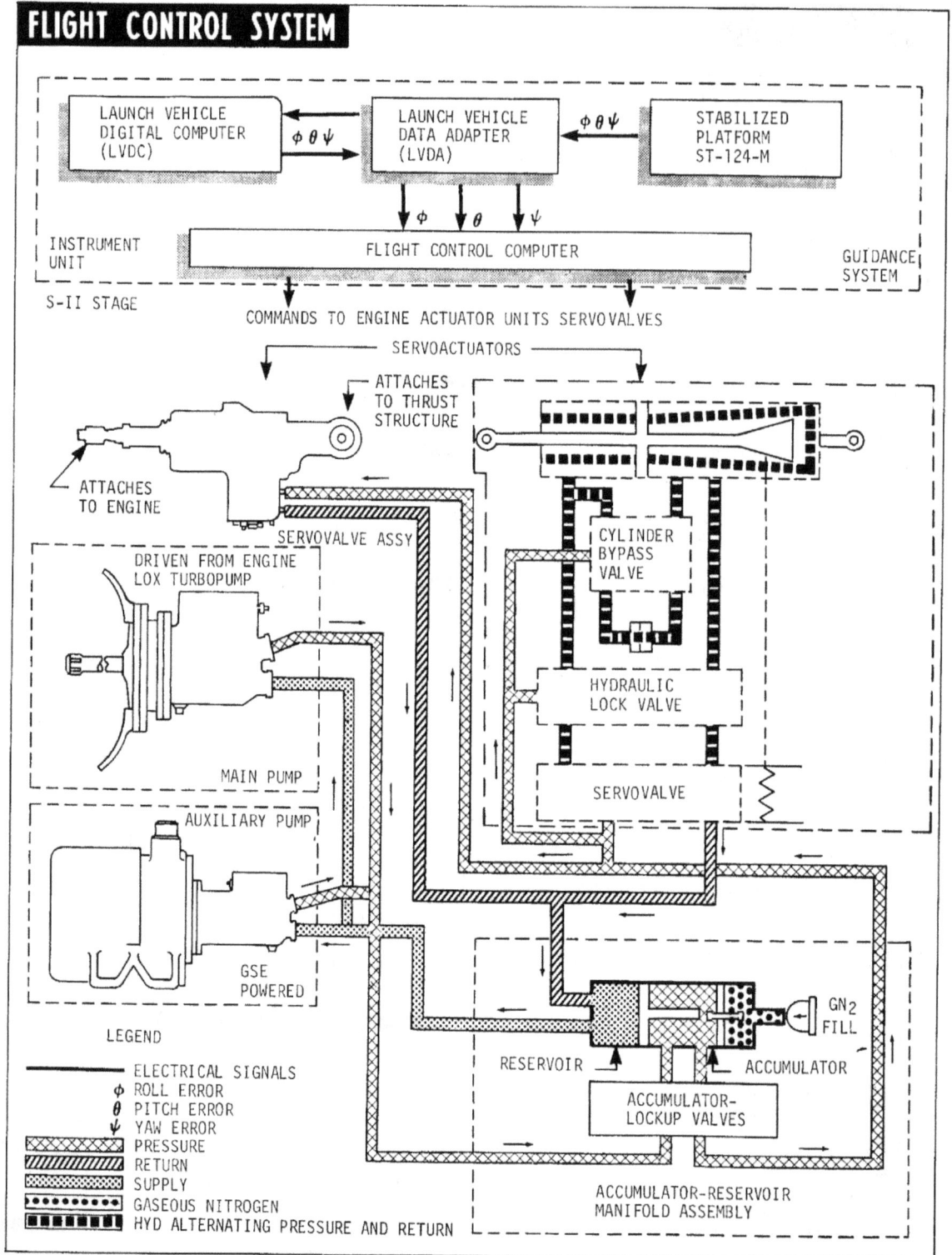

Figure 5-10

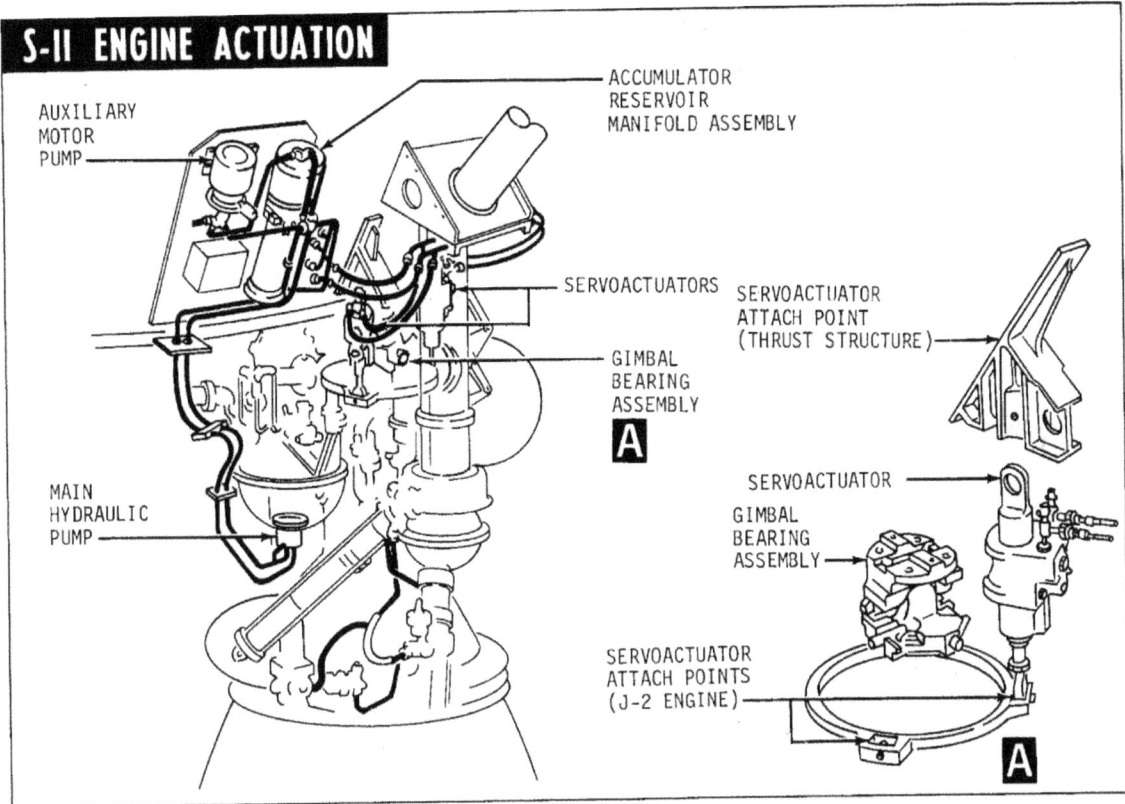

# S-II ENGINE ACTUATION

AUXILIARY
MOTOR
PUMP

ACCUMULATOR
RESERVOIR
MANIFOLD ASSEMBLY

SERVOACTUATORS

SERVOACTUATOR
ATTACH POINT
(THRUST STRUCTURE)

GIMBAL
BEARING
ASSEMBLY

**A**

MAIN
HYDRAULIC
PUMP

SERVOACTUATOR

GIMBAL
BEARING
ASSEMBLY

SERVOACTUATOR
ATTACH POINTS
(J-2 ENGINE)

**A**

Figure 5-11

Actuator return fluid is routed to the reservoir which stores hydraulic fluid at sufficient pressure to supply a positive pressure at the main pump inlet.

## PREFLIGHT OPERATION

During and following propellant loading, the hydraulic system fluid is intermittently recirculated by the electrically driven auxiliary pump in order to prevent the fluid from freezing. Recirculation is terminated just prior to S-IC ignition command. Recirculation is not necessary during S-IC burn, due to the short duration of the burn.

At approximately T-42 minutes, fluid is stored under high pressure in the accumulator by closing both hydraulic lockup valves contained in the accumulator/reservoir manifold assembly (figure 5-10).

## INFLIGHT OPERATION

After S-IC/S-II stage separation, an S-II switch selector command unlocks the accumulator lockup valves, releasing high pressure fluid to each of the two servoactuators. The accumulator stored fluid provides gimbaling power prior to main hydraulic pump operation. During S-II mainstage operation the main hydraulic pump supplies high pressure fluid to the servoactuators for gimbaling.

## PNEUMATIC CONTROLS

The pneumatic control system (figure 5-12) consists of the ground pneumatic control system and the onboard pneumatic control system. The ground system utilizes helium supplied directly from a ground source and the onboard system utilizes helium from onboard storage spheres.

## GROUND PNEUMATICS

Ground supplied helium controls and actuates various valves during preflight operations. These include the vent valves, fill and drain valves, recirculation return line valves, and main propellant line prevalves.

## ONBOARD PNEUMATICS

The onboard pneumatic control systems consist of a stage propellant valve control system and an engine pneumatic actuation and purge system.

### Stage Propellant Valve Control System

The stage onboard pneumatic control system is supplied from the helium receiver. It is pressurized to 3000 psig at approximately T-31 minutes. Pneumatic pressure from the helium receiver is regulated to 750 psig by the control regulator and is used during flight to actuate the prevalves and recirculation valves.

# PNEUMATIC CONTROL SYSTEM

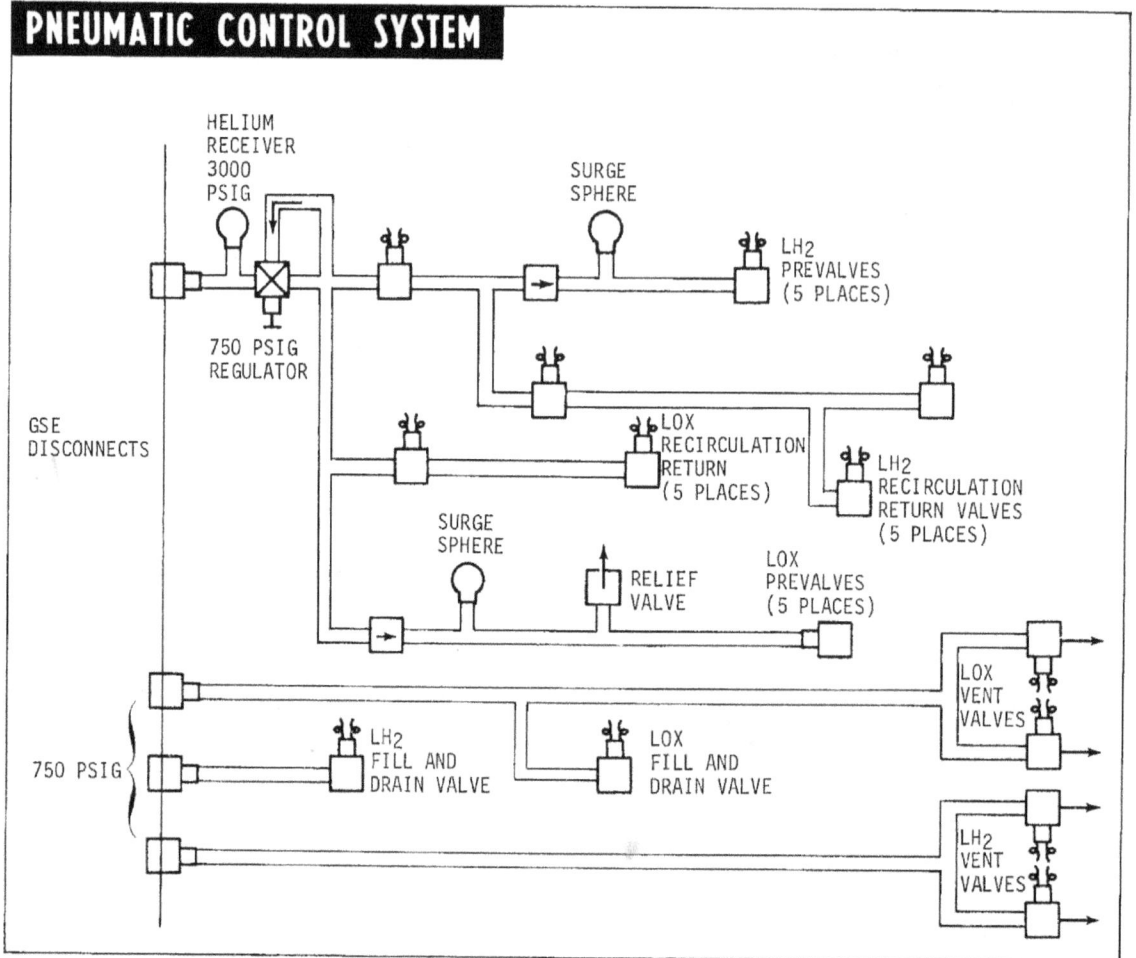

Figure 5-12

## PROPELLANTS

The propellant systems supply fuel and oxidizer to the five J-2 rocket engines. This is accomplished by the propellant management components and the servicing, conditioning, and delivery subsystems.

## PROPELLANT SERVICING SYSTEM

Pad servicing operations include the filling, draining, and purging of propellant tanks and lines, as required during launch preparations.

Ground interface is through the umbilicals, to the fill and drain valves, and into the propellant tanks. propellants then enter the engine feed lines, stopping at the closed main valves. Refer to figure 5-13 for propellant loading data. The tanks are vented by opening the tank vent valves, two per propellant tank, to allow ullage gas to escape from the tanks. Actuation pressure for the propellant tanks vent valves is provided by two separate 750-psig ground-supplied helium systems. One system actuates the lox tank vent valves, and the other system actuates the LH2 tank vent valves. The vent valves are open during propellant loading operations and closed for tank pressurization.

If the launch is aborted, draining of the propellant tanks can be accomplished by pressurizing the tanks, opening the fill valves, and reversing the fill operation.

## RECIRCULATION SYSTEM

Propellant recirculation is accomplished in order to maintain uniform cryogenic density and temperature and to preclude the formation of gas in propellant plumbing. This gas could cause cavitation in the turbopumps during J-2 engine start, or result in a slow engine start, slow thrust buildup or power surges.

### Lox Recirculation

Lox conditioning by natural convection (figure 5-14) is initiated shortly after start of lox fill and continues until T-30 minutes. At that time, helium is injected into the lox recirculation return line to boost recirculation. Helium boost is continuous until just prior to S-II ignition. After launch helium is supplied from a pressurized sphere. During recirculation, lox prevalves and recirculation return valves

| PROPELLANT LOADING DATA | | | | |
|---|---|---|---|---|
| PROPELLANT | TYPE FILL | RATE (GALLONS/MINUTE) | LEVEL (TANK % LEVEL) | COUNT TIME (HR: MIN: SEC) |
| LOX | PRECOOL FAST SLOW REPLENISH | 500 5,000 1,000 0 to 100 | 0 to 5 5 to 98 98 to 100 | T-7:04:00 to T-6:49:00 to T-6:32:00 to T-5:24:00 to T-185 seconds |
| LH₂ | PRECOOL FAST SLOW REPLENISH | 1,000 10,000 1,000 0 to 500 | 0 to 5 5 to 98 98 to 100 | T-5:21:00 to T-5:00:00 to T-4:34:34 to T-4:29:00 to T-36 seconds |

Figure 5-13

remain open. Return line valves are closed at termination of recirculation.

Lox conditioning is accomplished by recirculating the lox down the engine feed ducts through the prevalves, the lox turbopump, into the return lines, through the engine bleed valves, and back into the lox tank.

### LH₂ Recirculation

LH₂ recirculation (figure 5-15) is initiated at approximately T-30 minutes and is terminated just prior to S-II ignition. Forced recirculation during launch and S-IC boost consists of closing the LH₂ feed line prevalves and starting the LH₂ recirculation pumps. A separate recirculation pump is provided for each of the five feed ducts.

LH₂ conditioning is accomplished by pumping the fuel through the recirculation bypass valve, into the LH₂ feed ducts downstream of the prevalves, through the LH₂ turbopump, through the LH₂ bleed valve, the recirculation return valve, and back into the fuel tank.

Recirculation is terminated by opening the prevalves, stopping the pumps, and closing the recirculation return valves.

### PREPRESSURIZATION

After loading has been completed, and shortly before liftoff, the vent valves are closed and the propellant tanks are pressurized to their required levels by helium from ground supplies. Pressurization of the propellant tanks is required prior to liftoff to provide the required net positive suction head (NPSH) at the turbopump inlets for engine start. It is accomplished from a ground regulated helium source. Pressurization is initiated by the terminal countdown sequencer at approximately T-3 minutes and 7 seconds for the lox tank and T-1 minute and 37 seconds for the LH₂ tank. Pressurization is terminated at T-30 seconds for the lox tank and the LH₂ tank.

Both propellant tanks are pressurized in the same manner by separate systems (figures 5-14 and 5-15). At initiation of prepressurization, the tank vent valves are closed and the

disconnect valve and ground prepressurization valves are opened to allow GHe at minus 275 degrees F to flow from the ground source through the prepressurization solenoid valve into the tank pressurization line. This line carries helium into the propellant tank through the tank gas distributor.

Each propellant tank has a fill overpressure switch for personnel safety. The switch sends a signal to the GSE and is used only during loading.

The vent valves act as relief valves allowing ullage gas to be vented directly overboard during flight. The LH₂ vent valves operate between 27.5 and 29.5 psig prior to S-II ignition and at 30.5 to 33 psig during S-II burn. The lox vent valves crack at 42 psia and reseat at 39.5 psia.

### PROPELLANT DELIVERY SUBSYSTEMS

The function of the engine feed systems is to transfer the liquid propellants from their tanks to the J-2 rocket engines. Each propellant tank is provided with five prevalves which provide open/close control of the flow of propellants through separate feedlines to each engine.

The prevalves are normally open, pneumatically actuated, electrically controlled, butterfly-gate type valves. Built-in four-way pneumatic control solenoids permit 750 ± 50 psig helium pressure to actuate the prevalves. Should a loss of pneumatic or electrical power occur, the prevalves are spring actuated to the open position. The prevalves remain open during S-II powered flight unless a signal is received from the engine shutdown system.

### LOX FEED SYSTEM

The lox feed system furnishes lox to the five engines. This system includes four 8-inch, vacuum jacketed feed ducts, one uninsulated feed duct, and five normally open prevalves. At engine start, lox flows from the tank, through the prevalves and feed lines, to each engine. Approximately 300 milliseconds after main valve closure, the lox prevalves are closed, providing a redundant shutoff for the lox feed system.

# LOX SYSTEM PRESSURIZATION FLOW AND CONDITIONING

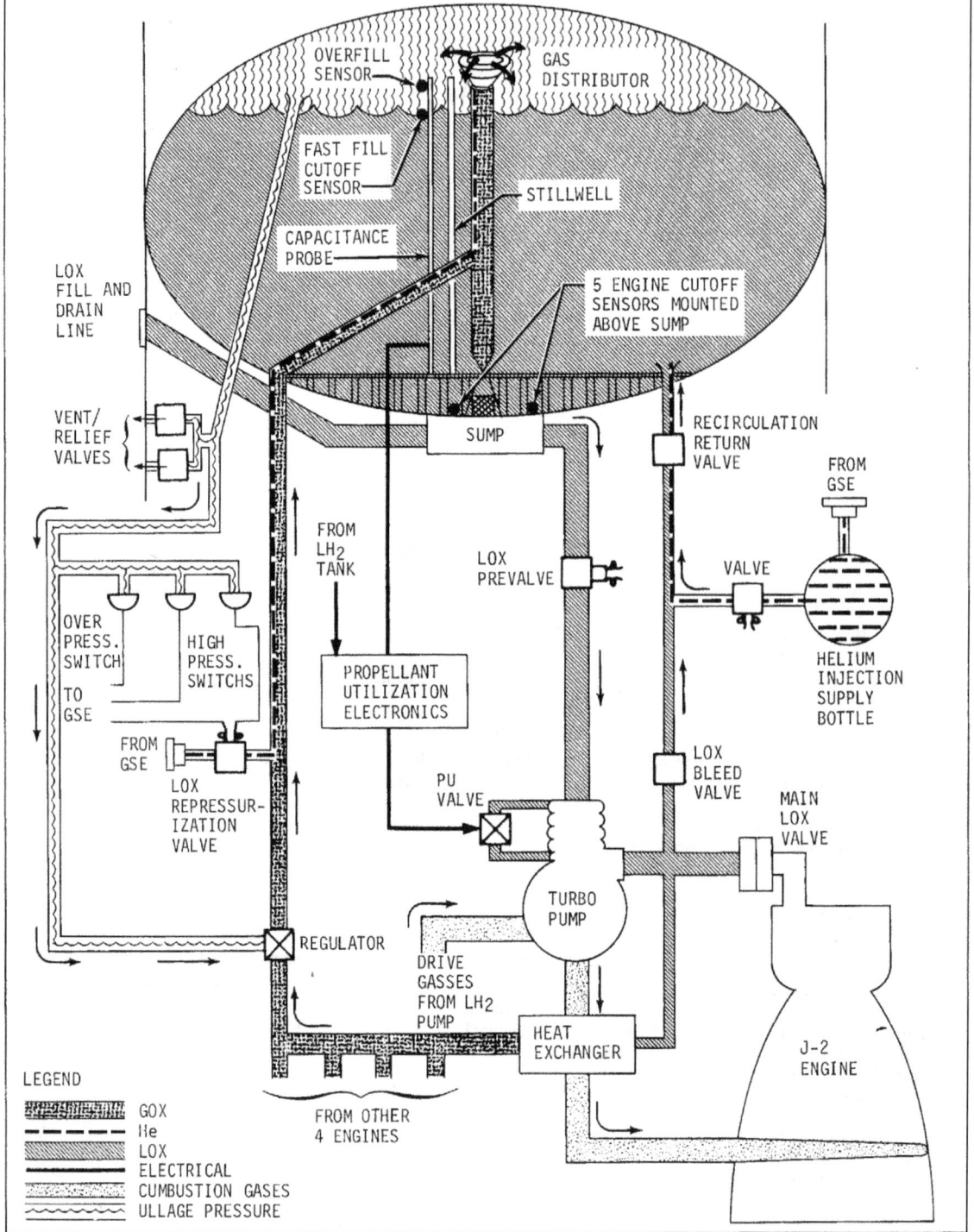

Figure 5-14

# LH₂ SYSTEM-PRESSURIZATION FLOW AND CONDITIONING

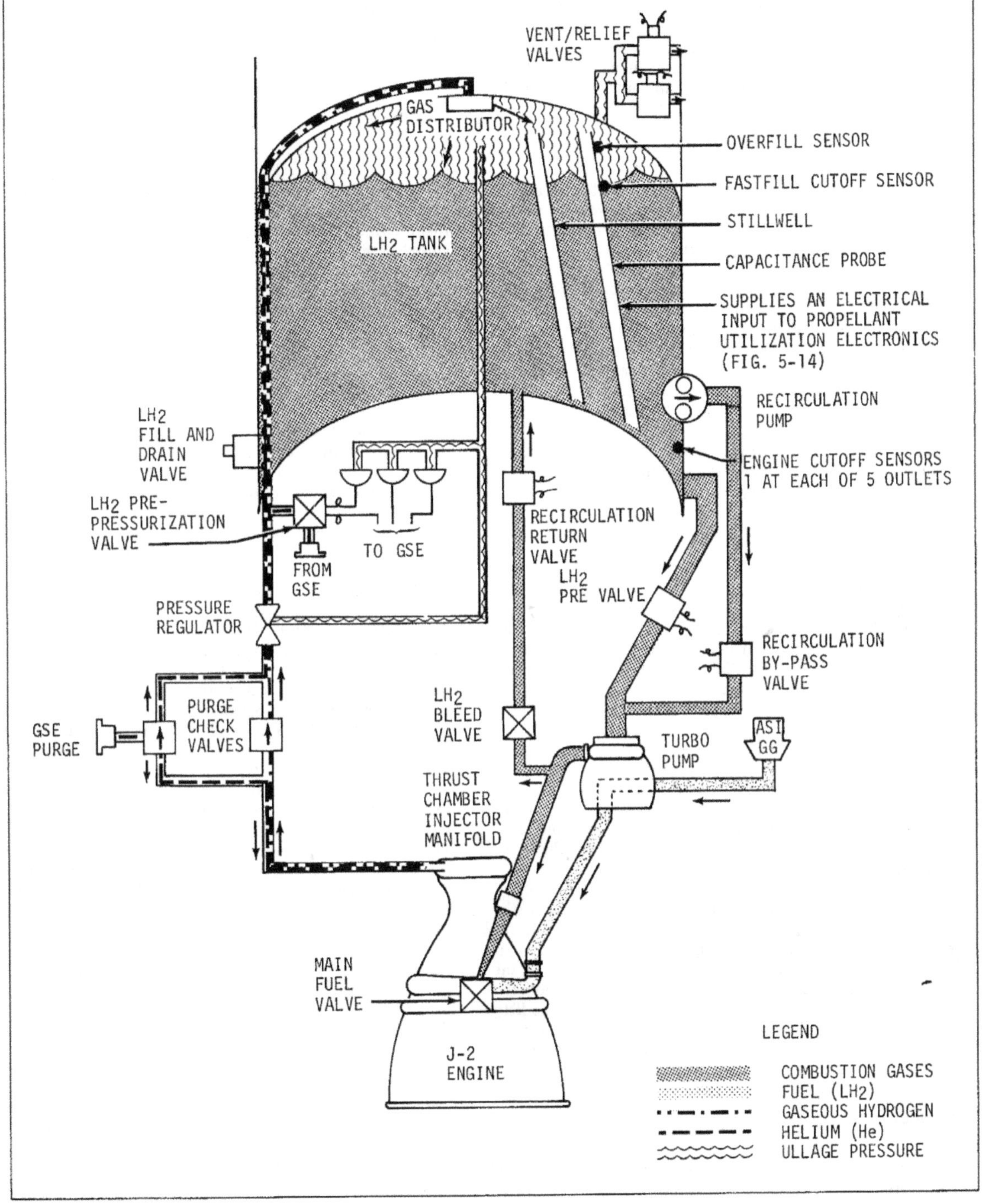

VENT/RELIEF VALVES

GAS DISTRIBUTOR

LH₂ TANK

OVERFILL SENSOR

FASTFILL CUTOFF SENSOR

STILLWELL

CAPACITANCE PROBE

SUPPLIES AN ELECTRICAL INPUT TO PROPELLANT UTILIZATION ELECTRONICS (FIG. 5-14)

RECIRCULATION PUMP

ENGINE CUTOFF SENSORS 1 AT EACH OF 5 OUTLETS

LH₂ FILL AND DRAIN VALVE

LH₂ PRE-PRESSURIZATION VALVE

TO GSE

FROM GSE

PRESSURE REGULATOR

RECIRCULATION RETURN VALVE

LH₂ PRE VALVE

RECIRCULATION BY-PASS VALVE

GSE PURGE

PURGE CHECK VALVES

LH₂ BLEED VALVE

TURBO PUMP

ASI GG

THRUST CHAMBER INJECTOR MANIFOLD

MAIN FUEL VALVE

J-2 ENGINE

LEGEND

COMBUSTION GASES
FUEL (LH2)
GASEOUS HYDROGEN
HELIUM (He)
ULLAGE PRESSURE

Figure 5-15

## LH2 FEED SYSTEM

The LH2 feed system furnishes LH2 to the five engines. This system includes five 8-inch vacuum-jacketed feed ducts and five normally open prevalves. The prevalves are closed following tank loading and remain closed until just prior to S-II ignition command. At engine start, LH2 flows from the tank, through the prevalves and feed lines, to each engine. Approximately 425 milliseconds after main valve closure, the prevalves are closed, providing a redundant shutoff for the LH2 feed system.

## LOX TANK PRESSURIZATION

Lox tank pressurization (figure 5-14) is initiated at S-II ignition and continues until engine cutoff. Pressurization is accomplished with gaseous oxygen obtained by heating lox bled from the lox turbopump outlet.

When the turbine discharge pressure reaches a pressure differential of 100 psi, a portion of the lox supplied to the engine is diverted into the heat exchanger where it is turned into gox. The gox flows from each heat exchanger into a common pressurization duct through the tank pressurization regulator, and into the tank through the gas distributor. The flowrate is varied according to the tank ullage pressure, which is sensed by the reference pressure line connecting the tank and the tank pressurization regulator. The tank pressurization regulator provides continuous regulation of the tank pressure throughout S-II powered flight.

## LH2 TANK PRESSURIZATION

During S-II powered flight gaseous hydrogen (GH2) for LH2 tank pressurization (figure 5-15) is bled from the thrust chamber hydrogen injector manifold of each of the four outboard engines. After S-II engine ignition, liquid hydrogen is preheated in the regenerative cooling tubes of the engine and tapped off from the thrust chamber injector manifold in the form of GH2 to serve as a pressurizing medium.

The GH2 passes from each injector manifold into a stage manifold, through the pressurization line and tank pressurization regulator, and into the tank through the LH2 tank gas distributor. The flowrate is varied according to the LH2 tank ullage pressure, which is sensed by the reference pressure line connecting the LH2 tank and the tank pressurization regulator.

At approximately 5 minutes after S-II engine ignition a step pressurization command from the stage switch selector activates the regulator to a fully open position, where it remains the rest of S-II boost. When the regulator is in the full open position, LH2 tank pressure increases to a nominal 33 psia. Pressure in excess of 33 psia is prevented by the LH2 tank vent valves. This step pressurization compensates for the loss of head pressure caused by the lowering of the fuel level in the tank.

## PROPELLANT MANAGEMENT

The propellant management systems provide a means of monitoring and controlling propellants during all phases of stage operation. Continuous capacitance probes and point level sensors in the LH2 and lox tanks monitor propellant mass. During the propellant loading sequence, the point level sensors are used to indicate to GSE the level of propellants in the tanks. In flight, the level sensors provide signals to the LVDC in order to accomplish a smooth engine cutoff at propellant depletion. The capacitance probes provide outputs used to operate the propellant utilization (PU) control valve. This rotary type valve controls the quantity of lox flowing to the engine. Figures 5-14 and 5-15 illustrate the components of the systems.

## PROPELLANT UTILIZATION SUBSYSTEM

The PU subsystem has been modified for the AS-503 mission. Normally, the PU valve follows the signal developed from inputs provided by the propellant monitoring capacitance probes. This provided for minimum propellant residuals at flight termination. The AS-503 modification has broken the closed loop and provides for LVDC controlled relays to furnish step control signals to the PU valve. See figure 6-22 for a representative block diagram of the PU system. The excursion effect caused by varying MR is illustrated in figure 5-8.

### Propellant Depletion

Five discrete liquid level sensors in each propellant tank provide initiation of engine cutoff upon detection of propellant depletion. The LH2 tank sensors are located above each feedline outlet while the lox tank sensors are located directly above the sump. The cutoff sensors will initiate a signal to shutdown the engines when two out of five engine cutoff signals from the same tank are received.

## ELECTRICAL

The electrical system is comprised of the electrical power and electrical control subsystems. The electrical power system provides the S-II stage with the electrical power source and distribution. The electrical control system interfaces with the IU to accomplish the mission requirements of the stage. The LVDC in the IU controls inflight sequencing of stage functions through the stage switch selector. The stage switch selector can provide up to 112 individual outputs in response to the appropriate commands. These outputs are routed through the stage electrical sequence controller or the separation controller to accomplish the directed operation. These units are basically a network of low power transistorized switches that can be controlled individually and, upon command from the switch selector, provide properly sequenced electrical signals to control the stage functions.

Figure 5-2 shows the relative location of the stage electrical system equipment.

## ELECTRICAL POWER SYSTEM

The electrical power system consists of six dc bus systems and a ground supplied ac bus system. In flight the electrical power system busses are energized by four zinc-silver oxide batteries. See figure 5-16 for battery characteristics. An integral heater and temperature probe are included in each battery. Power for battery heaters and for auxiliary hydraulic pump motors is supplied by GSE and is available only during prelaunch operations. Stage-mounted motor driven power transfer switches are employed to remotely disconnect all batteries from busses until just before launch. Approximately 50 seconds prior to liftoff, a power transfer sequence is initiated which changes the source power over to the stage mounted batteries. During the prelaunch checkout period all electrical power is supplied from GSE.

The motorized power transfer switches have a make-before-break (MBB) action to prevent power interruption during transfer from ground power to onboard battery power.

Each power source has an independant distribution system. There are no provisions for switching between the primary power sources or their associated distribution systems. No electrical failure of any type in one system can cause a failure in the other systems.

Distribution

Figure 5-17 illustrates the electrical system distribution. The loads have been distributed between the various busses in accordance with the following criteria:

1. Inflight loads, critical to mission continuance without performance degradation, are supplied by the main dc bus.

2. All instrumentation loads are supplied by the instrumentation dc bus.

3. All loads operational only on the ground are isolated from flight loads and supplied from ground power.

4. Two independent power sources supply the propellant dispersion and emergency detection systems.

| S-II BATTERY CHARACTERISTICS | |
|---|---|
| Type | Map 4301 Dry Charge |
| Material | Alkaline Silver-zinc |
| Electrolyte | Potassium Hydroxide (KOH) in demineralized water |
| Cells | 20 with taps to reduce voltage as required |
| Nominal Voltage | 28 ± 2 vdc |
| Current Rating | 35 Amp Hours |
| Gross Weight | 165 Pounds |

Figure 5-16

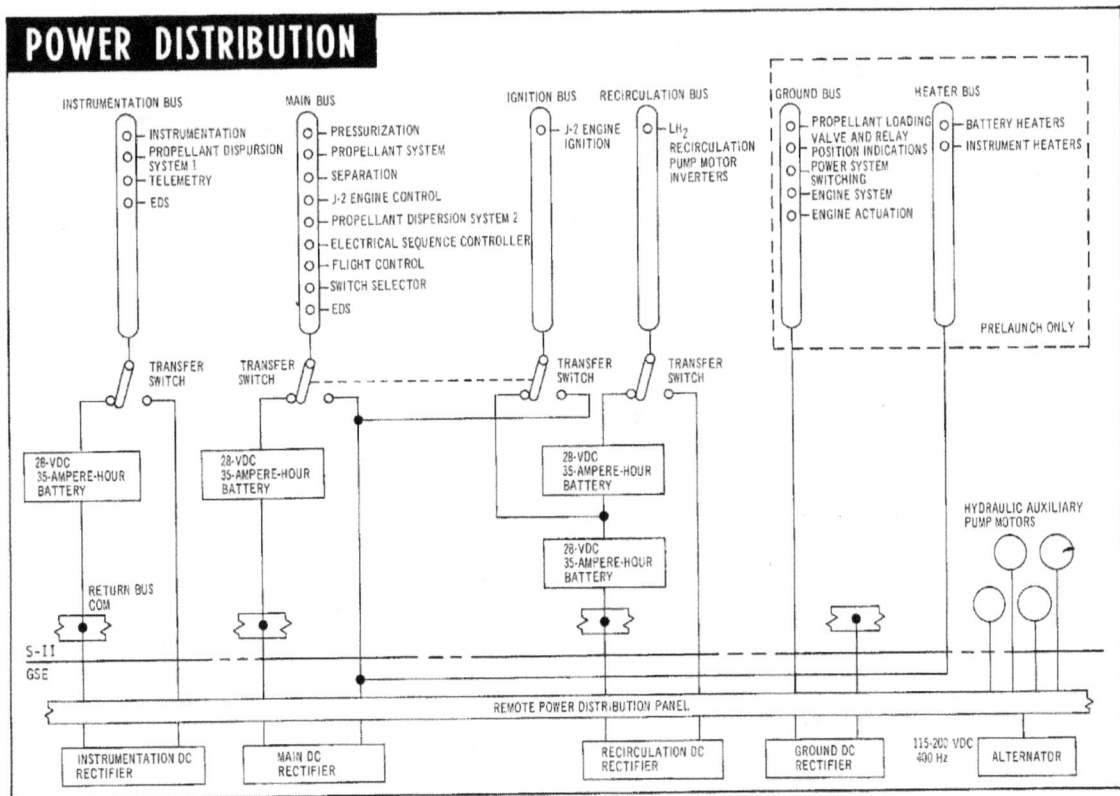

Figure 5-17

5. The recirculation pump motor system is supplied from a 56 volt dc system (two 28 volt batteries in series).

The division of loads between the main dc bus and the instrumentation dc bus leads to several advantages: closer voltage regulation and freedom from voltage variations is obtained; the number of loads on the main dc bus may be minimized and thus potential failure modes for the bus system minimized; instrumentation of most stage systems is still maintained after partial or total failure of the main dc bus system so that failure analysis capability is maintained.

Primary power is fed to high current capacity busses in the power distributor. Power is then routed to auxiliary and control distributors, or to measuring distributors for instrumentation power. Components which require high current levels are supplied directly from the main power distributor busses.

## ELECTRICAL CONTROL SYSTEM

The electrical control system provides electrical control for the various stage-mounted systems to implement normal flight operations, or prelaunch operations and checkout functions. The electrical control system contains most of the electrical and electronic components that are installed on the stage and required by the various mechanical systems for normal stage operation.

The primary stimuli for the electrical control system are provided by the IU and/or the GSE. Through the switch selector and sequence controller, various subsystems and functions are controlled. They include:

1. The propellant feed system which controls the fill and drain valves and the recirculation and conditioning valves.

2. The propellant management system which helps control the fill operation and the propellant utilization system during flight.

3. The pressurization system which controls the tank prepressurization valves, inflight pressurization valves, and the vent/relief valves.

4. The separation system which provides for S-IC/S-II separation, S-II ullage rocket ignition, S-II skirt separation, S-II/S-IVB separation, and S-II retrorocket ignition.

5. The propellant dispersion system which provides for thrust termination through engine cutoff and for explosively rupturing the propellant tanks.

6. The Emergency Detection System.

7. The starting and cutoff of the J-2 engine and the monitoring of certain engine system conditions.

8. The operation of the propellant prevalves.

## INSTRUMENTATION

The S-II instrumentation system consists of both operational and R&D measurement and telemetry systems. The measurement system monitors and measures conditions on the S-II stage while the telemetry system transmits this information to ground stations.

## MEASUREMENT SYSTEM

The measurement system consists of transducers, signal conditioners, and distribution equipment necessary to provide the required measurement ranges and to present suitably scaled signals to the telemetry system.

The measurement system monitors numerous stage conditions and characteristics. This data is processed and conditioned into a form acceptable to the telemetry systems.

Measurements fall into a number of basic categories depending upon the type of measured variable, the variable rate of change with time, and other considerations. Because the stage engines are ignited in flight, a large number of engine and environmental control measurements are required. Figure 5-18 presents a summary of stage instrumentation data.

## TELEMETRY SYSTEM

The telemetry system accepts the signals produced by the measuring portion of the instrumentation system and transmits them to the ground stations. Telemetry equipment includes signal multiplexers, subcarrier oscillators, amplifiers, modulators, transmitters, RF power amplifiers, RF multiplexers and an omnidirectional system of four antennae.

| STAGE MEASUREMENTS SUMMARY | |
|---|---|
| TYPE | QTY |
| Acceleration | 11 |
| Acoustics | 5 |
| Discrete Signals | 225 |
| Flow Rate | 10 |
| Liquid Level | 4 |
| Miscellaneous | 4 |
| Position | 36 |
| Pressure | 192 |
| RPM | 10 |
| Strain | 16 |
| Temperature | 309 |
| Vibration | 72 |
| Voltage, Current, Frequency | 60 |
| TOTAL | 954 |

Figure 5-18

This equipment also includes tape recorders for recording critical data during separation telemetry blackout (separation period) for later play back during stage free fall. The telemetry subsystems use multiplex techniques (signal mixing and time sharing) to transmit large quantities of measurement data over a relatively small number of basic RF links (figure 5-19).

Inflight data is transmitted in the form of frequency-modulated RF carriers in the 225 to 260-MHz band, through the common omnidirectional antenna system.

Several telemetry subsystems are provided in the S-II stage. Telemetry data is grouped in three general categories: low frequency data, medium frequency data, and high frequency

# INSTRUMENTATION AND TELEMETRY SYSTEMS

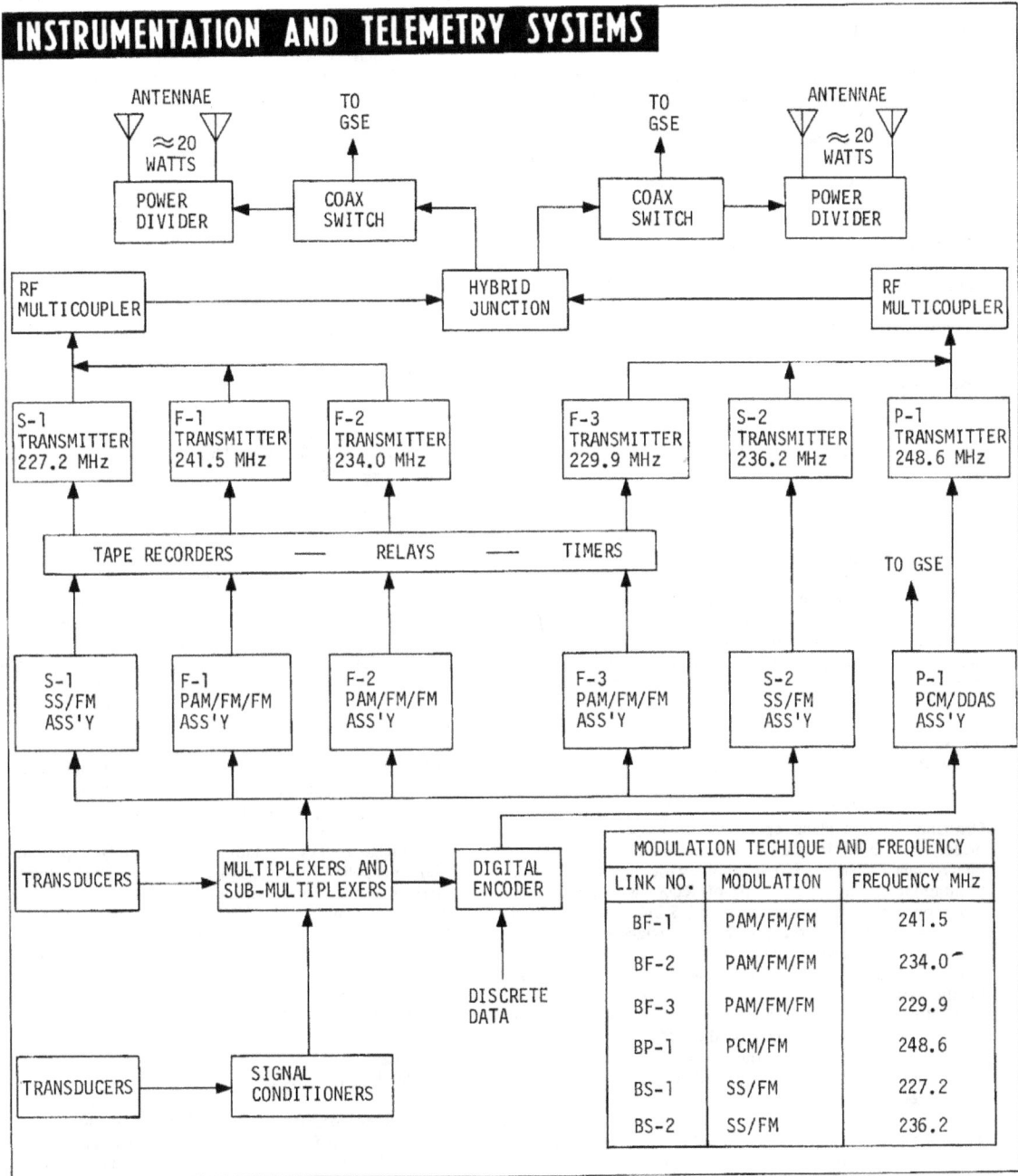

| MODULATION TECHIQUE AND FREQUENCY | | |
|---|---|---|
| LINK NO. | MODULATION | FREQUENCY MHz |
| BF-1 | PAM/FM/FM | 241.5 |
| BF-2 | PAM/FM/FM | 234.0 |
| BF-3 | PAM/FM/FM | 229.9 |
| BP-1 | PCM/FM | 248.6 |
| BS-1 | SS/FM | 227.2 |
| BS-2 | SS/FM | 236.2 |

Figure 5-19

data. Several different modulation techniques are employed in the telemetry systems to facilitate both quality and quantity of measured parameters. These modulation techniques include: pulse amplitude modulation/frequency modulation/frequency modulation (PAM/FM/FM) and pulse code modulation/frequency modulation (PCM/FM) for low-frequency data; and single sideband/frequency modulation (SS/FM) for high-frequency information. A pulse code modulation/digital data acquisition system (PCM/DDAS) transmits measurements by coaxial cable for automatic ground checkout of the stage.

The PCM/DDAS assembly converts analog transducer signals into digital representations, combines these representations with direct inputs, such as those from the guidance computer, and arranges this information into a format for transmission to the ground station on a 600 KHz carrier signal by means of coaxial cable for ground checkout of the stage.

## ANTENNAE

Four antennae, installed at 90 degree intervals (see figure 5-2), are employed to provide omnidirectional coverage. The Antennae are linear cavity-backed slot antennae which are fed from a hybrid junction ring and power dividers.

## ORDNANCE

The S-II ordnance systems include the separation, ullage rocket, retrorocket, and propellant dispersion (flight termination) systems.

## SEPARATION SYSTEM

The Saturn V launch vehicle system provides for separation of an expended stage from the remainder of the vehicle. For S-IC/S-II separation, a dual plane separation technique is used wherein the structure between the two stages is severed at two different planes (figure 5-20). The S-II/S-IVB separation occurs at a single plane (figure 5-20). All separations are controlled by the launch vehicle digital computer (LVDC) located in the instrument unit (IU).

A sequence of events for S-IC/S-II/S-IVB separations and block diagram of the separation systems is contained on figure 5-21.

Ordnance for first plane separation consists of two exploding bridgewire (EBW) firing units, two EBW detonators, and one linear shaped charge (LSC) assembly, which includes the LSC (containing 25 grains per foot of RDX) with a detonator block on each end (figures 5-20 and 5-21). The EBW firing units are installed on the S-IC/S-II interstage slightly below the S-II first separation plane. The leads of the EBW firing units are attached to the EBW detonators which are installed in the detonator blocks of the LSC assembly. The LSC detonator blocks are installed on adjustable mounts to provide for length variations of the LSC assembly and the circumference tolerances of the interstage. The LSC is routed from the detonator blocks around the periphery of the interstage.

The LSC is held in place by retaining clips and encased by covers which are secured by clips and sealed to environmentally protect the LSC. The two EBW firing units provide redundant signal paths for initiation of the LSC

assembly. The storage capacitor in each of the EBW firing units is charged by 28 vdc power during the latter part of S-IC boost. The trigger signal causes the storage capacitor to discharge into an EBW detonator which explodes the bridgewire to release energy to detonate the explosive charge in the detonator. The output of the detonators initiates each end of the LSC assembly.

Detonation of the LSC assembly severs the tension members attaching the S-IC/S-II interstage at station 1564.

The second plane separation ordnance is similar in composition and function to that of the first plane separation. The EBW firing units are installed on the S-IC/S-II interstage slightly below the separation plane. Detonation of the LSC assembly severs the tension members attaching the S-IC/S-II interstage at station 1760.

No heat-sensitive primary explosives are used and the detonators are not sensitive to accidental application of vehicle or gound power, static discharge, or RF energy. A spark gap in one pin of the firing circuitry prevents burnout of the bridgewire if power is accidentally applied.

S-II/S-IVB third plane separation is discussed in Section VI.

## ULLAGE ROCKET SYSTEM

To ensure stable flow of propellants into the J-2 engines, a small forward acceleration is required to settle the propellants in their tanks. This acceleration is provided by ullage rockets (figure 5-22).

The S-II ullage rocket system consists of two EBW firing units, two EBW detonators, two CDF manifolds, nine CDF assemblies, eight CDF initiators and four ullage rockets. CDF assemblies connect the two CDF manifolds together and both manifolds to each of the four ullage rockets (see block diagram on figure 5-21). The ullage rockets are mounted parallel to vehicle centerline 90 degrees apart on the periphery of the S-IC/S-II interstage at its aft end (figure 5-20). The rocket nozzles are just above the first separation plane and are canted outward 10 degrees to reduce the moment that would result from one or more rockets malfunctioning and to reduce exhaust plume impingement. With any one ullage rocket inoperative, the remaining rockets are capable of maintaining a minimum vehicle acceleration necessary for proper S-II engine ignition.

Each ullage rocket contains approximately 336 pounds of solid propellant, cast-in-place, in a four point star configuration. Ammonium perchlorate composes 82 percent of the propellant weight. The case is 4130 steel. The rocket is approximately 89 inches long by 12-1/2 inches in diameter and develops 22,700 pounds of thrust in a burn-time of 3.7 seconds.

## RETROROCKET SYSTEM

To separate and retard the S-II stage, a deceleration is provided by the retrorocket system.

The system consists of two EBW firing units, two EBW detonators, two CDF manifolds, nine CDF assemblies, eight pyrogen initiators, and four retrorockets (figure 5-22). The components are connected to each other in a manner similar to that of the ullage rocket system (see block diagram on

# STAGE SEPARATION SYSTEMS

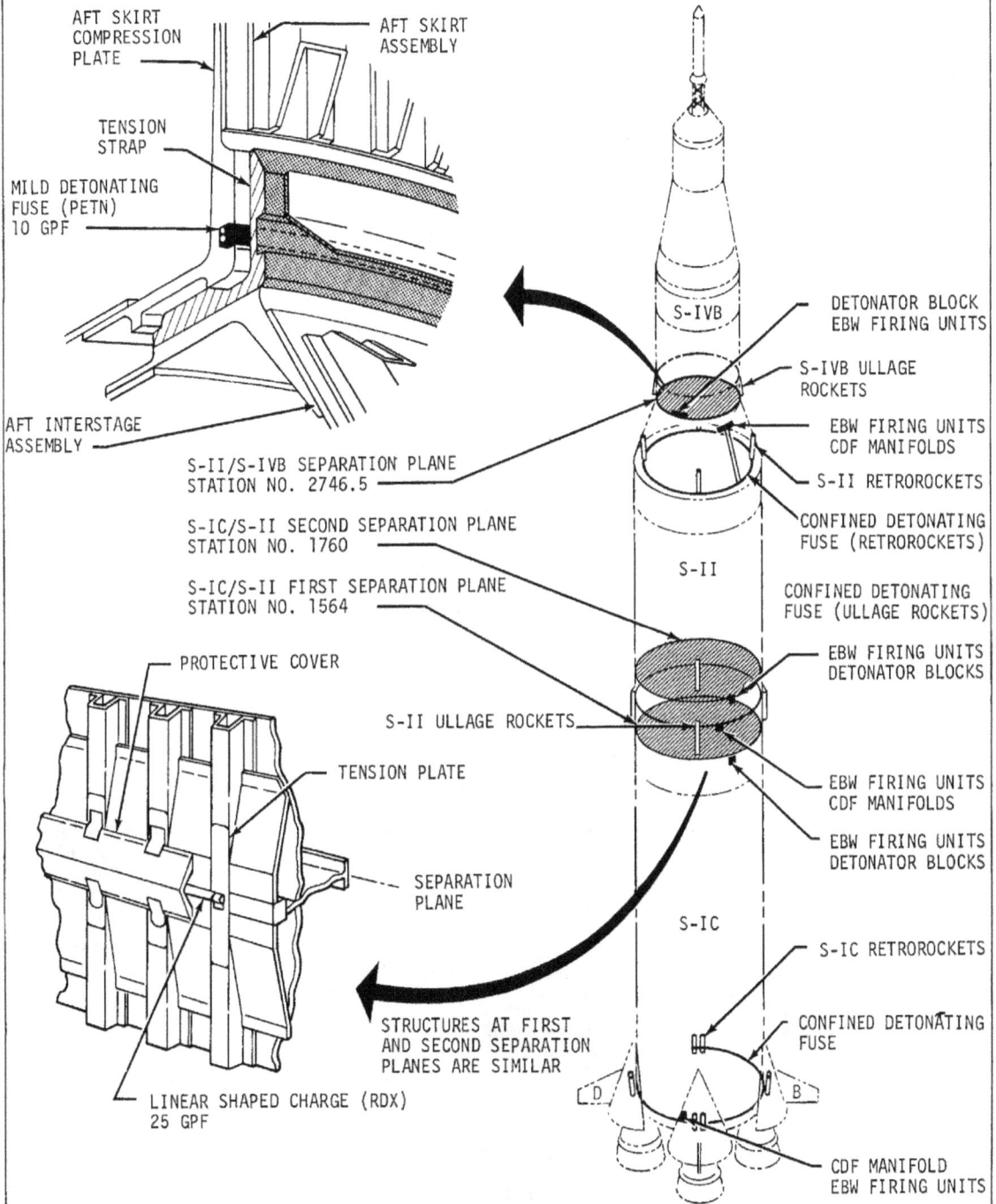

AFT SKIRT
COMPRESSION
PLATE

AFT SKIRT
ASSEMBLY

TENSION
STRAP

MILD DETONATING
FUSE (PETN)
10 GPF

AFT INTERSTAGE
ASSEMBLY

S-IVB

DETONATOR BLOCK
EBW FIRING UNITS

S-IVB ULLAGE
ROCKETS

EBW FIRING UNITS
CDF MANIFOLDS

S-II RETROROCKETS

CONFINED DETONATING
FUSE (RETROROCKETS)

S-II/S-IVB SEPARATION PLANE
STATION NO. 2746.5

S-IC/S-II SECOND SEPARATION PLANE
STATION NO. 1760

S-IC/S-II FIRST SEPARATION PLANE
STATION NO. 1564

S-II

CONFINED DETONATING
FUSE (ULLAGE ROCKETS)

PROTECTIVE COVER

EBW FIRING UNITS
DETONATOR BLOCKS

S-II ULLAGE ROCKETS

TENSION PLATE

EBW FIRING UNITS
CDF MANIFOLDS

EBW FIRING UNITS
DETONATOR BLOCKS

SEPARATION
PLANE

S-IC

STRUCTURES AT FIRST
AND SECOND SEPARATION
PLANES ARE SIMILAR

S-IC RETROROCKETS

CONFINED DETONATING
FUSE

D

B

LINEAR SHAPED CHARGE (RDX)
25 GPF

CDF MANIFOLD
EBW FIRING UNITS

Figure 5-20

# S-IC/S-II AND S-II/S-IVB SEPARATION

S-IC/S-II separation

> 1

EBW firing units enabled

A ground-latched interlock renders all the EBW firing units on the Saturn V inoperative while the vehicle is on the launch pad. The interlock is released with umbilical disconnect during liftoff, and the subsystem is reset to flight conditions.

> 2

S-IC/S-II separation ordnance arm

The ordnance-arm command is routed through the S-II switch selector to both the S-IC stage electrical circuitry to supply 28 vdc to the EBW units for first-plane separation and retrorocket ignition, and to the S-II stage electrical circuitry to supply 28 vdc to the EBW units for ullage rocket ignition and second-plane separation.

> 3

S-IC outboard engine cutoff followed by S-II ullage rocket ignition

> 4

First plane separation

Second plane separation is enabled by the removal of an electrical interlock during first plane separation.

> 5

Second plane separation

The second plane separation command is generated by the IU approximately thirty seconds after first plane separation.

This delay permits the transient vehicle motion, associated with first plane separation, to dampen out.

The separation command is routed to the S-II switch selector to trigger the ordnance train and ignite the LSC for second plane separation. The LSC detonates, severing the S-II interstage from the S-II stage. The combined effect of vehicle acceleration and the reaction caused by the J-2 engine exhaust plume impingement retards the interstage .

S-II/S-IVB separation

Physical separation is initiated by the IU at the end of the S-II boost phase following shutdown of the five J-2 engines. Separation requires the performance of the following major functions in the sequence described:

> 6

S-II/S-IVB separation ordnance arm

The ordnance-arm command is routed through the S-II switch selector to both the S-II and S-IVB stage electrical circuitry and carries 28 vdc to the EBW firing units for S-II/S-IVB separation and retrorocket ignition.

> 7

S-II/S-IVB separation

Four solid propellant S-II retrockets. (figure 5-22) are mounted at equal intervals on the periphery of the S-II/S-IVB interstage structure and are used to retard the S-II stage after separation.

Figure 5-21 (Sheet 1 of 2)

figure 5-21). The retrorockets are mounted 90 degrees apart in the aft end of S-II/S-IVB interstage between stations 2519 and 2633 (figure 5-22). The retrorockets are canted out from the vehicle centerline approximately three degrees with the nozzles canted out nine and one-half degrees from the centerline.

Each retrorocket contains approximately 268.2 pounds of case-bonded, single-grain, solid propellant with a tapered, five-point star configuration. The 4130 steel case is 9 inches in diameter and 90.68 inches long. The approximate length and weight of the rocket are 104.68 inches and 377.5 pounds, respectively. Each produces a thrust of 34,810 pounds in 1.52 seconds of burning time.

PROPELLANT DISPERSION SYSTEM

The S-II propellant dispersion system (PDS) provides for termination of vehicle flight during the S-II boost phase if the vehicle flight path varies beyond its prescribed limits or if continuation of vehicle flight creates a safety hazard. The S-II

PDS may be safed after the launch escape tower is jettisoned. The system is installed in compliance with Air Force Eastern Test Range (AFETR) Regulation 127-9 and AFETR Safety Manual 127-1.

The S-II PDS is a dual channel, redundant system composed of two segments (figure 5-23). The radio frequency segment receives, decodes, and controls the propellant dispersion commands. The ordnance train segment consists of two EBW firing units, two EBW detonators, one safety and arming (S&A) device (shared by both channels), six CDF assemblies, two CDF tees, one $LH_2$ tank LSC assembly, two lox tank destruct charge adapters and one lox tank destruct charge assembly.

Should emergency flight termination become necessary, two coded radio frequency commands are transmitted to the launch vehicle by the range safety officer. The first command arms the EBW firing units (figure 5-23) and initiates S-II stage engine cutoff. The second command, which is delayed

# S-IC/S-II AND S-II/S-IVB SEPARATION

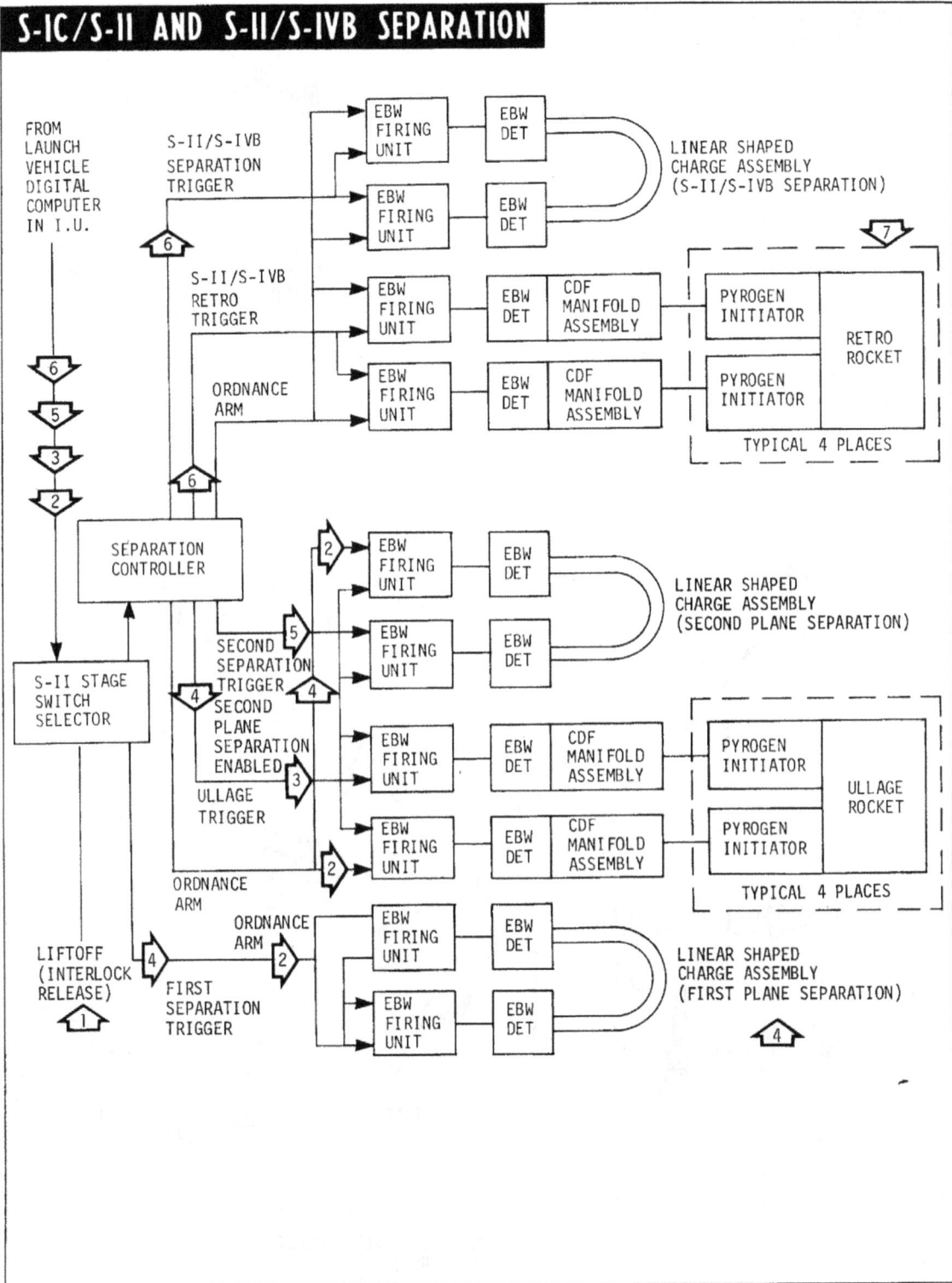

Figure 5-21 (Sheet 2 of 2)

# S-II ULLAGE AND RETROROCKETS

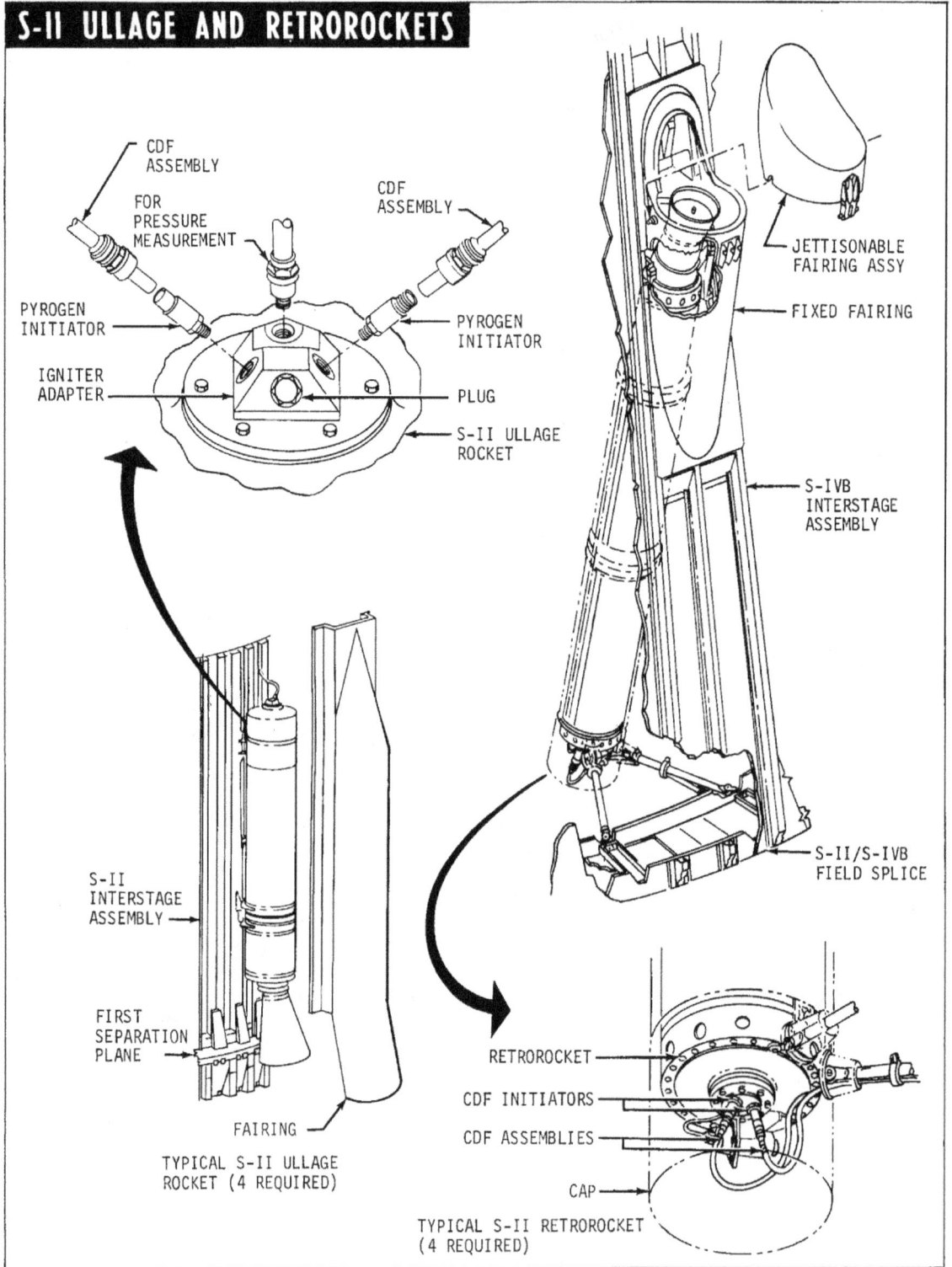

CDF
ASSEMBLY

FOR
PRESSURE
MEASUREMENT

CDF
ASSEMBLY

JETTISONABLE
FAIRING ASSY

FIXED FAIRING

PYROGEN
INITIATOR

PYROGEN
INITIATOR

IGNITER
ADAPTER

PLUG

S-II ULLAGE
ROCKET

S-IVB
INTERSTAGE
ASSEMBLY

S-II
INTERSTAGE
ASSEMBLY

S-II/S-IVB
FIELD SPLICE

FIRST
SEPARATION
PLANE

RETROROCKET

CDF INITIATORS

CDF ASSEMBLIES

FAIRING

CAP

TYPICAL S-II ULLAGE
ROCKET (4 REQUIRED)

TYPICAL S-II RETROROCKET
(4 REQUIRED)

Figure 5-22

# PROPELLANT DISPERSION

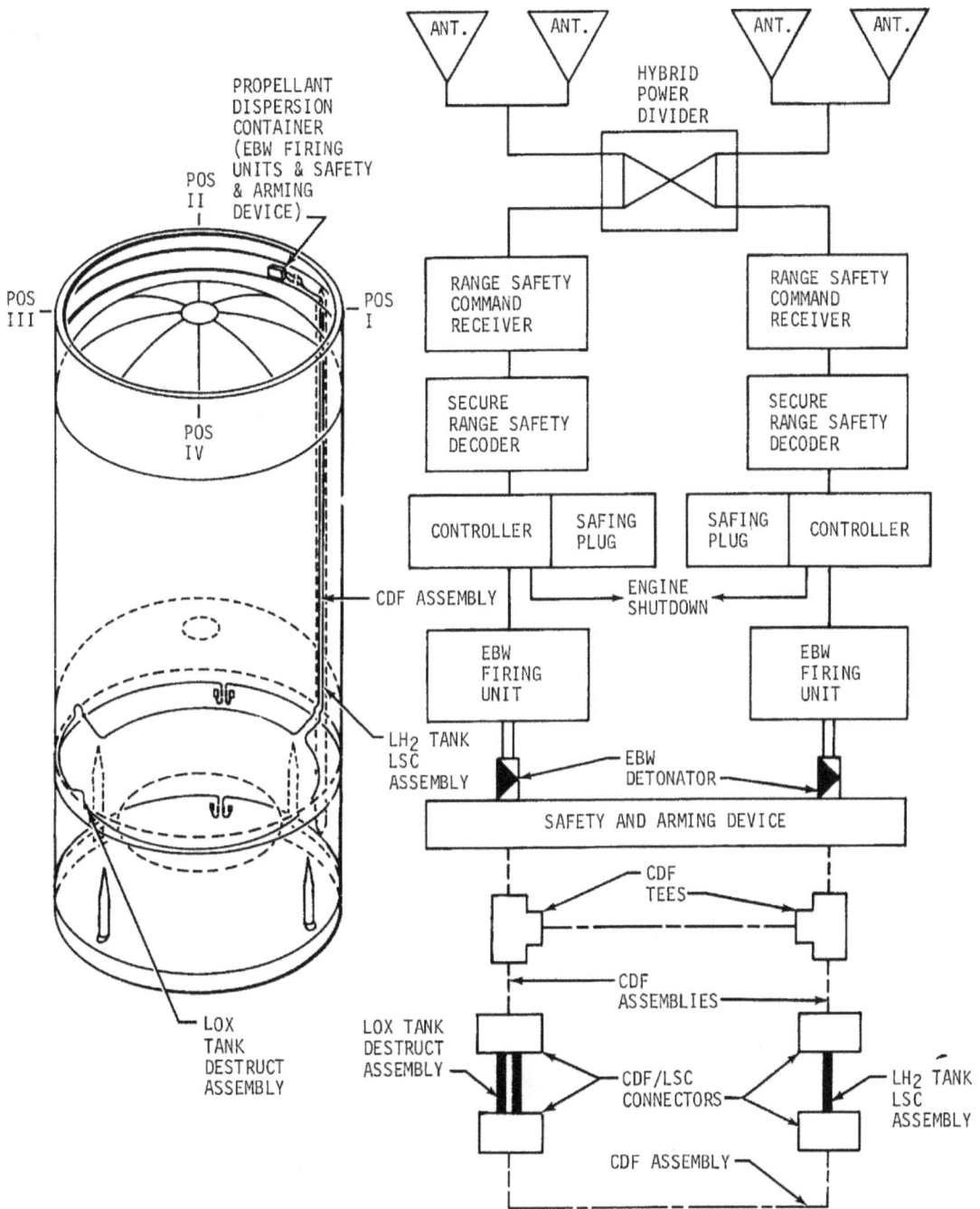

Figure 5-23

to permit charging of the EBW firing units, discharges the storage capacitors in the EBW firing units across the exploding bridgewire in the EBW detonators mounted on the S&A device. The resulting explosive wave propagates through the S&A device inserts to the CDF assemblies and to the CDF tees. The CDF tees, installed on the S-II forward skirt, propagate the wave to two CDF assemblies which detonate to their respective destruct assemblies. The destruct assemblies are connected by a CDF assembly to provide redundancy to the system.

A description of the S&A device is included in the PDS discussion in Section IV.

The LH$_2$ tank linear shaped charge when detonated cuts a 30-foot vertical opening in the tank. The LSC assembly consists of two 15-foot sections of RDX loaded at 600 grains per foot.

The lox tank destruct charges cut 13-foot lateral openings in the lox tank and the S-II aft skirt simulataneously. The destruct assembly consists of two linear explosive charges of RDX loaded at 800 grains per foot. The destruct charges are installed in a figure-eight tube mounted on the inside of the aft skirt structure near station number 1831.0.

# SECTION VI

# S-IVB STAGE

## TABLE OF CONTENTS

## INTRODUCTION

The Saturn S-IVB (figure 6-1) is the third booster stage. Its single J-2 engine is designed to boost the payload into a circular orbit on the first burn, then boost the payload to a proper position and velocity for lunar intercept with a second burn. The stage dry weight (including interstage) is approximately 33,142 pounds. The total weight at ignition is approximately 262,300 pounds. The major systems of the stage are: structures, environmental control, propulsion, flight control, pneumatics, propellants, electrical, instrumentation and ordnance systems.

## STRUCTURE

The basic S-IVB stage airframe, illustrated in figure 6-1, consists of the following structural assemblies: the forward skirt, propellant tanks, aft skirt, thrust structure, and aft interstage. These assemblies, with the exception of the propellant tanks, are all of a skin/stringer type aluminum alloy airframe construction. In addition, there are two longitudinal tunnels which house wiring, pressurization lines, and propellant dispersion systems. The tunnel covers are made of aluminum stiffened by internal ribs. These structures do not transmit primary shell loads but act only as fairings.

### FORWARD SKIRT ASSEMBLY

Cylindrical in shape, the forward skirt (figure 6-1) extends forward from the intersection of the liquid hydrogen ($LH_2$) tank sidewall and the forward dome, providing a hard attach point for the Instrument Unit (IU). It is the load supporting member between the $LH_2$ tank and the IU. An access door in the IU allows servicing of the equipment in the forward skirt. The five environmental plates which support and thermally condition various electronic components, such as the transmitters and signal conditioning modules, are attached to the inside of this skirt. The forward umbilical plate, antennae, $LH_2$ tank flight vents and the tunnel fairings are attached externally to this skirt.

### PROPELLANT TANK ASSEMBLY

The propellant tank assembly (figure 6-1) consists of a cylindrical tank with a hemispherical shaped dome at each end, and a common bulkhead to separate the lox from the $LH_2$. This bulkhead is of sandwich type construction consisting of two parallel hemispherical shaped aluminum alloy (2014-T6) domes bonded to and separated by a fiberglass-phenolic honeycomb core. The internal surface of the $LH_2$ tank is machine milled in a waffle pattern to obtain required tank stiffness with minimum structural weight. To minimize $LH_2$ boil off polyurethane insulation blocks, covered with a fiberglass sheet and coated with a sealant, are bonded into the milled areas of the waffle patterns.

The walls of the tank support all loads forward of the forward skirt attach point and transmit the thrust to the payload. Attached to the inside of the $LH_2$ tank are a 34 foot propellant utilization (PU) probe, nine cold helium spheres, brackets with temperature and level sensors, a chilldown pump, a slosh baffle, a slosh deflector, and fill, pressurization and vent pipes. Attached to the inside of the lox tank are slosh baffles, a chilldown pump, a 13.5 foot PU probe, temperature and level sensors, and fill, pressurization and vent pipes. Attached externally to the propellant tank are helium pipes, propellant dispersion components, and wiring which passes through two tunnel fairings. The forward edge of the thrust structure is attached to the lox tank portion of the propellant tank.

### THRUST STRUCTURE

The thrust structure assembly (figure 6-1) is an inverted, truncated cone attached at its large end to the aft dome of the lox tank and attached at its small end to the engine mount. It provides the attach point for the J-2 engine and distributes the engine thrust over the entire tank circumference. Attached external to the thrust structure are the engine piping, wiring and interface panels, eight ambient helium spheres, hydraulic system, oxygen/hydrogen burner, and some of the engine and lox tank instrumentation.

### AFT SKIRT ASSEMBLY

The cylindrical shaped aft skirt assembly is the load bearing structure between the $LH_2$ tank and aft interstage. The aft skirt assembly is bolted to the tank assembly at its forward edge and connected to the aft interstage. A frangible tension tie separates it from the aft interstage at S-II separation.

### AFT INTERSTAGE ASSEMBLY

The aft interstage is a truncated cone that provides the load supporting structure between the S-IVB stage and the S-II stage (figure 6-1). The interstage also provides the focal point for the required electrical and mechanical interface between the S-II and S-IVB stages. The S-II retrorocket motors are attached to this interstage and at separation the interstage remains attached to the S-II stage.

## ENVIRONMENTAL CONTROL

There are three general requirements for environmental control during checkout and flight operations of the S-IVB stage. The first is associated with ground checkout and prelaunch operations and involves thermal conditioning of the environments around the electrical equipment, auxiliary propulsion system (APS), and hydraulic accumulator reservoir. In addition, there is a requirement for aft skirt and interstage purging. The second involves forward skirt area purging, while the third concerns inflight heat dissipation for the electrical/electronic equipment.

### AFT SKIRT AND INTERSTAGE THERMOCONDITIONING

During countdown, air/$GN_2$ is supplied by the environmental control system, which is capable of switching from air to $GN_2$ purge. Air or $GN_2$ is supplied at the rate of

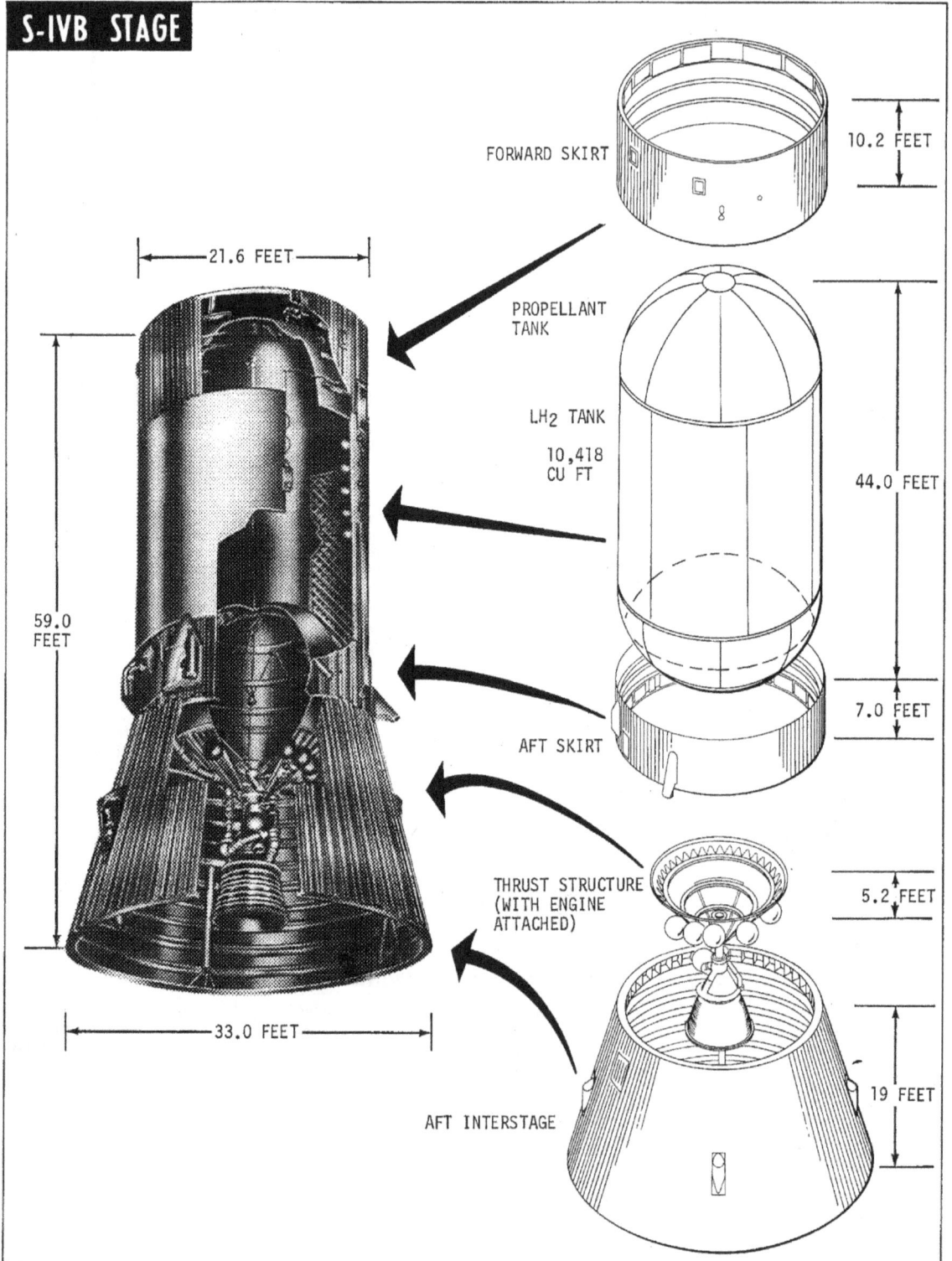

# S-IVB STAGE

FORWARD SKIRT

10.2 FEET

21.6 FEET

PROPELLANT
TANK

LH2 TANK

10,418
CU FT

44.0 FEET

59.0
FEET

7.0 FEET

AFT SKIRT

THRUST STRUCTURE
(WITH ENGINE
ATTACHED)

5.2 FEET

33.0 FEET

19 FEET

AFT INTERSTAGE

Figure 6-1

approximately 3600 scfm. The air purge is initiated when electrical power is applied to the vehicle. $GN_2$ flow is initiated 20 minutes prior to lox chilldown (at T-8 hrs.) and continued until liftoff. During periods of hold, $GN_2$ purge is continued. The aft skirt and interstage thermoconditioning and purge subsystem provides the following:

1. Thermal conditioning of the atmosphere around electrical equipment in the aft skirt during ground operations.

2. Thermal conditioning of the APS, hydraulic accumulator reservoir, and ambient helium bottle.

3. Purging of the aft skirt, aft interstage and thrust structure, and the forward skirt of the S-II stage of oxygen, moisture and combustible gases.

The subsystem consists of a temperature-controlled air or $GN_2$ distribution system (figure 6-2). The purging gas passes over electrical equipment below the ring frame and flows into the interstage. A duct from the skirt manifold directs air or $GN_2$ to a thrust structure manifold. Another duct directs the gas to a shroud covering the ambient helium bottle used to purge lox and $LH_2$ pump shaft seal cavities. From the thrust structure manifold supply duct a portion of air or $GN_2$ is directed to a shroud covering the hydraulic accumulator reservoir.

Temperature control is accomplished by two dual element thermistor assemblies located in the gaseous exhaust stream of each of the APS modules. One element from each thermistor assembly is wired in series to sense average temperature. One series is used for temperature control, the other for temperature recording.

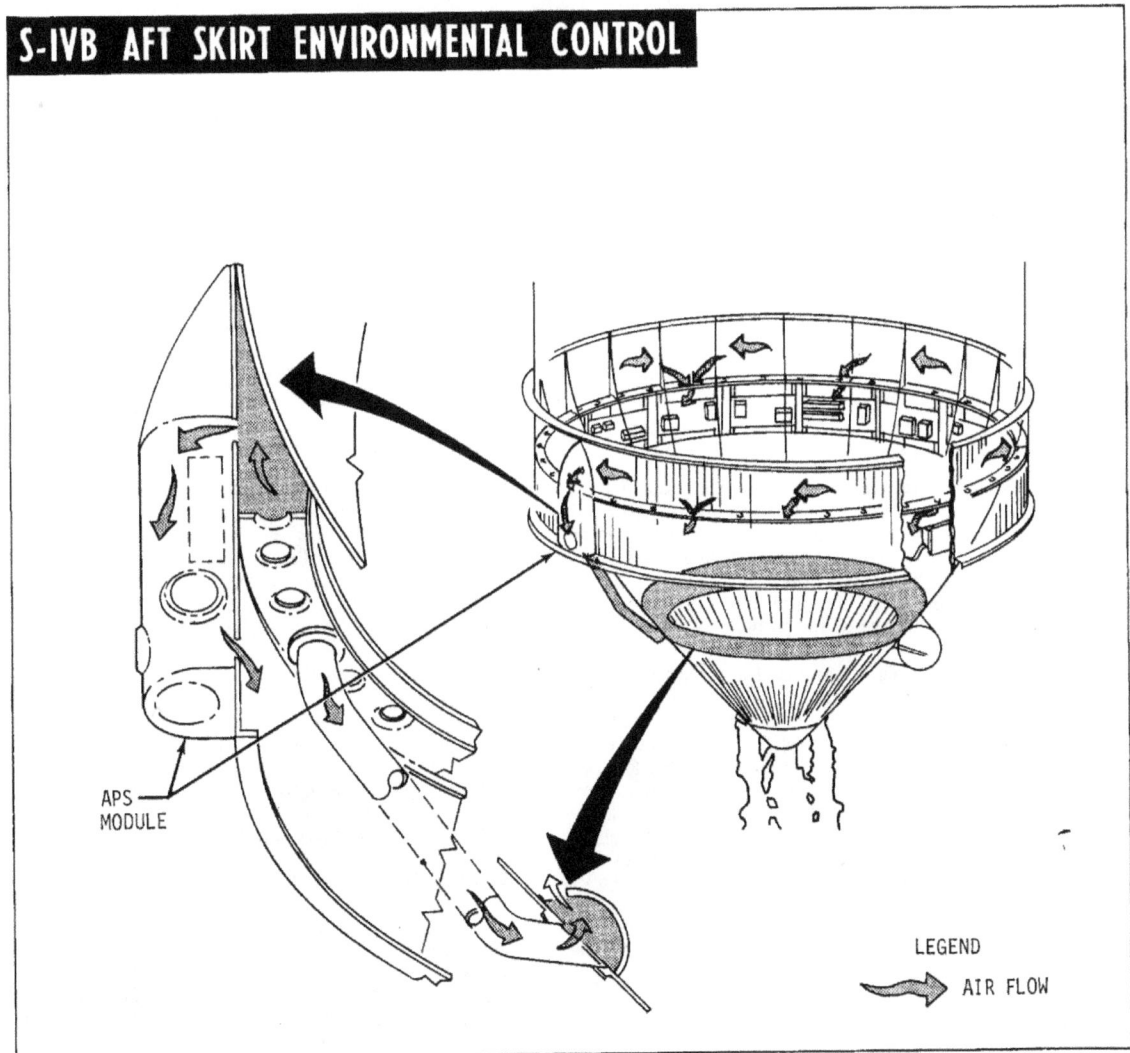

# S-IVB AFT SKIRT ENVIRONMENTAL CONTROL

APS
MODULE

LEGEND

➤ AIR FLOW

Figure 6-2

## FORWARD SKIRT THERMOCONDITIONING

The electrical/electronic equipment in the S-IVB forward skirt area is thermally conditioned by a heat transfer subsystem using a circulating coolant for the medium. Principal components of the system, located in the S-IVB stage forward skirt area, are a fluid distribution subsystem and cold plates. The coolant is supplied to the S-IVB by the IU thermoconditioning system starting when electrical power is applied to the vehicle and continuing throughout the mission. For a description of this system refer to Section VII.

## FORWARD SKIRT AREA PURGE

The forward skirt area is purged with $GN_2$ to minimize the danger of fire and explosion while propellants are being loaded or stored in the stage, or during other hazardous conditions. The purge is supplied by the IU purge system which purges the entire forward skirt/IU/adapter area. The total flow rate into this area is approximately 3500 scfm.

## PROPULSION

This stage provides vehicle propulsion twice during this mission. The first burn occurs immediately after S-II/S-IVB separation and lasts long enough to insert the vehicle into earth orbit. The second burn injects the spacecraft into a high apogee elliptical orbit.

At J-2 engine first burn cutoff, the auxiliary propulsion system (APS) ullage engines are ignited and burn until $T_5$ + 1 minute and 28 seconds, providing stabilization and settling of the propellants. At APS engine ignition, the APS yaw and pitch control modes are enabled (roll already active) for the required attitude control of the stage payload during coast. $LH_2$ continuous venting is activated at $T_5$ + 59.0 seconds and continues until the oxygen/hydrogen ($O_2/H_2$) burner second start.

Prior to second burn, the systems are again readied for an engine start. Approximately 4 minutes before restart, the chilldown systems are reactivated to condition the lines by removing gases collected in the propellant supply system. The $O_2/H_2$ burner is started approximately 9 minutes prior to second burn to pressurize the propellant tanks ullage space and to provide thrust to settle the propellants.

$LH_2$ continuous venting is terminated immediately after the $O_2/H_2$ burner start. Approximately one minute before engine start, the APS ullage engines are fired and the $O_2/H_2$ burner is shut down. The recirculation system is deactivated and engine restart is initiated. With this, the APS ullage engines are shut off, and the APS yaw and pitch control modes are deenergized. The roll control mode remains active throughout the second burn.

At the end of J-2 engine second burn the APS pitch and yaw modes are again enabled to provide attitude control of the stage and payload during terminal coast. In addition, the lox and $LH_2$ pressures are vented through the stage nonpropulsive vents.

## J-2 ROCKET ENGINE

The J-2 rocket engine (figure 5-9) is a high performance, multiple restart engine utilizing liquid oxygen and liquid hydrogen as propellants. The engine attains a thrust of 203,000 pounds during first burn and 203,000 pounds during second burn. The only substances used in the engine are the propellants and helium gas. The extremely low operating temperature of the engine prohibits the use of lubricants or other fluids. The engine features a single, tubular-walled, bell-shaped thrust chamber and two independently driven direct drive turbopumps for liquid oxygen and liquid hydrogen. Both turbopumps are powered in series by a single gas generator, which utilizes the same propellants as the thrust chamber. The main hydraulic pump is driven by the oxidizer turbopump turbine. The ratio of fuel to oxidizer is controlled by bypassing liquid oxygen from the discharge side of the oxidizer turbopump to the inlet side through a servovalve.

The engine valves are controlled by a pneumatic system powered by gaseous helium which is stored in a sphere inside the start bottle. An electrical control system, which uses solid state logic elements, is used to sequence the start and shutdown operations of the engine. Electrical power is supplied from aft battery No. 1.

During the burn periods, the lox tank is pressurized by flowing cold helium through the heat exchanger in the oxidizer turbine exhaust duct. The heat exchanger heats the cold helium, causing it to expand. The $LH_2$ tank is pressurized during burn periods by $GH_2$ from the thrust chamber fuel manifold.

During burn periods in the pitch and yaw planes, thrust vector control is achieved by gimbaling the main engine. Hydraulic pressure for gimbal actuation is provided by the main hydraulic pump. During coast mode the APS engines give the pitch and yaw thrust vector control. Roll control during both the burn periods and the coast modes is achieved by firing the APS engines.

### Start Preparations

Preparations for an engine start include ascertaining the positions and status of various engine and stage systems and components. The J-2 engine electrical control system controls engine operation by means of electrical signals. The heart of the engine electrical control system is the electrical control package (17, figure 5-9). It sequences and times the functions required during engine start or cutoff.

Each cutoff automatically causes the electrical control package circuitry to reset itself, ready for restart, providing that all reset conditions are met. The LVDC issues an engine ready bypass signal just prior to each engine start attempt. This bypass signal acts in the same manner as a cutoff would act. The reset signals engine ready and this allows the LVDC to send its start command. Receipt of the start command initiates the engine start sequence.

## ENGINE START SEQUENCE

When engine start is initiated (3, figure 6-3), the spark exciters in the electrical control package provide energy for the gas generator (GG) and augmented spark igniter (ASI) spark plugs (4). The helium control and ignition phase control valves, in the pneumatic control package (1), are simultaneously energized, allowing helium from the helium tank (2) to flow through the pneumatic regulator to the pneumatic control system. The helium is routed through the internal check valve in the pneumatic control package (1) to

# S-IVB ENGINE START

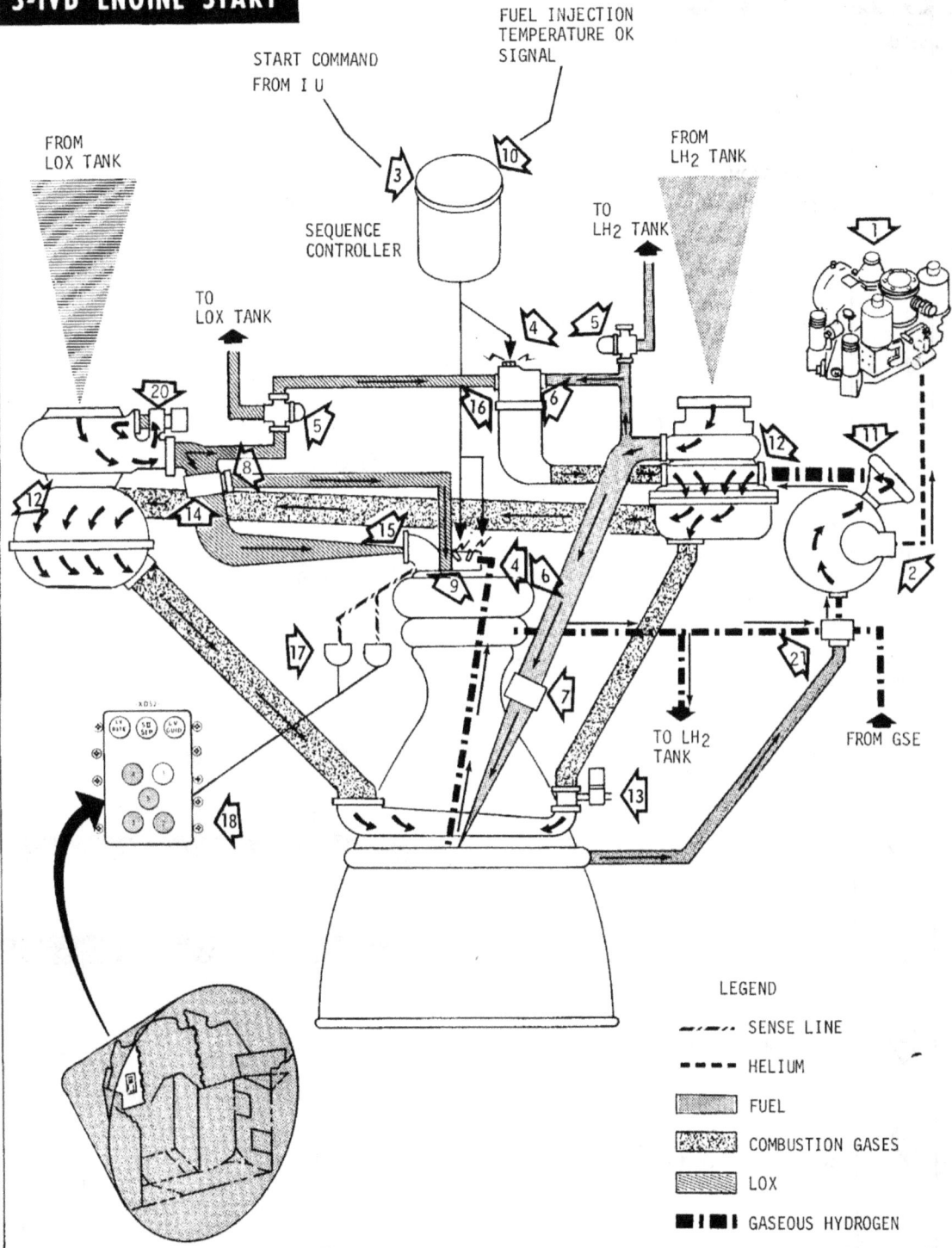

FROM
LOX TANK

START COMMAND
FROM I U

FUEL INJECTION
TEMPERATURE OK
SIGNAL

FROM
LH₂ TANK

SEQUENCE
CONTROLLER

TO LOX TANK

TO
LH₂ TANK

TO
LOX TANK

FROM GSE

TO LH₂
TANK

LEGEND

—··—··—  SENSE LINE

- - - -  HELIUM

▨  FUEL

▨  COMBUSTION GASES

▨  LOX

▮▯▮▯▮  GASEOUS HYDROGEN

Figure 6-3 (Sheet 1 of 3)

# S-IVB ENGINE START

| SEQUENCE | EVENT | TIME IN SECONDS |
|---|---|---|

| | | S-IVB 1ST BURN | T4 + 1.0 | 3.0 | 5.0 | 7.0 | 9.0 | 11.0 | 13.0 | 15.0 |
| | | S-IVB 2ND BURN | T6 + 570.0 | 572.0 | 574.0 | 576.0 | 578.0 | 580.0 | 582.0 | 584.0 |

LEGEND
████ S-IVB 1ST BURN
▨▨▨▨ S-IVB 2ND BURN

**PRE-IGNITION**

NOTE

ALL PNEUMATIC VALVES ARE CONTROLLED BY THE PNEUMATIC CONTROL PACKAGE ⟨1⟩ WHICH IS LOCATED ON THE ENGINE. HELIUM FOR THE PNEUMATIC SYSTEM IS SUPPLIED BY THE CONTROL SPHERE INSIDE THE GH$_2$ START TANK ⟨2⟩.

⟨3⟩ Start Command (Occurs at T$_4$ + 1.0 and T$_6$ + 570.0) Engine Ready → [AND gate] → Engine Start Sequence

⟨4⟩ Augmented Spark Igniter (ASI) Spark Plugs and gas generator (GG) spark plugs fire.

⟨5⟩ Bleed valves stop return flow to propellant tanks.

⟨6⟩ Oxidizer dome and gas generator oxidizer injector are purged.

⟨7⟩ Main fuel valve allows LH$_2$ to flow into engine thrust chamber and into ASI.

⟨8⟩ ASI oxidizer valve allows lox flow to ASI.

⟨9⟩ Sparks ignite the propellants in the ASI.

⟨10⟩ Fuel injection temperature OK signal received causing the

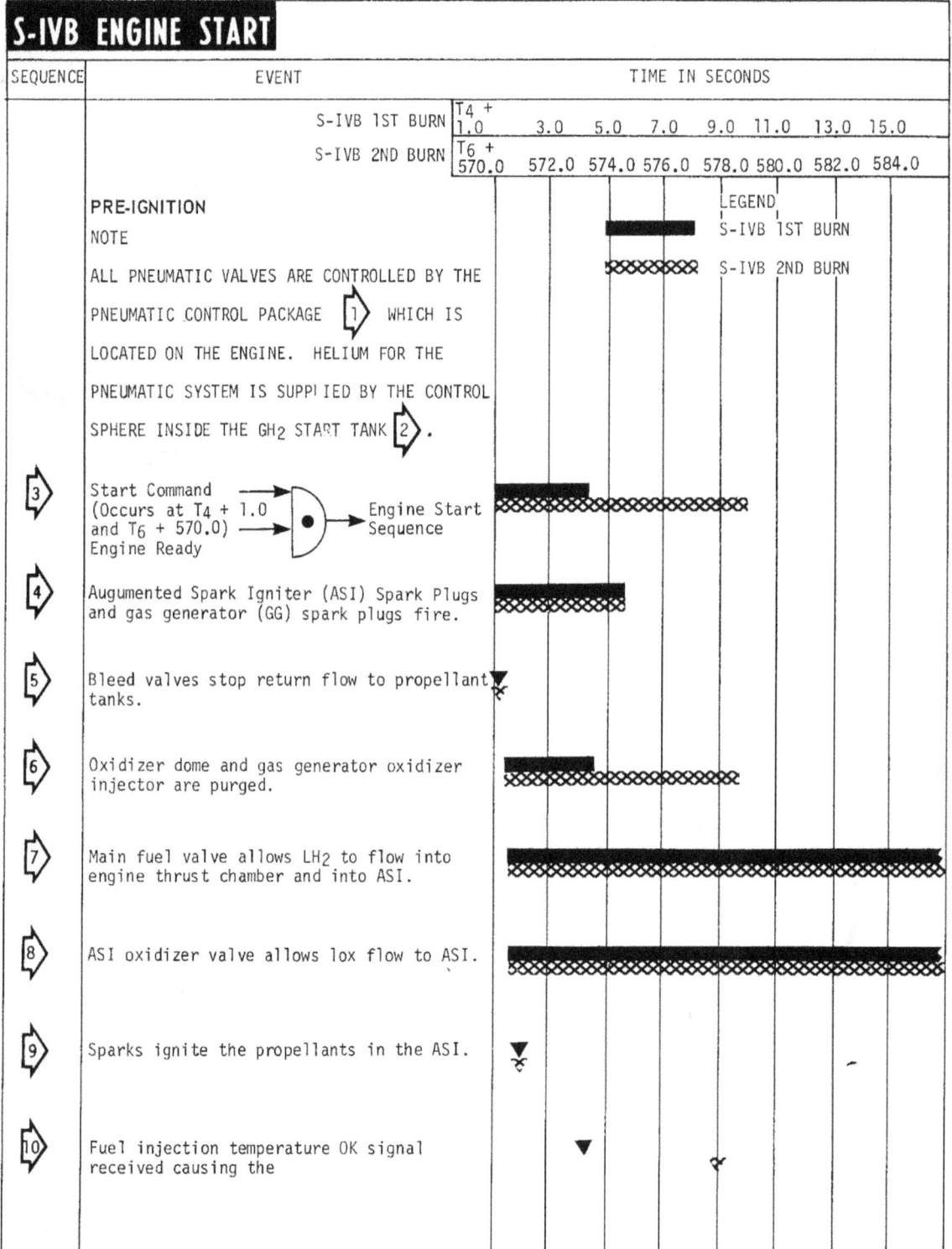

Figure 6-3 (Sheet 2 of 3)

# S-IVB ENGINE START

| SEQUENCE | EVENT | TIME IN SECONDS | | | | | | | |
|---|---|---|---|---|---|---|---|---|---|
| | S-IVB 1ST BURN | $T_4 +$ 1.0 | 3.0 | 5.0 | 7.0 | 9.0 | 11.0 | 13.0 | 15.0 |
| | S-IVB 2ND BURN | $T_6 +$ 570.0 | 572.0 | 574.0 | 576.0 | 578.0 | 580.0 | 582.0 | 584.0 |

11  start tank discharge valve to open.

12  GH2 spins LH2 and lox turbopumps causing propellant pressure to buildup.

13  Lox turbopump bypass valve open to control lox pump speed.

14  Main oxidizer valve opens allowing.

15  lox to be injected into thrust chamber.

16  G G valves admit propellants. (Spark ignites propellants causing pressure build up.)

**MAIN STAGE**

17  OK pressure switches send mainstage OK signal to CM.

18  No. 1 engine out light goes out. (Remaining 4 lights not operative).

19  Engine reaches and maintains 90% thrust or more.

20  GH2 Start tank is refilled with GH2 and LH2.

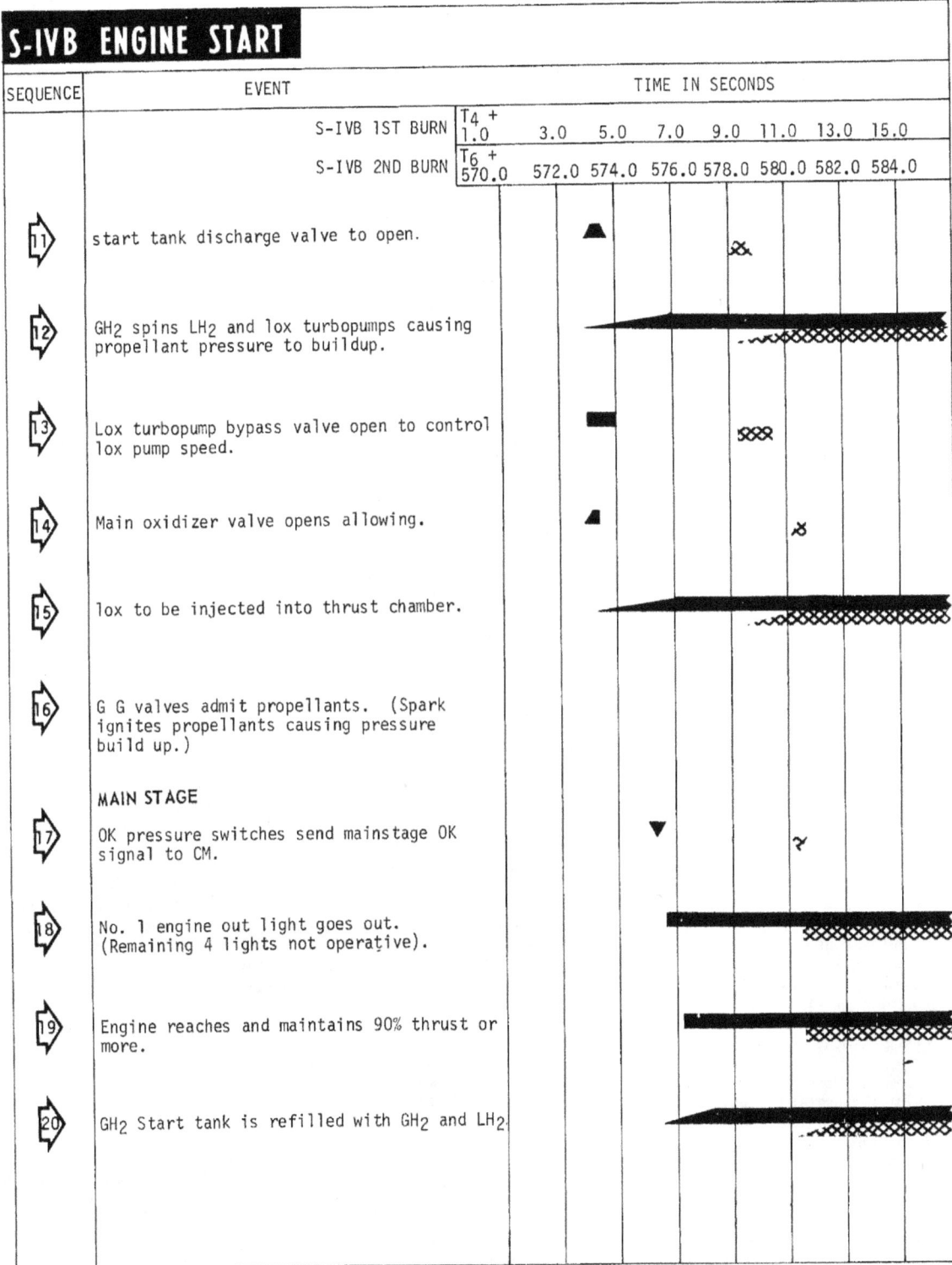

Figure 6-3 (Sheet 3 of 3)

ensure continued pressure to the engine valves in the event of helium supply failure. The regulated helium fills a pneumatic accumulator, closes the propellant bleed valves (5), and purges (6) the oxidizer dome and gas generator oxidizer injector manifold. The oxidizer turbopump (12) intermediate seal cavity is continuously purged. The mainstage control valve holds the main oxidizer valve closed, and opens the purge control valve, which allows the oxidizer dome and gas generator oxidizer injector to be purged (6). The mainstage control valve also supplies opening control pressure to the oxidizer turbine bypass valve (13). The ignition phase control valve, when actuated, opens the main fuel valve (7) and the ASI oxidizer valve (8), and supplies pressure to the sequence valve located within the main oxidizer valve (14). Fuel is tapped from downstream of the main fuel valve for use in the ASI (4). Both propellants, under tank pressure, flow through the stationary turbopumps (12).

The sequence valve, in the main fuel valve (7), opens when the fuel valve reaches approximately 90% open, and routes helium to the start tank discharge valve (STDV) (11) control valve. Simultaneously with engine start, the STDV delay timer is energized. Upon expiration of the STDV timer and the receipt of a fuel injection temperature OK signal, the STDV control valve and ignition phase timer are energized. As the STDV control valve energizes, the discharge valve opens allowing gaseous hydrogen under pressure to flow through the series turbine drive system. This accelerates both turbopumps (12) to the proper operating levels to allow subsequent ignition and power buildup of the gas generator (16). The relationship of fuel to lox turbopump speed buildup is controlled by an orifice in the oxidizer turbine bypass valve (13). During the start sequence, the normally open oxidizer bypass valve (13) permits a percentage of the gas to bypass the oxidizer turbine.

During this period, ASI combustion is detected by the ASI ignition monitor. (Absence of the ignition detection signal, or a start tank depressurized signal, will cause cutoff at the expiration of the ignition phase timer.) With both signals present at ignition phase timer expiration, the mainstage control valve energizes. Simultaneously, the sparks deenergize timer is energized and the STDV control valve is deenergized, causing the STDV to close. Helium pressure is vented from the main oxidizer valve (14) and from the purge control valve through the mainstage control valve. The purge control valve closes, terminating the oxidizer dome and gas generator oxidizer injector manifold purges (6). Pressure from the mainstage control valve is routed to open the main oxidizer valve (14). A sequence valve, operated by the main oxidizer valve (14), permits gaseous helium to open the gas generator control valve (4) and to close the oxidizer turbine bypass valve (13). Flow to close the oxidizer turbine bypass valve (13) is restricted as it passes through an orifice.

Propellants flowing into the gas generator (16) are ignited by the spark plugs (4). Combustion of the propellants causes the hot gases to drive the turbopumps (12). The turbopumps force propellant into the thrust chambers where it is ignited by the torch from the ASI.

Transition into mainstage occurs as the turbopumps (12) accelerate to steadystate speeds. As the oxidizer pressure increases, a thrust OK signal is generated by either of the two thrust OK pressure switches (17). (Cutoff occurs if no signal is received before expiration of the sparks deenergized timer.) The ASI and GG spark exciters are deenergized at expiration

of the sparks deenergized timer. Cutoff occurs if both pressure switch actuated signals (thrust OK) are lost during mainstage operation.

Steadystate operation is maintained until a cutoff signal is initiated. During this period, gaseous hydrogen is tapped from the fuel injection manifold to pressurize the LH$_2$ tank. The lox tank is pressurized by gaseous helium heated by the heat exchanger in the turbine exhaust duct. Gaseous hydrogen is bled from the thrust chamber fuel injection manifold, and liquid hydrogen is bled from the ASI fuel line to refill start tank for engine restart. (Approximately 50 seconds of mainstage engine operation is required to recharge the start tank.) During engine mainstaging propellant utilization is accomplished as described in paragraphs on propellant management and propellant utilization subsystems.

### Engine Cutoff

The J-2 engine may receive cutoff signals from the following sources: EDS No.'s 1 and 2, range safety systems No.'s 1 and 2, thrust OK pressure switches, propellant depletion sensors, and an IU programmed command (velocity or timed) via the switch selector.

The switch selector, range safety system No. 2, EDS No. 2, and the propellant depletion sensors cutoff commands are tied together (but diode isolated) and sent to the electrical control package cutoff circuit. The dropout of the thrust OK pressure switches removes a cutoff inhibit function in the electrical control package cutoff circuit. EDS No. 1 and range safety system No. 1 cutoff commands will indirectly transfer the engine control power switch to the OFF position, causing the engine to shut down due to power loss.

### Cutoff Sequence

The engine cutoff sequence is shown graphically in figure 6-4. When the electrical control package receives the cutoff signal (1), it deenergizes the mainstage and ignition phase control valves in the pneumatic control package (2), while energizing the helium control deenergize timer. The mainstage control valve closes the main oxidizer valve (3), and opens the purge control valve and the oxidizer turbine bypass valve (8). The purge control valve directs a helium purge (11) to the oxidizer dome and GG oxidizer injector. The ignition phase control valve closes the ASI oxidizer valve (4) and the main fuel valve (5), while opening the fast shutdown valve. The fast shutdown valve now rapidly vents the return flow from the GG control valve. All valves, except the ASI oxidizer valve (4) and oxidizer turbine bypass valve (8), are spring loaded closed. This causes the valves to start moving closed as soon as the pressure to open them is released. GG combustion pressure aids closing of the GG control valve.

Expiration of the helium control deenergize timer causes the helium control valve to close. When the helium control valve closes, it causes the oxidizer dome and GG oxidizer injector purges (11) to stop. An orifice in the locked up lines bleeds off pressure from the propellant bleed valves (13). This loss of pressure allows springs to open the valves. When open, the propellant bleed valves allow propellants to flow back to the propellant tanks.

# S-IVB ENGINE CUTOFF

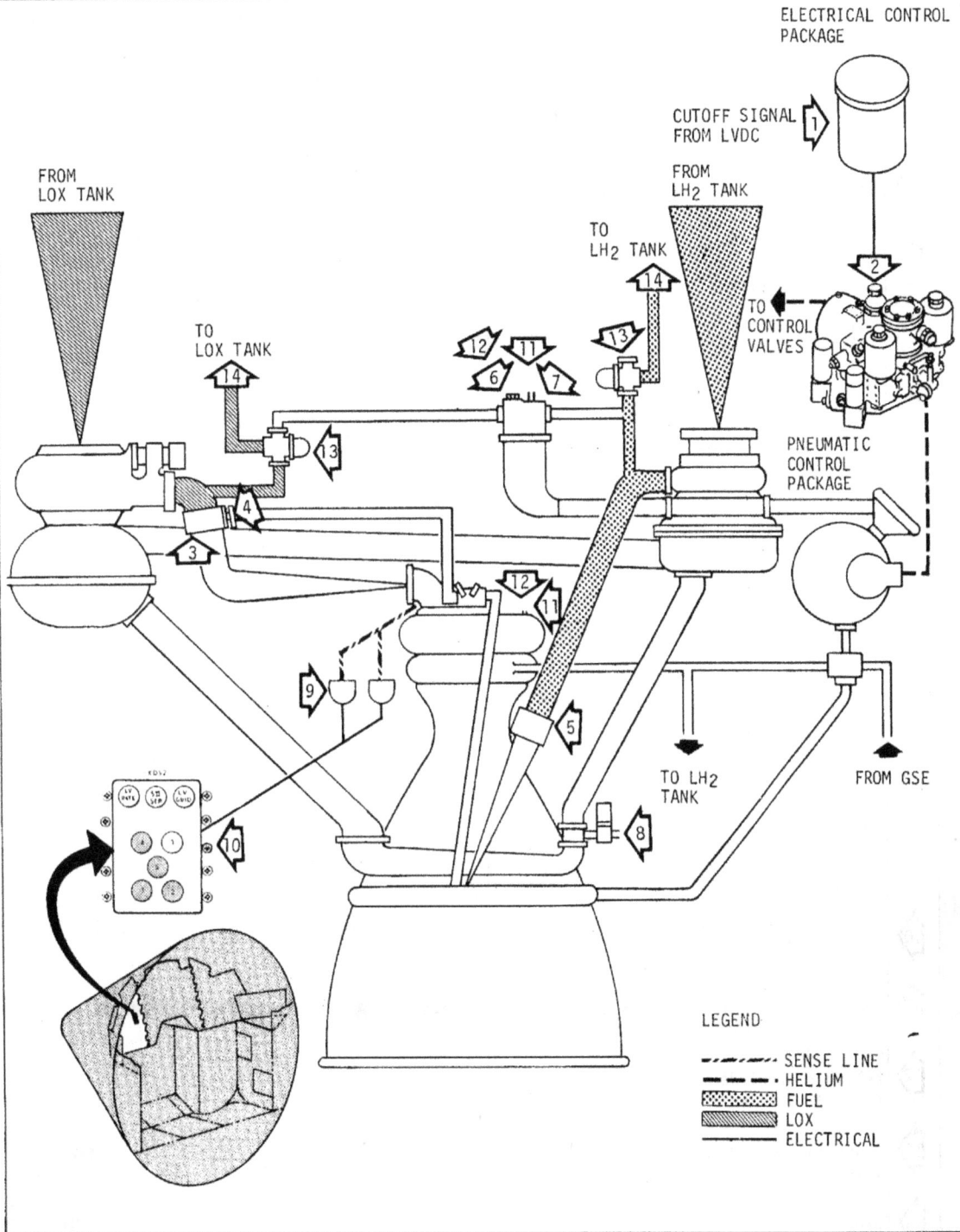

Figure 6-4 (Sheet 1 of 2)

| SEQUENCE | EVENTS | TIME IN SECONDS FROM CUTOFF SIGNAL |
|---|---|---|
| | | 0.0 0.1 0.2 0.3 0.4 0.5 0.6 0.7 0.8 0.9 1.0 1.1 1.2 |
| 1 | Cutoff signal from LVDC triggers | |
| 2 | valves in pneumatic control package. | |
| 3 | Pneumatic pressure closes main oxidizer valve; | |
| 4 | ASI oxidizer valve; | |
| 5 | main fuel valve; | |
| 6 | GG oxidizer valve; | |
| 7 | GG fuel valve and | |
| 8 | Opens oxidizer turbine bypass valve. (Within 10 seconds.) | |
| 9 | Thrust OK pressure switches sense loss of thrust. (This event starts time base No. 5 and No. 7.) | |
| | NOTE | |
| | Loss of Thrust could have been the cause of the cutoff signal 1 and have preceeded previous events: | |
| 10 | No. 1 engine out light illuminates. | |
| 11 | Helium flow purges oxidizer dome and GG oxidizer injector. | |
| 12 | Helium control deenergize timer stops oxidizer dome and GG oxidizer purge. | |
| 13 | Propellant bleed valves open allowing | |
| 14 | propellant flow. | |

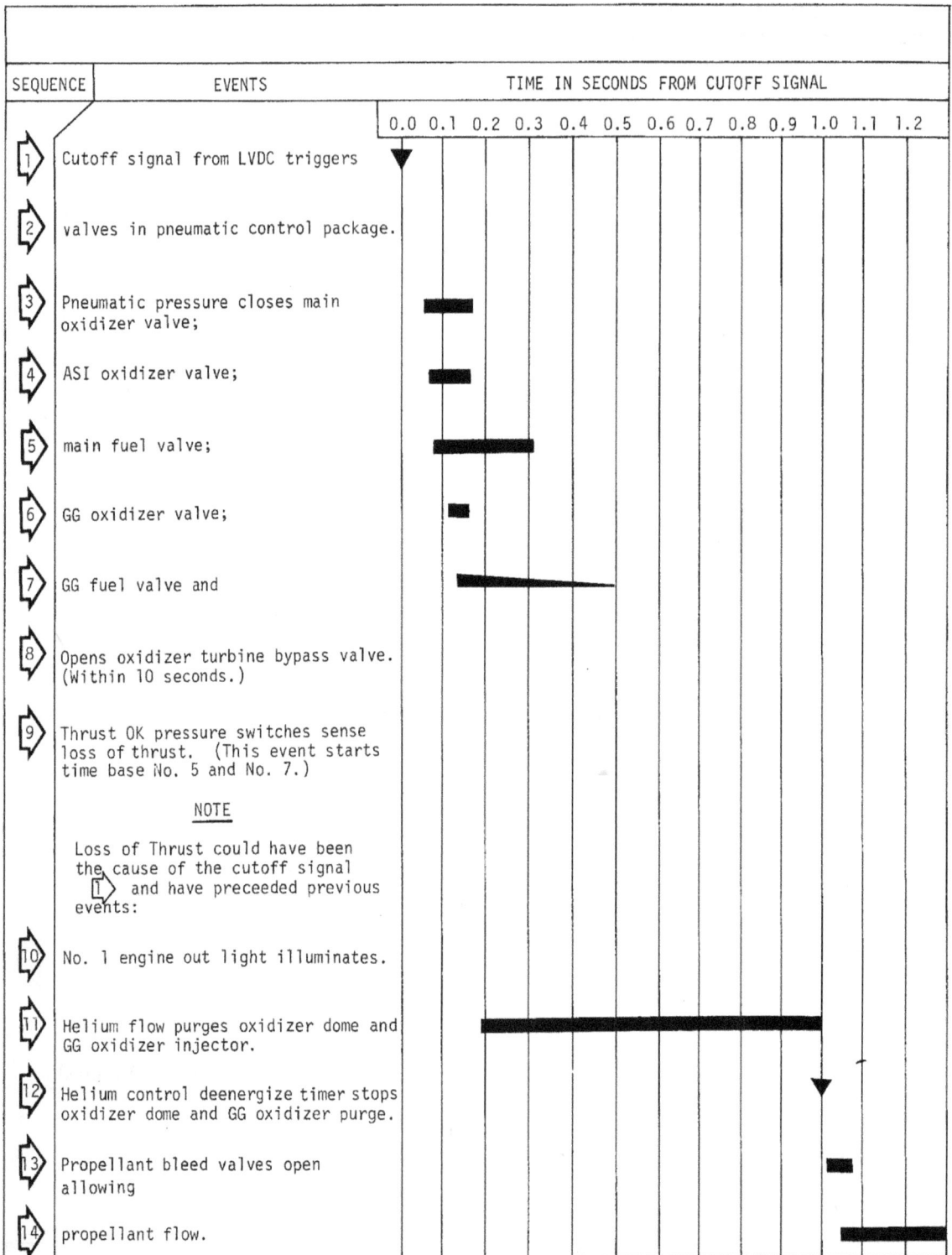

Figure 6-4 (Sheet 2 of 2)

## Restart

The restart of the J-2 engine is identical to the initial start except for the fill procedure of the start tank. The start tank is filled with $LH_2$ and $GH_2$ during the first burn period.

To ensure that sufficient energy will be available for spinning the $LH_2$ and lox pump turbines, a waiting period of approximately 90 minutes to 6 hours is permitted. The minimum time is required to build sufficient pressure by warming the start tank through natural means. The minimum wait is also needed to allow the hot gas turbine exhaust system to cool. Prolonged heating will cause a loss of energy in the start tank. This loss occurs when the $LH_2$ and $GH_2$ warms and raises the gas pressure to the relief valve setting. If this venting continues over a prolonged period, the total stored energy will be depleted. This limits the waiting period prior to a restart attempt to 6 hours.

## PROPELLANTS

The propellant tank assembly accepts, stores and supplies the lox and $LH_2$ used for J-2 engine operation.

Separate pressurization and repressurization systems (figure 6-5) are provided for each propellant tank to monitor and control tank pressures, in order to assure the engine propellant turbopump a net positive suction head (NPSH) of 42 feet for lox and 150 feet for $LH_2$ (minimum) during engine 1st and 2nd burn. Each tank is equipped with a pressure sensing switch which commands the opening or closing of the respective tank pressurization or repressurization control module valves. The control modules, in turn, control the passage of pressurizing gases to the tanks as required. Prior to flight the ambient and cold helium storage bottles are filled, and the propellant tanks are

pressurized utilizing a common ground source of cold each tank.

## OXYGEN/HYDROGEN BURNER

The oxygen/hydrogen $(O_2/H_2)$ burner (figure 6-6) uses stage onboard propellants as an energy source to heat cold helium. The $O_2/H_2$ burner use spark plugs to ignite onboard propellants in its combustion chamber. The chamber is surrounded by three sets of coils. Tank propellants flow through one set of coils where they are vaporized and injected into the combustion chamber. The energy of combustion heats the cold helium in the other two sets of coils, causing it to vaporize. Gaseous helium flow is controlled by the $LH_2$/lox tank repressurization control module (dual valves). The burner produces 15 to 35 pounds thrust through the center of gravity as shown in figure 6-6.

## PROPELLANT CONDITIONING

During filling operations, the prevalves are allowed to stay in the open position to provide a deadhead type chilldown of the feed system hardware (low pressure feed duct and engine pump), allowing temperature stabilization of this hardware prior to activation of the recirculation chilldown system. Approximately five minutes before liftoff, the prevalves are closed and recirculating chilldown flow is initiated. It continues for approximately 7-1/2 minutes, until J-2 engine prestart (figure 6-7). Since lox is already a subcooled liquid (no two- phase flow in the return line), prepressurization has negligible effect on the flowrate. The $LH_2$, however, becomes a subcooled liquid at prepressurization (eliminating two-phase flow in the return line), resulting in increased $LH_2$ chilldown flow.

Chilldown conditioning of the engine pumps, inlet ducting

| Pressurization and Venting Systems | | | |
|---|---|---|---|
| Time | Tank | Pressurization Source | Venting System |
| Prior to liftoff | LOX<br>LH2 | Helium ground source<br>Helium ground source | Through aft skirt<br>ground vent line |
| 1st & 2nd stage boost | LOX<br>LH2 | Cold helium*<br>Ambient helium** | Non-propulsive (pressure relief)<br>No venting required |
| During S-IVB 1st burn | LOX<br>LH2 | Cold helium heated in heat exchanger<br>(uses GH2 bled from J-2 engine) | No venting required<br>No venting required |
| Earth parking orbit | LOX<br><br>LH2 | Ambient helium** or cold helium heated<br>by oxygen/hydrogen burner | Non-propulsive<br><br>Propulsive (8-15 pounds thrust)<br>through forward skirt |
| During 2nd burn | LOX<br>LH2 | Cold helium heated in heat exchanger | No venting required |
| Trans Lunar Injection | LOX<br>LH2 | None required<br>None Required | Non-propulsive (Command)<br>Non-propulsive (Command) |
| NOTE: See Figure 2-1 for specific time sequence.<br>*Cold helium is stored in 9 spheres in the LH2 tank.<br>**Ambient helium is stored in 8 spheres on the thrust structure. | | | |

Figure 6-5

# OXYGEN/HYDROGEN BURNER

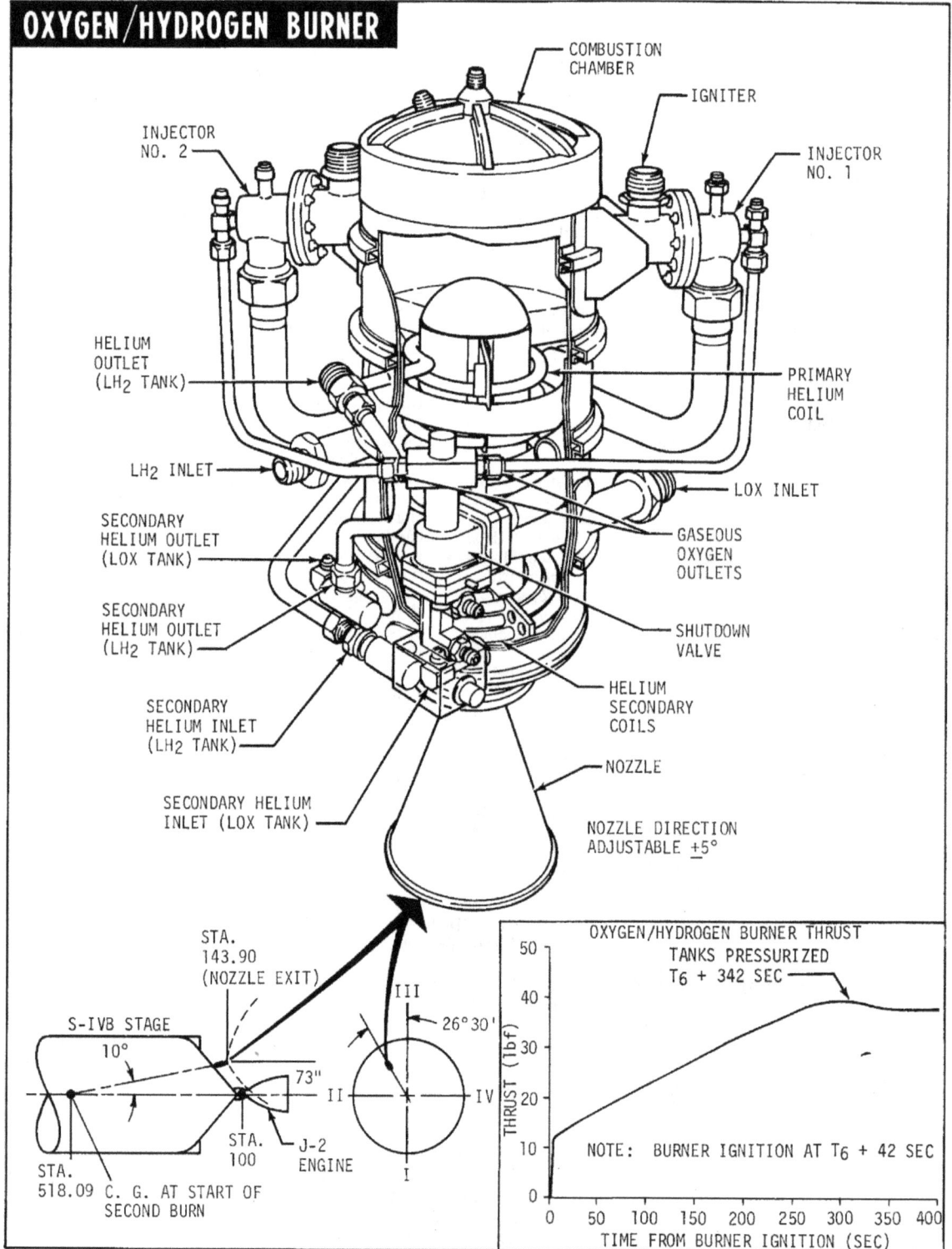

COMBUSTION CHAMBER

IGNITER

INJECTOR NO. 2

INJECTOR NO. 1

HELIUM OUTLET (LH$_2$ TANK)

PRIMARY HELIUM COIL

LH$_2$ INLET

LOX INLET

SECONDARY HELIUM OUTLET (LOX TANK)

GASEOUS OXYGEN OUTLETS

SECONDARY HELIUM OUTLET (LH$_2$ TANK)

SHUTDOWN VALVE

SECONDARY HELIUM INLET (LH$_2$ TANK)

HELIUM SECONDARY COILS

SECONDARY HELIUM INLET (LOX TANK)

NOZZLE

NOZZLE DIRECTION ADJUSTABLE ±5°

STA. 143.90 (NOZZLE EXIT)

S-IVB STAGE

10°

73"

26° 30'

III

II

IV

I

STA. 100

J-2 ENGINE

STA. 518.09 C. G. AT START OF SECOND BURN

OXYGEN/HYDROGEN BURNER THRUST TANKS PRESSURIZED T$_6$ + 342 SEC

NOTE: BURNER IGNITION AT T$_6$ + 42 SEC

THRUST (lbf)

TIME FROM BURNER IGNITION (SEC)

Figure 6-6

# RECIRCULATION SYSTEMS

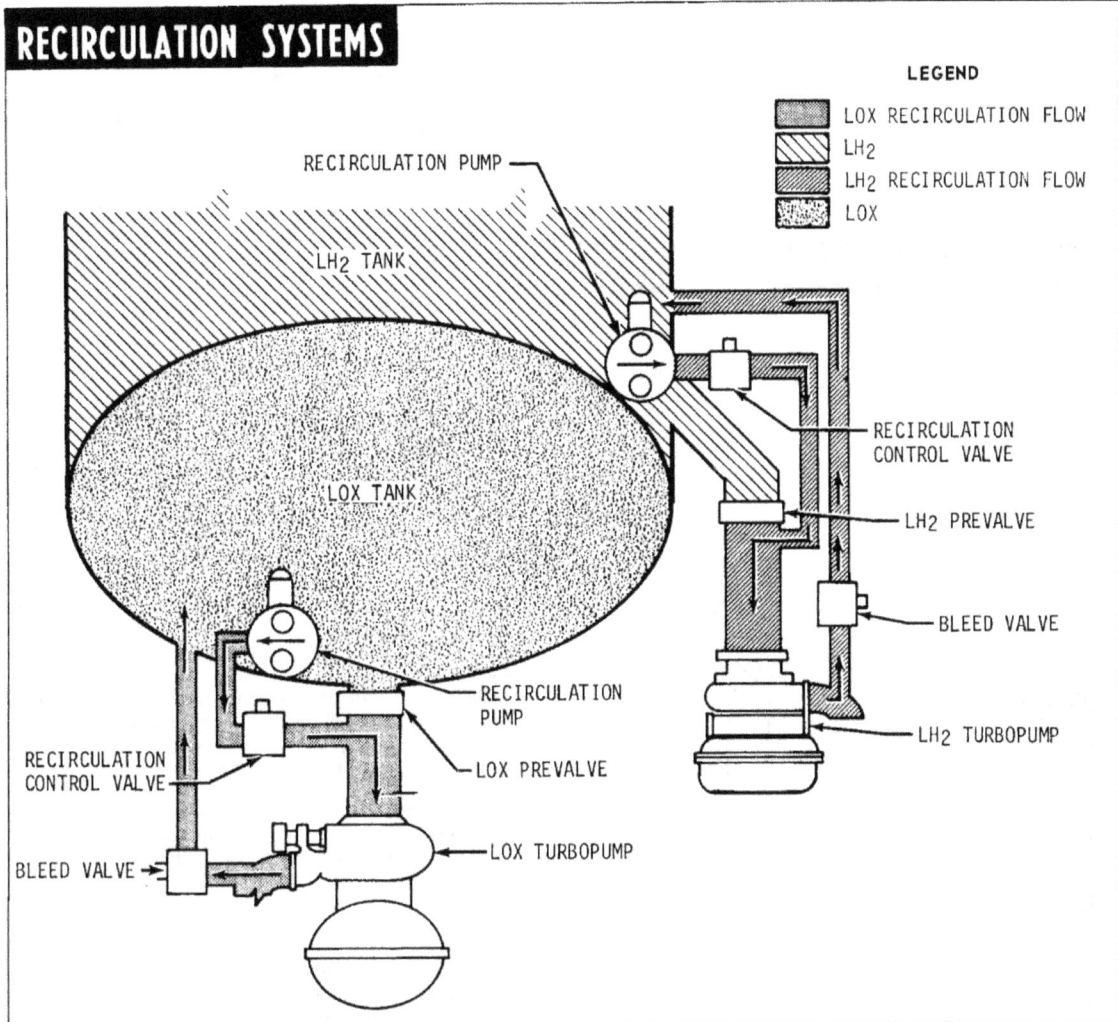

Figure 6-7

and the engine hardware for both engine starts is accomplished by separate lox and $LH_2$ chilldown systems. The purpose of the chilldown is to condition the ducting and engine to the proper temperature level, and to eliminate bubbles (two-phase flow) prior to pressurization. The chilldown, along with the net positive suction head which is obtained by the proper pressure levels, provides the proper starting conditions.

Propellants from each tank are recirculated through the feed systems and return bleed lines by chilldown pumps. Check valves, prevalves, shutoff valves, and ducts control and route the fluids to perform the chilldown. Pneumatic pressure for operating the shutoff valves and prevalves is supplied by the stage pneumatic helium control bottle.

## PROPELLANT VENTING

The vent-relief subsystems (figure 6-9) on the stage protect the propellant tanks against overpressurization and enable command venting at any time that controlled venting of tank pressure is required. The lox and $LH_2$ tanks each have a command relief and venting subsystem.

Both lox and $LH_2$ tank venting sequences provide for propulsive and non-propulsive venting (figure 6-5). The lox tank is vented through its propulsive vent duct, located in the aft skirt, during loading. During flight, lox venting is normally a relief valve function and escaping vapors are routed through the non-propulsive vents. Specific command relief non-propulsive venting is accomplished by flight program or by ground command (figure 6-5). During fill, $LH_2$ venting is routed through ground vent lines to the $GH_2$ burn pond. At T-40 seconds, the vent valves are commanded closed and the relief valves provide venting through the non-propulsive vents. Non-propulsive venting is continued until after the J-2 engine first burn, when, as a part of second burn preparations, the $LH_2$ tank is command vented through the propulsive vents to provide propellant settling. Specific venting sequences are noted in figure 2-1 and figure 6-5.

## LOX SYSTEM

Lox is stored in the aft tank of the propellant tank structure (figure 6-8) at a temperature of -29 degrees F. Total volume of the tank is approximately 2830 cubic feet with an ullage volume of approximately 108 cubic feet. The tank is prepressurized between 38 and 41 psia and is maintained at that pressure during boost and engine operation. Gaseous helium is used as the pressurizing agent.

# LOX TANK SUPPLY AND VENT

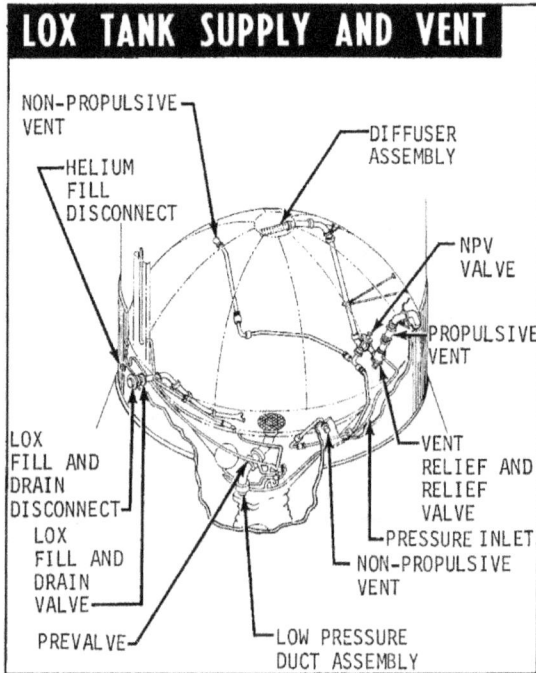

Figure 6-8

### Lox Fill and Drain

The lox fill and drain valve is capable of allowing flow in either direction for fill or drain operations. During tank fill, the valve is capable of flowing 1000 gpm of lox at -297 degrees F at an inlet pressure of 51 psia. Pneumatic pressure for operating the fill and drain valve is supplied by the stage pneumatic control bottle.

Loading begins with a precooling flowrate of 500 gpm. When the 5% load level is reached, fast fill (1000 gpm) is initiated. At the 98% load level, fast fill stops and a slow fill at 300 gpm begins. A fast fill emergency cutoff sensor has been provided to compensate for a primary control cutoff failure. Slow fill is terminated at the 100% load level and this level is then maintained by a replenish flowrate of 0 to 30 gpm, as required. The replenish flow is maintained through the complete lox tank prepressurization operation and the 100% lox load operation. Liquid level during fill is monitored by means of the lox mass probes.

Pressure sensing switches are used to control the tank pressure during fill and flight. In the event of lox tank overpressurization (41 ± 0.5 psia) the pressure switch for the lox tank ground fill valve signals the ground automatic

sequencer to close the lox ground fill valve.

### Lox Engine Supply

A six-inch low pressure supply duct supplies lox from the tank to the engine. During engine burn, lox is supplied at a nominal flowrate of 392 pounds per second, and at a transfer pressure above 37 psia. The supply duct is equipped with bellows to provide compensating flexibility for engine gimbaling, manufacturing tolerances, and thermal movement of structural connections.

## LOX TANK PRESSURIZATION

Lox tank pressurization is divided into three basic procedures. These procedures are called prepressurization, pressurization, and repressurization. The term prepressurization is used for that portion of the pressurization performed on the ground prior to liftoff. The term pressurization is used to indicate pressurization during engine burn periods, and lastly, repressurization indicates pressurization just before a burn period.

The pressurant used during the three lox tank pressurization procedures is gaseous helium. Cold helium from a ground source (1, figure 6-9) is used during the prepressurization period. This ground source of cold helium is also used to charge the nine cold helium storage spheres (2). The cold helium storage spheres, located in the LH₂ tank, supply cold helium for both the pressurization and repressurization periods. The ambient helium storage spheres (3), filled by ground support equipment (26), supply an alternate source of helium for use during the repressurization period.

The lox tank pressure is controlled by the flight control pressure switches (4) (dual redundant), regardless of the pressurization procedure used. These switches control solenoid shutoff valves in each of the supply subsystems.

### Prepressurization

At T-167 seconds, the lox tank is pressurized (figure 6-9) by ground support equipment. Pressure regulated cold helium (1) passes through the lox tank pressure control module (5) and flows (6) into the lox tank. When the lox tank pressure increases to 41 psia, the pressurization is completed and the flight control pressure switch shuts off the ground supply of cold helium.

### Pressurization

After S-II/S-IVB separation at $T_4$ + 0.8, the lox tank flight pressure system is activated. When this system is activated, cold helium from cold helium storage spheres (2, figure 6-9) is routed through the lox tank pressure control module (5) to the J-2 engine heat exchanger (7). When the cold helium passes through the lox tank pressure control module (5), its pressure is reduced to approximately 385 psia. As it passes through the heat exchanger it is expanded and routed to the lox tank. A small portion of cold helium bypasses the heat exchanger through a control orifice, and mixes with the hot gas prior to entering the lox tank. The flight control pressure switch (4) controls the operation of a solenoid valve in the lox tank pressure control module (5) to control lox tank pressure to 38-41 psia. The flow (8) of the helium to the lox tank is monitored by a backup pressure switch (9). If normal pressure regulation fails, this switch (9) will control lox tank

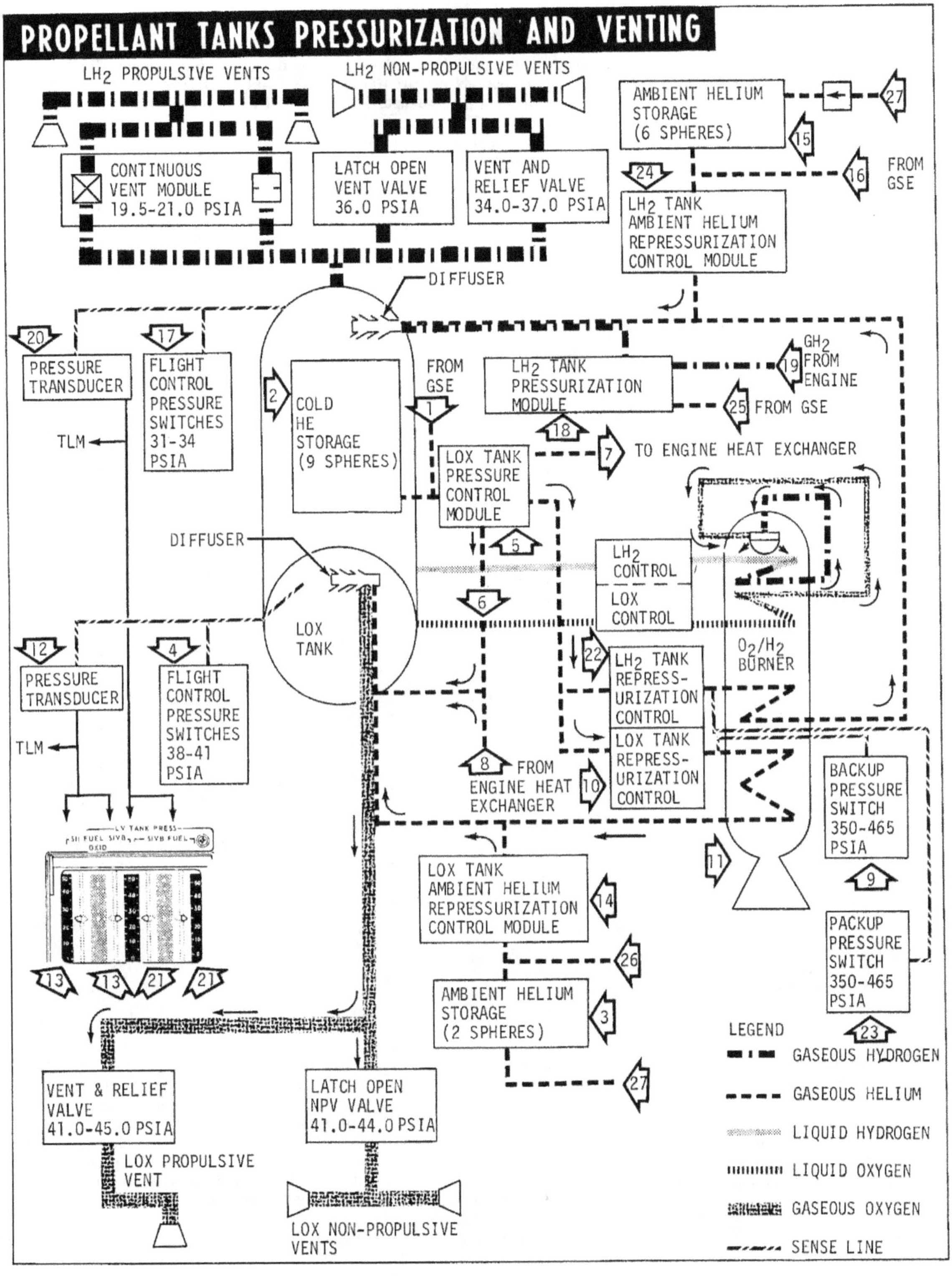

# PROPELLANT TANKS PRESSURIZATION AND VENTING

LH₂ PROPULSIVE VENTS          LH₂ NON-PROPULSIVE VENTS

CONTINUOUS
VENT MODULE
19.5-21.0 PSIA

LATCH OPEN
VENT VALVE
36.0 PSIA

VENT AND
RELIEF VALVE
34.0-37.0 PSIA

AMBIENT HELIUM
STORAGE
(6 SPHERES)

15

24

16   FROM
     GSE

LH₂ TANK
AMBIENT HELIUM
REPRESSURIZATION
CONTROL MODULE

DIFFUSER

20          17

PRESSURE
TRANSDUCER

FLIGHT
CONTROL
PRESSURE
SWITCHES
31-34
PSIA

TLM

2   COLD
    HE
    STORAGE
    (9 SPHERES)

FROM
GSE
1

LH₂ TANK
PRESSURIZATION
MODULE

GH₂
19  FROM
    ENGINE

25  FROM GSE

18

7   TO ENGINE HEAT EXCHANGER

LOX TANK
PRESSURE
CONTROL
MODULE

5

6

LH₂
CONTROL

LOX
CONTROL

O₂/H₂
BURNER

DIFFUSER

LOX
TANK

12          4

PRESSURE
TRANSDUCER

FLIGHT
CONTROL
PRESSURE
SWITCHES
38-41
PSIA

TLM

22

LH₂ TANK
REPRESS-
URIZATION
CONTROL

LOX TANK
REPRESS-
URIZATION
CONTROL

8   FROM
    ENGINE HEAT
    EXCHANGER

10

11

BACKUP
PRESSURE
SWITCH
350-465
PSIA

9

LV TANK PRESS
SII FUEL SIVB   SIVB FUEL
OXID

LOX TANK
AMBIENT HELIUM
REPRESSURIZATION
CONTROL MODULE

14

PACKUP
PRESSURE
SWITCH
350-465
PSIA

13   13   21   21

26

AMBIENT HELIUM
STORAGE
(2 SPHERES)

3

23

LEGEND

VENT & RELIEF
VALVE
41.0-45.0 PSIA

LATCH OPEN
NPV VALVE
41.0-44.0 PSIA

27

GASEOUS HYDROGEN

GASEOUS HELIUM

LIQUID HYDROGEN

LIQUID OXYGEN

GASEOUS OXYGEN

SENSE LINE

LOX PROPULSIVE
VENT

LOX NON-PROPULSIVE
VENTS

Figure 6-9

pressure through operation of solenoid valves in the lox tank pressure control module (5). In this manner, lox tank pressurization is maintained during all engine burn periods. An S-IVB lox tank pressure reading becomes available in the command module (CM) at S-II/S-IVB separation. This pressure is sensed by the pressure transducer (12) and is relayed to the S-II FUEL/S-IVB OXID gauges (13) in the CM and, via telemetry, to the ground.

### Repressurization

The normal repressurization procedure is initiated at $T_6$ + 48.1 seconds. It uses cold helium from the cold helium storage spheres (2, figure 6-9). The cold helium pressure is reduced to approximately 385 psia as it flows through the lox tank pressure control module (5). It next flows through the lox tank repressurization control (10), and into the $O_2/H_2$ burner (11). Should the regulator in the lox tank pressure control module (5) fail, the backup pressure switch (9) will maintain a pressure of 350-465 psia at the $O_2/H_2$ burner. The backup pressure switch controls the pressure by opening or closing valves in the lox tank repressurization module (10). As the cold helium is heated in the $O_2/H_2$ burner it expands and is routed to the lox tank. Pressure in the lox tank increases and is sensed by the flight control pressure switch (4) and the pressure transducer (12). The pressure switch (4) maintains lox tank pressure between 38-41 psia by opening and closing solenoid shutoff valves in the lox tank repressurization control module (10). The pressure transducer (12) transmits a continuous pressure reading to telemetry and to the LV TANK PRESS gauges (13) in the CM. At $T_6$ + 496.6 seconds, cryogenic repressurization is switched off. Ambient repressurization is turned on at $T_6$ + 497.6 seconds. Ambient helium from the ambient helium storage spheres (3) flows through the lox tank ambient helium repressurization control module (14) to the lox tank. Here the pressure is sensed by the flight control pressure switches (4). The pressure switches (4) control lox tank pressure by opening or closing the control valves in the lox tank ambient helium, repressurization control module (14). Just before the engines second burn, ambient repressurization is terminated ($T_6$ + 520.0 seconds)

### NOTE

The ambient repressurization portion of the repressurization procedure is in reality a backup procedure. Should the $O_2/H_2$ burner fail, ambient repressurization ensures lox tank pressure for engine start requirements.

### Lox Venting

The lox tank vent subsystem provides for controlled lox tank venting during normal stage operation and for pressure relief venting when tank overpressures occur. The lox tank venting subsystem operates through a ground controlled combination vent and relief valve. This valve is pneumatically operated upon receipt of the ground command. Prior to loading, the vent and relief valve is placed in the open position and the boiloff of lox during loading is directed through one propulsive vent in the aft skirt of the stage (figure 6-8). When LOX tank prepressurization commences, the valve is closed and placed in the relief position. Over-pressures from the lox tank will then be vented as necessary.

A ground command non-propulsive function is available in the lox tank vent subsystem. The lox vent and relief valve is commanded to the open position permitting non-propulsive venting through two non-propulsive vents placed 180 degrees apart. See figures 2-1 and 6-5 for the sequence of operation.

### LH₂ SYSTEM

The $LH_2$ is stored in an insulated tank at less than -423°F. Total volume of the tank is approximately 10,500 cubic feet with an ullage volume of approximately 300 cubic feet. The $LH_2$ tank is prepressurized to 28 psia minimum and 31 psia maximum.

### LH₂ Low Pressure Fuel Duct

$LH_2$ from the tank is supplied to the J-2 engine turbopump through a vacuum jacketed low pressure 10-inch duct. This duct is capable of flowing 80-pounds per second at -423°F and at a transfer pressure of 28 psia. The duct is located in the aft tank side wall above the common bulkhead joint. Bellows in this duct compensate for engine gimbaling, manufacturing tolerances, and thermal motion.

### LH₂ Fill and Drain

Prior to loading, the $LH_2$ tank is purged with helium gas. At the initiation of loading, the ground controlled combination vent and relief valve is opened, and the directional control valve is positioned to route $GH_2$ overboard to the burn pond.

Loading begins with precool at a flow of 500 gpm. When the 5% load level is reached fast fill is initiated at a flow of 3000 gpm. At the 98% load level, fast fill stops and a slow fill at 500 gpm begins. A fast fill emergency cutoff sensor has been provided to compensate for a primary control cutoff failure. Slow fill is terminated at the 100% load level, and this level is then maintained by a replenish flowrate of 0 to 300 gpm, as required. The replenish flow is maintained through the $LH_2$ tank prepressurization operation.

Liquid level during fill is monitored by means of the $LH_2$ mass probes. A backup overfill sensor is provided to terminate flow in the event of a 100% load cutoff failure.

An $LH_2$ vent system provides command venting of the $LH_2$ tank plus overpressure relief capability. Pressure sensing switches are provided to control tank pressure during fill and flight.

### LH₂ TANK PRESSURIZATION

$LH_2$ tank pressurization is divided into three basic procedures. These procedures are called prepressurization, pressurization, and repressurization. The term prepressurization is used for that portion of the pressurization performed on the ground prior to liftoff. The term pressurization is used to indicate pressurization during engine burn periods, and lastly, repressurization indicates pressurization just before the second burn period.

The pressurants used during the three $LH_2$ tank pressurization procedures are gaseous hydrogen ($GH_2$) gaseous helium. Cold helium from a ground source (25, figure 6-9) is used during the prepressurization period. The cold helium storage spheres (2), located in the $LH_2$ tank, supply cold helium for use during the repressurization period. The five ambient helium storage spheres (15), filled by ground support equipment (16), supply an alternate source of helium for use during the repressurization period.

The LH2 tank pressure is controlled by the flight control pressure switches (17) (dual redundant) regardless of the pressurization procedure used. These switches control solenoid shutoff valves in each of the supply subsystems.

## Prepressurization

At $T_1$ –97 seconds, the LH2 tank is prepressurized (figure 6-9) by ground support equipment. Cold helium (25) flows through the LH2 tank pressurization module (18) and into the LH2 tank. When the LH2 tank pressure increases to 31 psia ($T_1$ –43 seconds), the flight control pressure switch (17) shuts off the ground supply of cold helium (25) to complete prepressurization.

## Pressurization

Pressurization is controlled by the flight control pressure switches (17, figure 6-9) which open or close solenoid valves in the LH2 tank pressurization module (18). Gaseous hydrogen (19) bled from the J-2 engine flows through the LH2 tank pressurization module (18) to the LH2 tank. As pressure in the LH2 tank increases to 31 psia, the flight control pressure switches (17) close valves in the LH2 tank pressurization module (18) to maintain tank pressure at 28-31 psia. This pressure is sensed by the pressure transducer (20) and is relayed to the S-IVB fuel gauges (21) in the CM and, via telemetry, to the ground. In this manner, LH2 tank pressurization is maintained during engine burn periods.

## Repressurization

The normal repressurization procedure is initiated at $T_6$ + 48.1 seconds. It uses cold helium from the cold helium storage spheres (2, figure 6-9). The cold helium pressure is reduced to approximately 385 psia as it flows through the lox tank pressure control module (5). The cold helium next flows through the LH2 tank repressurization control (22), and into the O2/H2 burner (11). Should the regulator in the lox tank pressure control module (5) fail, the backup pressure switch (23) will maintain a pressure of 350-465 psia at the O2/H2 burner. The backup pressure switch controls the pressure by opening or closing valves in the LH2 tank repressurization module (22). As the cold helium is heated in the O2/H2 burner (11), it expands and is routed to the LH2 tank. Pressure in the LH2 tank increases and is sensed by the flight control pressure switch (17) and the pressure transducer (20). The pressure switch (17) maintains LH2 tank pressure between 28-31 psia by opening and closing solenoid shutoff valves in the LH2 tank repressurization control module (22). The pressure transducer (20) transmits a continuous pressure reading to telemetry and to the LV TANK PRESS gauges (21) in the CM. At $T_6$ + 496.6 seconds, cryogenic repressurization is switched off. Ambient repressurization is turned on at $T_6$ + 497.6 seconds. Ambient helium from the ambient helium storage spheres (15) flows through the LH2 tank ambient helium repressurization control module (24) to the LH2 tank. Here the pressure is sensed by the flight control pressure switches (17). The pressure switches (17) control LH2 tank pressure by opening or closing, the control valves in the LH2 tank ambient helium repressurization control module (24). Just before the engines second burn, ambient repressurization is terminated ($T_6$ + 520.0 seconds).

## NOTE

The ambient repressurization portion of the repressurization procedure is in reality a backup procedure. Should the O2/H2 burner fail, ambient repressurization ensures LH2 tank pressure for engine start requirements.

## LH2 Venting

The LH2 tank vent subsystem (figure 6-9) is equipped to perform either a propulsive or non-propulsive venting function. The non-propulsive venting is the normal mode used.

The non-propulsive function is performed through the use of a ground controlled combination vent and relief valve which permits the option of routing the GH2 through either the ground vent lines or through non-propulsive relief venting. The valve is in the ground vent line open position until T-40 seconds at which time it is positioned to the in-flight non-propulsive relief function. The non-propulsive vents are located 180 degrees to each other so as to cancel any thrust effect.

The propulsive venting function is a command function which operates through two control valves upstream of the non-propulsive directional control valve. This mode vents the GH2 through two propulsive vents located axial to the stage. Propulsive venting provides a small additional thrust, prior to second J-2 engine burn, for propellant settling. Figure 2-1 and figure 6-5 illustrate the sequential operation of the venting subsystems.

## PNEUMATIC CONTROL

The pneumatic control system (figure 6-10) provides pressure for all pneumatically operated valves on the stage and for the engine start tank vent valve on the J-2 engine. The pneumatic control system is filled with gaseous helium from ground support equipment to 3200 ± 100 psig. The onboard pneumatic control system consists of the helium fill module, an ambient helium bottle and a pneumatic power control module.

The helium fill module regulates and reduces the incoming supply to 490 ± 25 psia for operation of control valves during preflight activities. The ambient helium storage bottle is initially pressurized to 750 psia, and is capable of supplying operating pressure to stage control valves at that pressure. After propellant loading has begun, and the cold helium bottles are chilled down, the pressure is raised to 3100 psia and both the ambient and cold helium bottles are then completely pressurized to their flight pressure of 3100 psia by the time the LH2 tank reaches a 92% load level.

The pneumatic power control module is set at 475 psig which is equivalent to 490 psia on the ground and 475 psia in orbit. These pressure levels are essential to the operation of the LH2 directional control valve, the propulsion vent shutoff valve, the lox and LH2 fill and drain valves, the lox and LH2 turbopump turbine purge module, the lox chilldown pump purge control, the lox and LH2 prevalves and chilldown and shutoff valves, the lox tank vent/relief valves, the LH2 propulsive vent valve, and the J-2 engine GH2 start system vent/relief valve. Each pneumatically operated component is attached to a separate actuation control module containing dual solenoids, which provide on-off control.

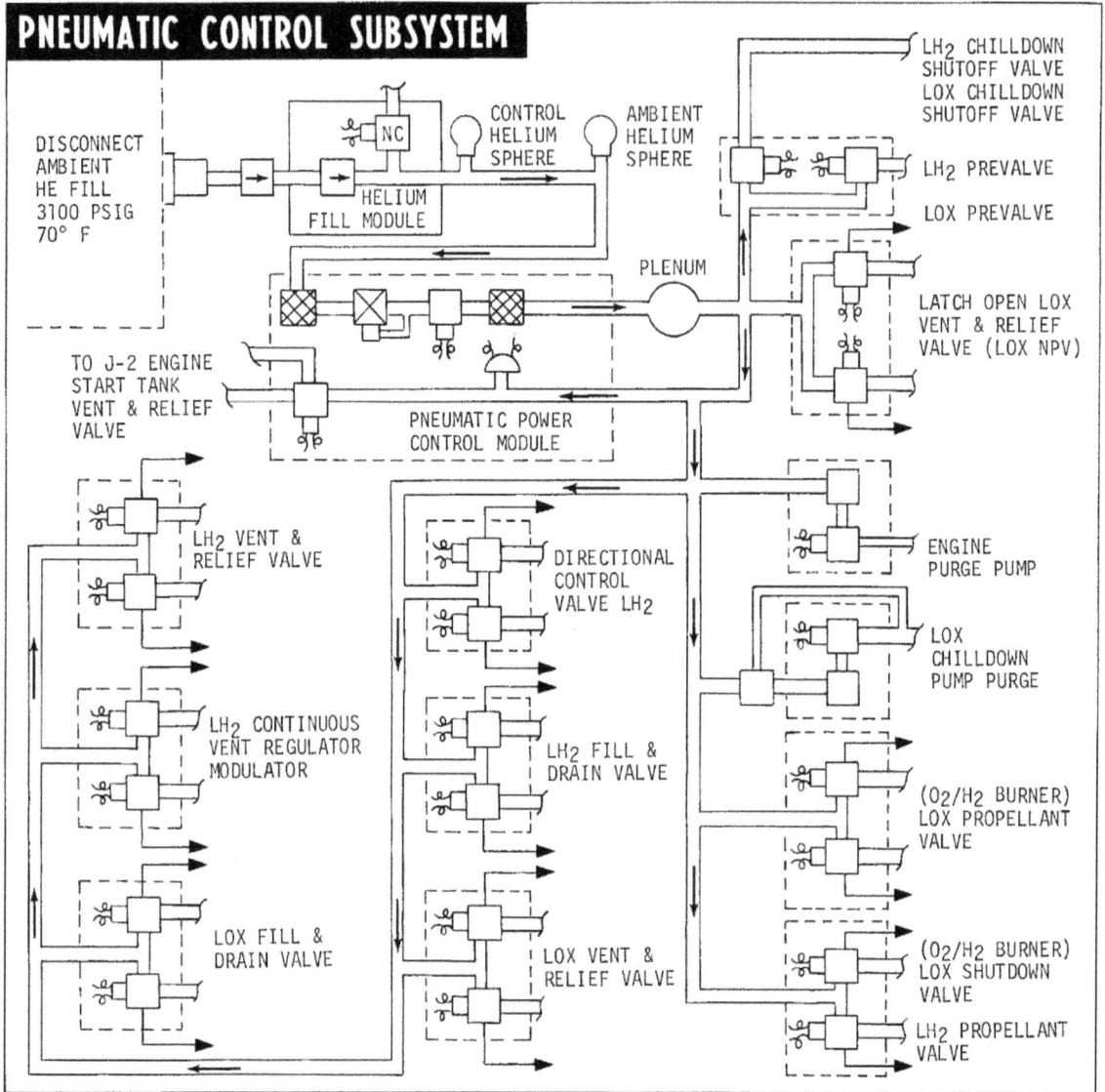

Figure 6-10

The pneumatic control system is protected from over-pressure by a normally open, pressure switch controlled, solenoid valve. This switch maintains system pressure between 490 - 600 psia.

**FLIGHT CONTROL**

The flight control system incorporates two systems for flight and attitude control. During powered flight, thrust vector steering is accomplished by gimbaling the J-2 engine for pitch and yaw control, and by operating the APS engines for roll control. Steering during coast flight is by use of the APS engines alone. (See Auxiliary Propulsion Systems subsection for coast flight steering.)

ENGINE GIMBALING

During the boost and separation phase, the J-2 engine is

commanded to the null position to prevent damage by shifting. The engine is also nulled before engine restart to minimize the possibility of contact between the engine bell and the interstage at S-II/S-IVB separation, and to minimize inertial effects at ignition. The engine is gimbaled (figure 6-11) in a 7.0 degrees square pattern by a closed loop hydraulic system. Mechanical feedback from the actuator to the servovalve completes the closed engine position loop.

When a steering command is received from the flight control computer, a torque motor in the servovalve shifts a control flapper to direct the fluid flow through one of two nozzles. The direction of the flapper is dependent upon signal polarity.

Two actuators are used to translate the steering signals into

# THRUST VECTOR CONTROL

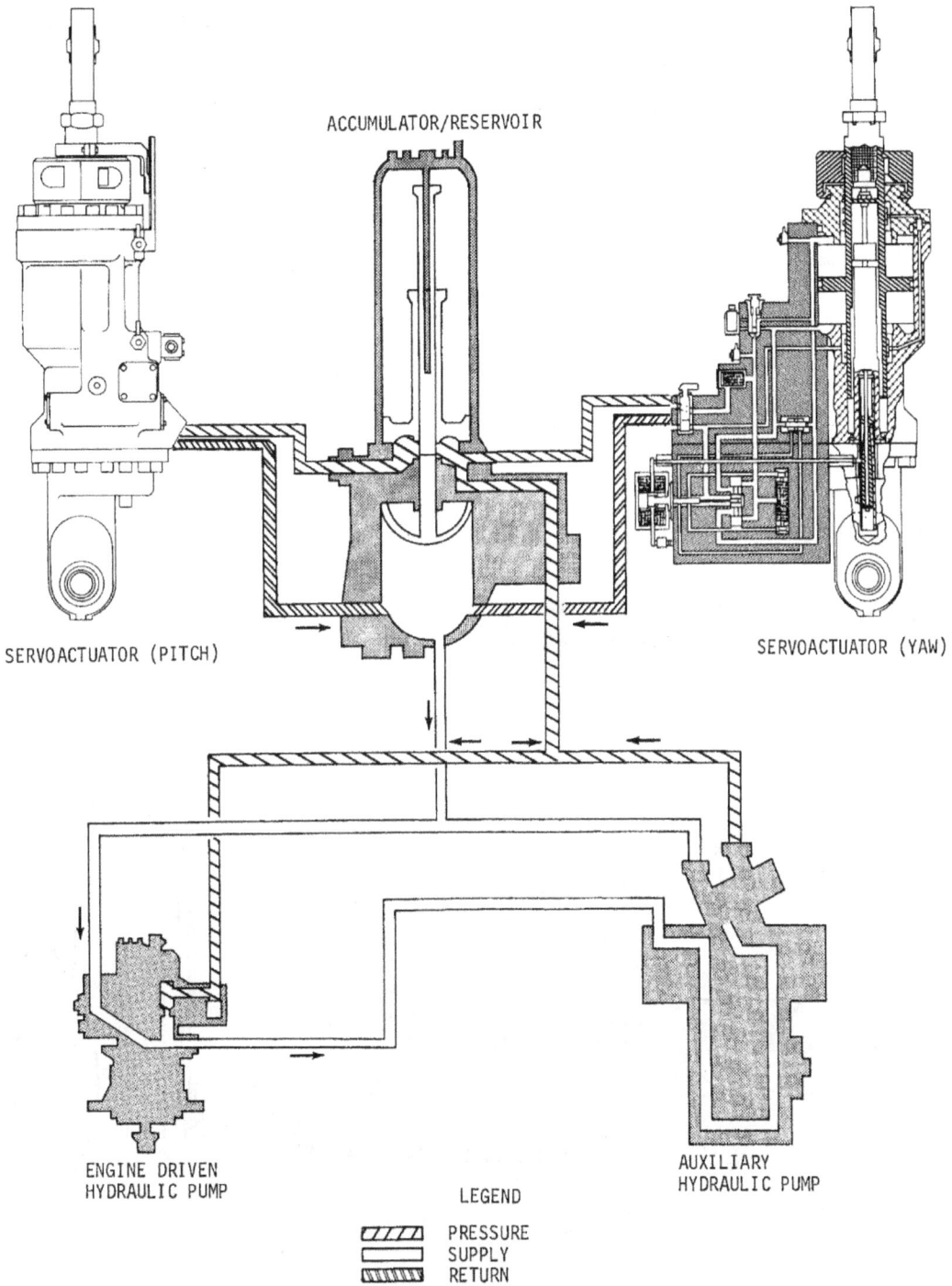

ACCUMULATOR/RESERVOIR

SERVOACTUATOR (PITCH)

SERVOACTUATOR (YAW)

ENGINE DRIVEN
HYDRAULIC PUMP

AUXILIARY
HYDRAULIC PUMP

LEGEND

| | |
|---|---|
| ▨▨ | PRESSURE |
| ☐ | SUPPLY |
| ▨▨ | RETURN |

Figure 6-11

vector forces to position the engine. The deflection rates are proportional to the pitch and yaw steering signals from the flight control computer.

## HYDRAULIC SYSTEM

Major components of the hydraulic system (figure 6-12) are an engine driven hydraulic pump, an electrically driven auxiliary hydraulic pump, two hydraulic actuator assemblies, and an accumulator/reservoir assembly.

### Hydraulic Pumps

The engine driven hydraulic pump is a variable displacement type driven directly from the engine oxidizer turbopump. In normal operation, the pump delivers up to 8 gpm under continuous working pressure.

The auxiliary hydraulic pump is an electrically driven pump which is capable of supplying a minimum of 1.5 gpm of fluid to the system. This pump supplies pressure for preflight checkout, to lock the J-2 engine in the null position during boost and separation phase, and as emergency backup. During orbit, the auxiliary pump, controlled by a thermal switch, circulates the hydraulic fluid to maintain it between +10°F and +40°F. The auxiliary pump is enabled before liftoff and during coast periods.

### Accumulator/Reservoir Assembly

The accumulator/reservoir assembly (figure 6-12) is an integral unit mounted on the thrust structure. The reservoir section is a storage area for hydraulic fluid and has a maximum volume of 167 cubic inches.

During system operation, between 60 and 170 psig is maintained in the reservoir (figure 6-12) by two pressure operated pistons contained in the accumulator section. In addition to maintaining pressure in the reservoir, the system accumulator supplies peak system demands and dampens high pressure surging.

### Hydraulic Actuators - Pitch and Yaw

The pitch and yaw actuators and servovalve (figure 6-12) are integrally mounted and are interchangeable. The actuators are linear and double acting. During powered flight, pitch and yaw control is provided by gimbaling the main engine, the two actuator assemblies providing deflection rates proportional to pitch and yaw steering signals from the flight control computer.

## AUXILIARY PROPULSION SYSTEM

The S-IVB auxiliary propulsion system provides three axis stage attitude control (figure 6-13) and main stage propellant control during coast flight.

## APS CONSTRUCTION

The APS engines are located in two modules 180 degrees apart on the aft skirt of the S-IVB stage (see figure 6-14). The

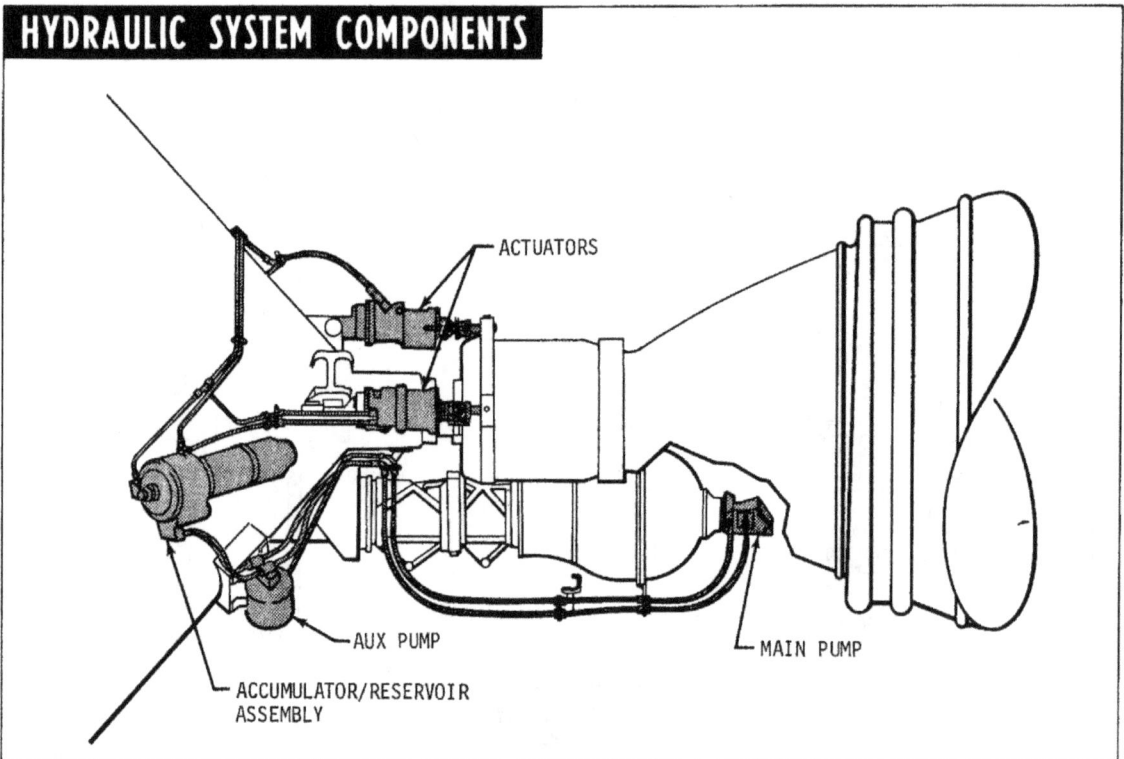

# HYDRAULIC SYSTEM COMPONENTS

ACTUATORS

AUX PUMP

ACCUMULATOR/RESERVOIR ASSEMBLY

MAIN PUMP

Figure 6-12

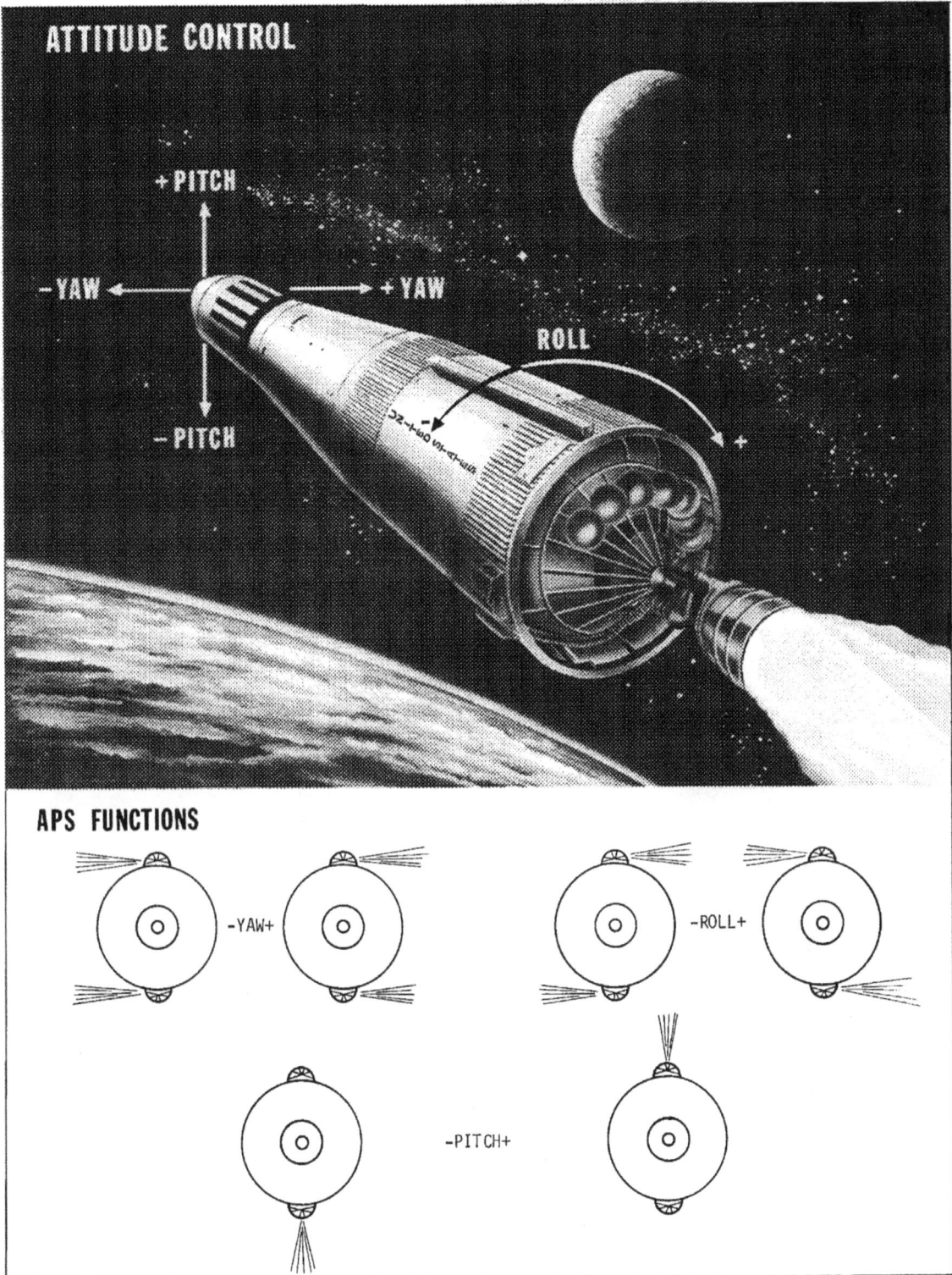

# ATTITUDE CONTROL

# APS FUNCTIONS

Figure 6-13

## 150-POUND THRUST APS ENGINE

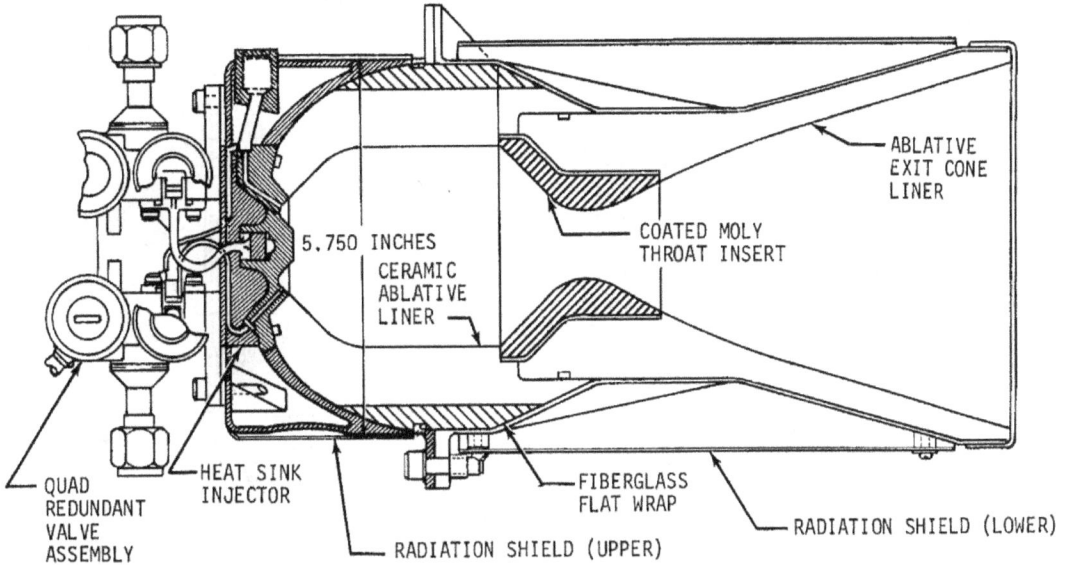

ABLATIVE
EXIT CONE
LINER

COATED MOLY
THROAT INSERT

5.750 INCHES
CERAMIC
ABLATIVE
LINER

QUAD
REDUNDANT
VALVE
ASSEMBLY

HEAT SINK
INJECTOR

FIBERGLASS
FLAT WRAP

RADIATION SHIELD (LOWER)

RADIATION SHIELD (UPPER)

Figure 6-15

## 70-POUND THRUST APS ENGINE

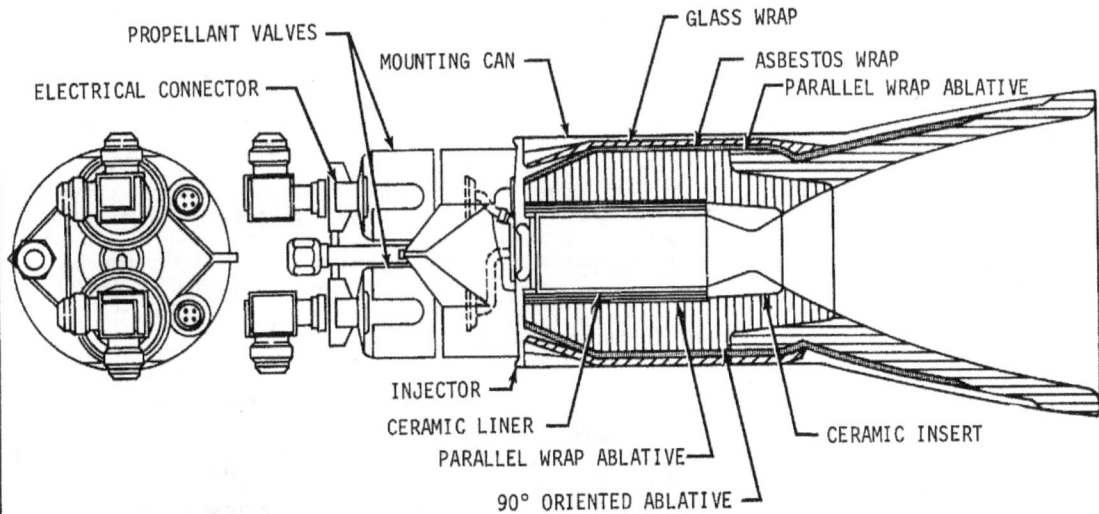

PROPELLANT VALVES

ELECTRICAL CONNECTOR

MOUNTING CAN

GLASS WRAP

ASBESTOS WRAP

PARALLEL WRAP ABLATIVE

INJECTOR

CERAMIC LINER

PARALLEL WRAP ABLATIVE

90° ORIENTED ABLATIVE

CERAMIC INSERT

Figure 6-16

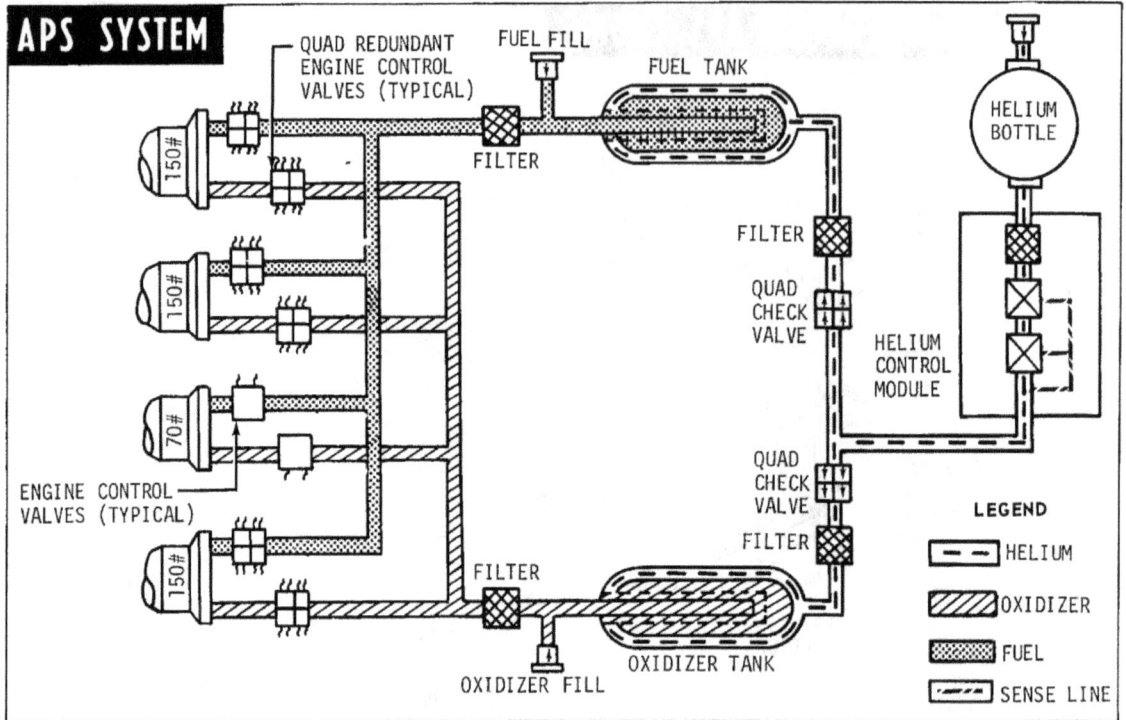

Figure 6-17

## BATTERIES

On board power is supplied by four zinc/silver-oxide batteries. Two are located in the forward equipment area and two in the aft equipment area. These batteries are activated and installed in the stage during the final prelaunch preparations. Heaters and instrumentation probes are an integral part of each battery. See figure 6-19 for a table of battery characteristics and figure 6-20 for systems drawing from the four batteries.

## POWER DISTRIBUTION

Two main forward busses and two main aft busses supply electrical power to all stage systems. Busses are electrically isolated from each other with each main bus utilizing a power transfer switch to switch from GSE power to stage mounted batteries.

Figure 6-21 illustrates the electrical power distribution for the S-IVB stage. The four main busses are configured to perform the following functions:

1. The function of forward bus No. 1 is to provide 28 vdc power to the telemetry transmitters, data acquisition system, transducers, level sensors, switch selector, range safety system No. 1, checkout measurement group, 5 vdc excitation modules and the forward battery No. 1 and No. 2 heaters.

2. Forward bus No. 2 supplies 28 vdc power to the

propellant utilization electronic assembly, the propellant utilization static inverter-converter, and range safety system No. 2. The requirements of redundant range safety and emergency detection systems dictated the use of a second battery in the forward skirt.

3. Aft bus No. 1 supplies 28 vdc power to the J-2 engine, sequencer, pressure switches, all propulsion system valves which are flight operational, attitude control system modules No. 1 and No. 2, ullage rocket motor ignition and jettison system, aft battery No. 1 and No. 2 heaters.

4. The aft bus No. 2 provides 56 vdc power to the LH$_2$ and lox chilldown inverters and to the auxiliary hydraulic pump motor. The high power requirements of the chilldown motors and hydraulic pump motor necessitated the 56-volt battery.

## CHILLDOWN INVERTERS

The chilldown inverters are three-phase, 1500 volt-amp, solid state power conversion devices. The purpose of the chilldown inverters is to provide electrical power to the motor driven pumps for circulation of the lox and LH$_2$, to ensure propellant temperature stabilization at the J-2 engine inlets.

## PROPELLANT UTILIZATION (PU) STATIC INVERTER CONVERTER

The static inverter-converter is a solid state power supply which provides all the regulated ac and dc voltages necessary

# ATTITUDE AND ULLAGE ENGINE CONTROL

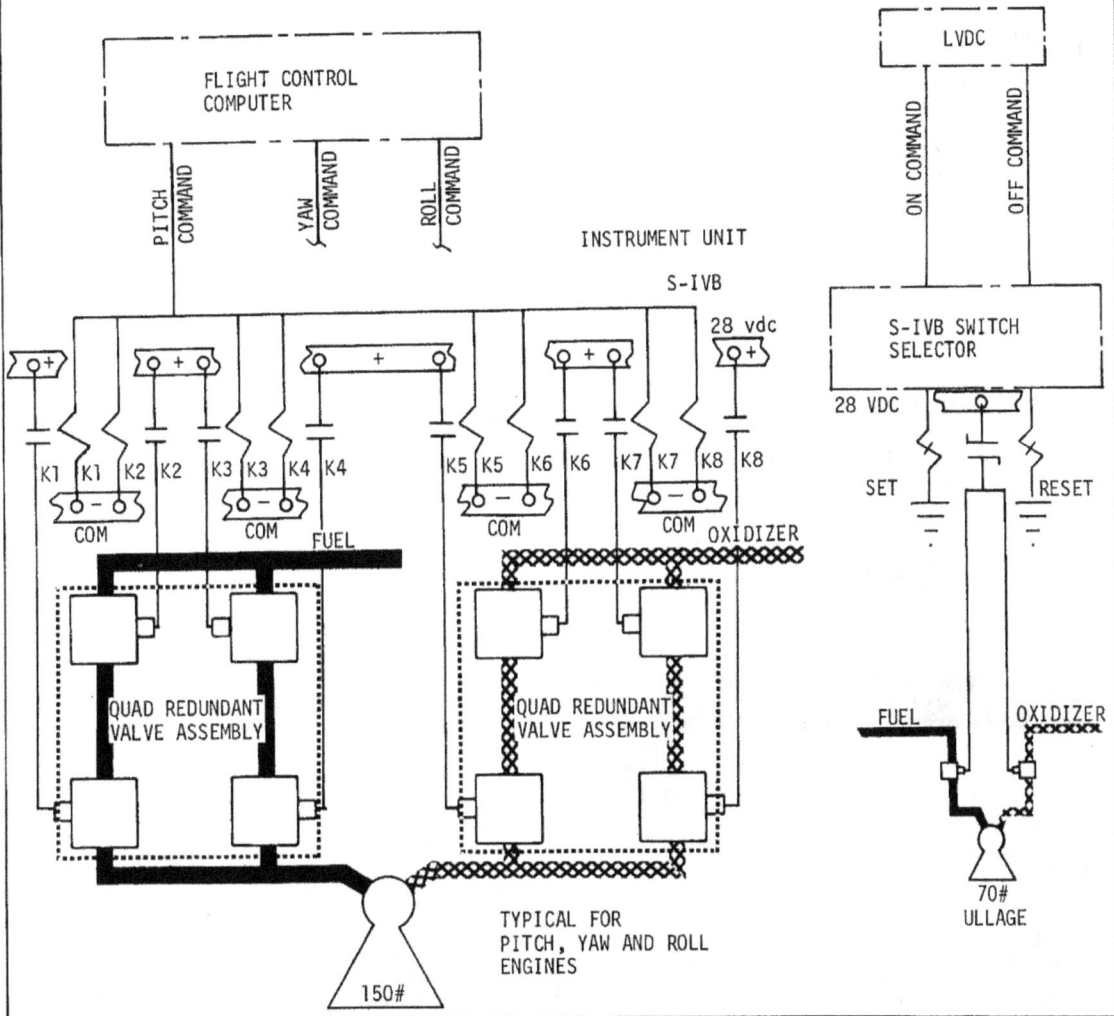

Figure 6-18

to operate the S-IVB stage propellant utilization electronics assembly.

The static inverter-converter converts a 28 vdc input into the following outputs:

1. 115 vac 400 Hz required to drive propellant utilization valve positioning motors and bridge rebalancing servomotors,

2. 2-vac, (peak-to-peak), square wave required to convert propellant utilization error signal to alternating current,

3. 5 vdc required to provide excitation for propellant utilization fine and coarse mass potentiometers,

4. regulated 22 vdc required for propellant utilization bridges,

5. 117.5 vdc floating supply used to provide propellant utilization summing potentiometer excitation,

6. 49 vdc floating supply required to provide excitation for valve feedback potentiometers.

## EXCITATION MODULES

The 5-volt and 20-volt excitation modules are transistorized power conversion devices which are used to convert the 28-vdc to the various regulated voltages required by the instrumentation, signal conditioning, and emergency detection system transducers.

| S-IVB BATTERY CHARACTERISTICS | | | | |
|---|---|---|---|---|
| TYPE | Dry charge | | | |
| MATERIAL | Zinc/silver-oxide | | | |
| ELECTROLYTE | Potassium hydroxide (KOH) in pure water | | | |
| CELLS | 20, with taps for selecting 18 or 19 to reduce output voltage as required | | | |
| NOMINAL VOLTAGE OUTPUT | 1.5 vdc per cell 28 (+2) vdc per 18 to 20 cell group Aft Battery No. 2 is made up of two regular 28 (+2) vdc batteries and has an output of 56 (+4) vdc | | | |
|  | FORWARD NO. 1 | FORWARD NO. 2 | AFT NO. 1 | AFT NO. 2 |
| CURRENT RATING | 179 AH | 12.2 AH | 179 AH | 49.6 AH |
| Gross Weight | 90 lbs | Two units: 20 lbs ea. | 90 lbs | 75 lbs |
| (Design target weight) | | | | |

Figure 6-19

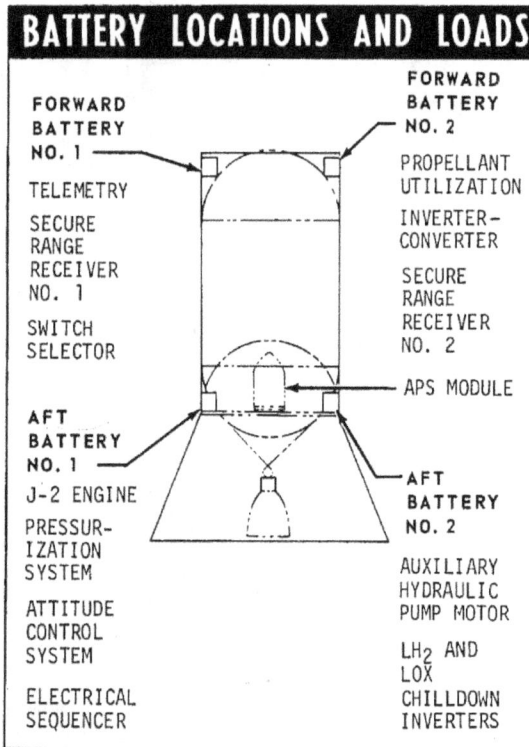

## BATTERY LOCATIONS AND LOADS

FORWARD BATTERY NO. 1
TELEMETRY
SECURE RANGE RECEIVER NO. 1
SWITCH SELECTOR

FORWARD BATTERY NO. 2
PROPELLANT UTILIZATION
INVERTER-CONVERTER
SECURE RANGE RECEIVER NO. 2

APS MODULE

AFT BATTERY NO. 1
J-2 ENGINE
PRESSURIZATION SYSTEM
ATTITUDE CONTROL SYSTEM
ELECTRICAL SEQUENCER

AFT BATTERY NO. 2
AUXILIARY HYDRAULIC PUMP MOTOR
LH$_2$ AND LOX CHILLDOWN INVERTERS

Figure 6-20

### ELECTRICAL CONTROL SUBSYSTEM

The electrical control subsystem function is to distribute the command signals required to control the electrical components of the stage. The major components of the electrical control subsystem are the power and control distributors, the sequencer assemblies, and the pressure sensing and control devices.

### SEQUENCER ASSEMBLIES

The two major elements in the stage sequencing system are the switch selector and the stage sequencer. During flight, sequencing commands are received from the IU. Each command is in digital form and consists of an 8-bit word accompanied by a "read" pulse. The commands are interpreted by the S-IVB stage at the switch selector.

#### Switch Selector

The switch selector is an electronic assembly utilized as the primary device for controlling the inflight sequencing of the stage. A switch selector is utilized in each stage of the launch vehicle.

The switch selector consists of relays, a diode matrix, and low-power transistor switches used as relay drivers controlled by binary-coded signals from the LVDC in the IU. The function of the switch selector is to operate magnetically latching relays in the sequencer and power distribution assemblies.

The switch selector provides electrical isolation between the IU and the S-IVB stage systems and decodes digital information from the IU into discrete output commands. Capability exists to provide 112 discrete commands to the stage sequencer.

#### Stage Sequencer

The stage sequencer operates upon receipt of discrete inputs from the switch selector (and other S-IVB stage subsystems) and initiates S-IVB flight functions by supplying or removing power from the appropriate equipment. Sequence circuits perform logistical gating of inputs necessary for sequencing control with as few timed commands from the IU as possible. It controls only those functions established as sequencing events.

### PRESSURE SWITCHES

Calibratable pressure switches (calips) are used on the S-IVB stage to perform various control functions. For example:

1.  LH$_2$ Tank System

    a.  Pressurization, ground fill, valve control,

    b.  LH$_2$ tank pressure control backup,

2.  Lox Tank System

    a.  Ground fill, valve control, pressurization,

    b.  Lox tank pressure control backup,

3.  Pneumatic Power System

# ELECTRICAL POWER DISTRIBUTION

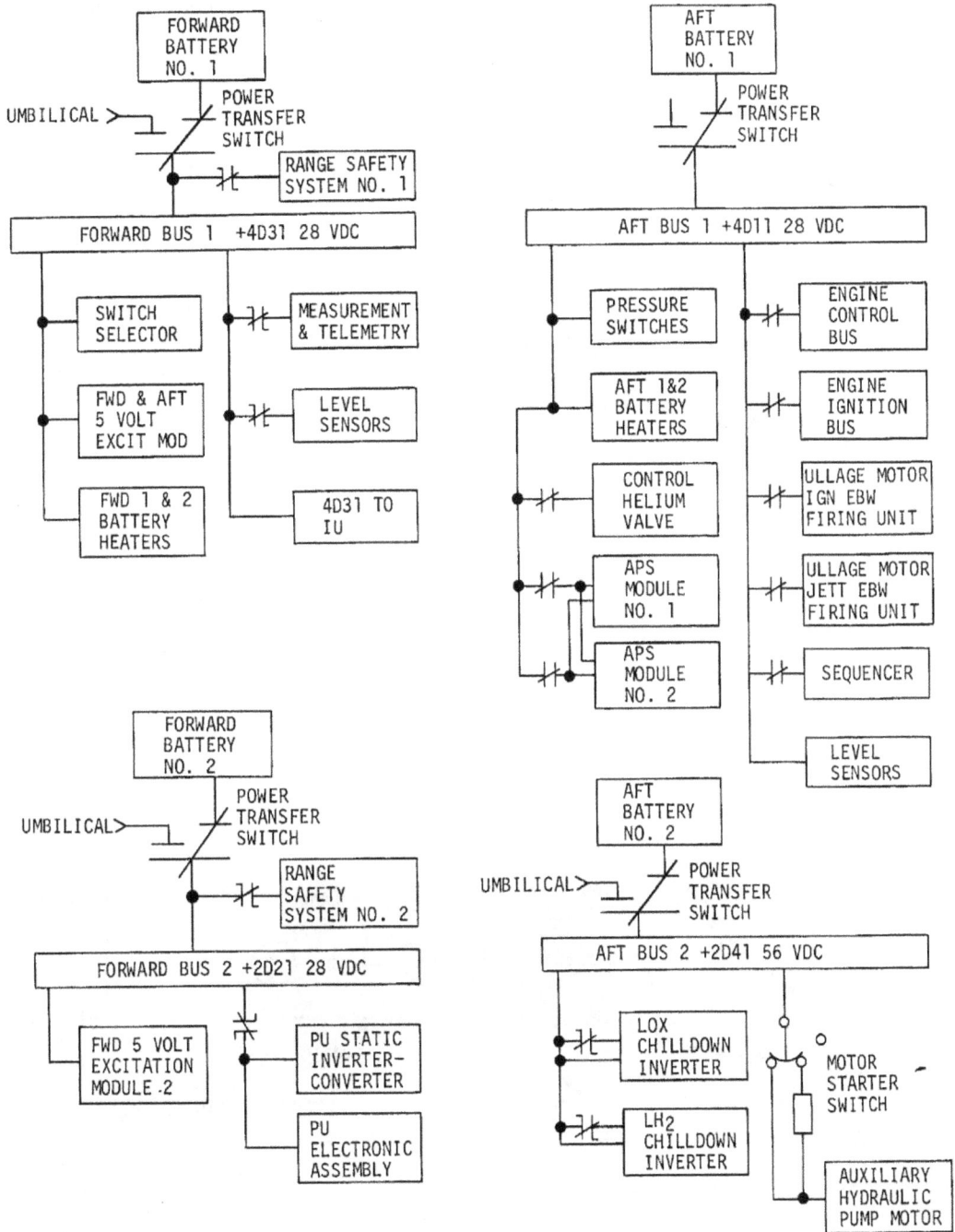

```
FORWARD
BATTERY
NO. 1
          POWER
UMBILICAL >  TRANSFER
             SWITCH
                      RANGE SAFETY
                      SYSTEM NO. 1

FORWARD BUS 1  +4D31  28 VDC

SWITCH              MEASUREMENT
SELECTOR            & TELEMETRY

FWD & AFT           LEVEL
5 VOLT              SENSORS
EXCIT MOD

FWD 1 & 2           4D31 TO
BATTERY             IU
HEATERS
```

```
AFT
BATTERY
NO. 1
          POWER
          TRANSFER
          SWITCH

AFT BUS 1  +4D11  28 VDC

PRESSURE            ENGINE
SWITCHES            CONTROL
                    BUS

AFT 1&2             ENGINE
BATTERY             IGNITION
HEATERS             BUS

CONTROL             ULLAGE MOTOR
HELIUM              IGN EBW
VALVE               FIRING UNIT

APS                 ULLAGE MOTOR
MODULE              JETT EBW
NO. 1               FIRING UNIT

APS                 SEQUENCER
MODULE
NO. 2

                    LEVEL
                    SENSORS
```

```
FORWARD
BATTERY
NO. 2
            POWER
UMBILICAL>   TRANSFER
             SWITCH
                      RANGE
                      SAFETY
                      SYSTEM NO. 2

FORWARD BUS 2  +2D21  28 VDC

FWD 5 VOLT          PU STATIC
EXCITATION          INVERTER-
MODULE .2           CONVERTER

                    PU
                    ELECTRONIC
                    ASSEMBLY
```

```
AFT
BATTERY
NO. 2
              POWER
UMBILICAL>    TRANSFER
              SWITCH

AFT BUS 2  +2D41  56 VDC

LOX
CHILLDOWN         MOTOR
INVERTER          STARTER
                  SWITCH
LH2
CHILLDOWN
INVERTER

                  AUXILIARY
                  HYDRAULIC
                  PUMP MOTOR
```

Figure 6-21

a. Regulator backup,

b. Engine purge,

c. Lox chilldown pump container purge,

These pressure switches are located in either the aft section (thrust structure) or the forward section (interstage), depending upon the pressure control system in which the switch is used.

The calips pressure switch employs two pressure ports, each isolated from the other. The test port provides for remote checkout without disconnecting or contaminating the primary pressure system. The test pressure settings are calibrated during manufacture of the switch to provide an accurate indication of the system pressure settings.

Calips pressure switches utilize a single Belleville spring which provides "snap" response to actuation or deactuation pressures. This response provides switching of 28 vdc power to relays in the stage sequencer for operation and control of propulsion system solenoid valves.

## PROPELLANT MANAGEMENT

The propellant management systems illustrated in figure 6-22 provide a means of monitoring and controlling propellants during all phases of stage operation. Continuous capacitance probes and point level sensors in both the $LH_2$ and lox tanks monitor propellant mass. Point level sensors are used during the propellant loading sequence to indicate to the GSE the level of propellants in the tanks. Level sensors signal the LVDC during flight so that engine cutoff can be accomplished smoothly at propellant depletion. The capacitance probes provide outputs which are used to operate a propellant utilization (PU) control valve. The PU valve is a rotary type valve which controls the quantity of lox flowing to the engine.

## PROPELLANT UTILIZATION SUBSYSTEM

For the AS-503 mission the PU subsystem has been modified so that it can be step commanded from the LVDC. Normally

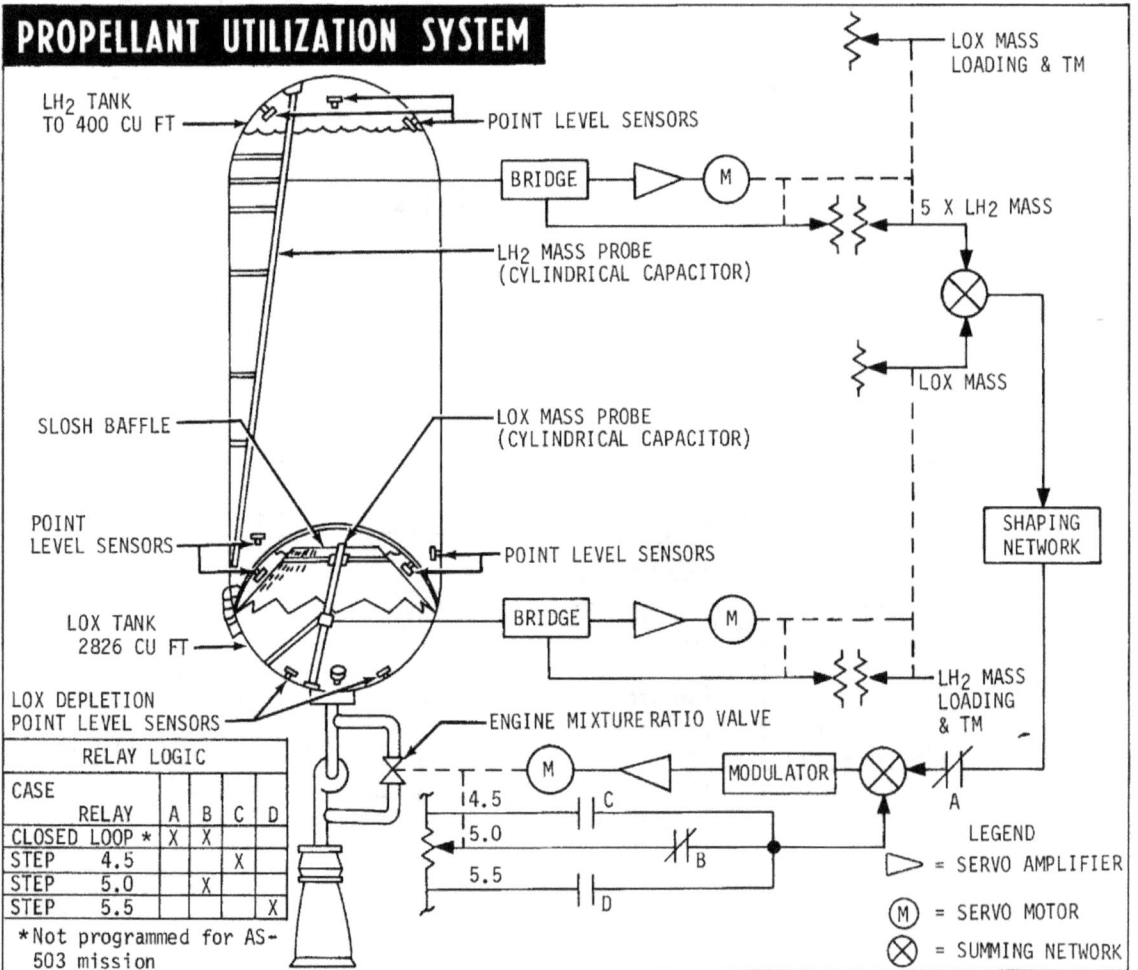

Figure 6-22

the PU valve would follow a signal developed from inputs provided by propellant monitoring capacitance probes. In this way, propellant utilization is managed so that propellant residuals at flight termination are at a minimum. This closed loop control has been broken for the AS-503 mission and LVDC controlled relays furnish step control signals to the PU valve.

Prior to S-IVB ignition for first burn, the PU valve is commanded to its null position. This will result in an engine mixture ratio (MR) of approximately 5:1 for start and throughout first burn.

Prior to S-IVB restart the PU valve is commanded to its full open position. This will result in a MR of approximately 4.5:1 during the start sequence. Approximately 2.5 seconds after mainstage the PU valve is commanded back to its null position for a MR of approximately 5:1 for the remainder of the second burn.

The excursion effect caused by varying the MR is illustrated in figure 5-8.

### INSTRUMENTATION

The S-IVB stage instrumentation monitors functional operations of stage systems. Before liftoff, measurements are telemetered by coaxial cable to ground support equipment.

During flight, radio frequency antennae convey data to ground stations. See figure 6-23 for a block diagram of the S-IVB measurement and telemetry systems.

### MEASUREMENT SYSTEM

Monitoring functional operations of stage systems is the purpose of the measurement subsystem. It acquires functional data, conditions it, and supplies it to the telemetry system for transmission to the ground stations.

Various parameters, as listed in figure 6-24, are measured by the types of transducers described in the following paragraphs.

1.  Temperature transducers are of two types: the platinum wire whose resistance changes with a change in temperature, and the thermocouple which shows an output voltage increase with an increase in temperature. Individual bridges and dc voltage amplifiers are employed for each temperature measurement. In cases where the temperature bridge connects directly to the remote analog submultiplexer (RASM), the external dc voltage amplifier is omitted.

2.  Pressure Tranducers. Two basic types of pressure transducers are used, the conventional potentiometer

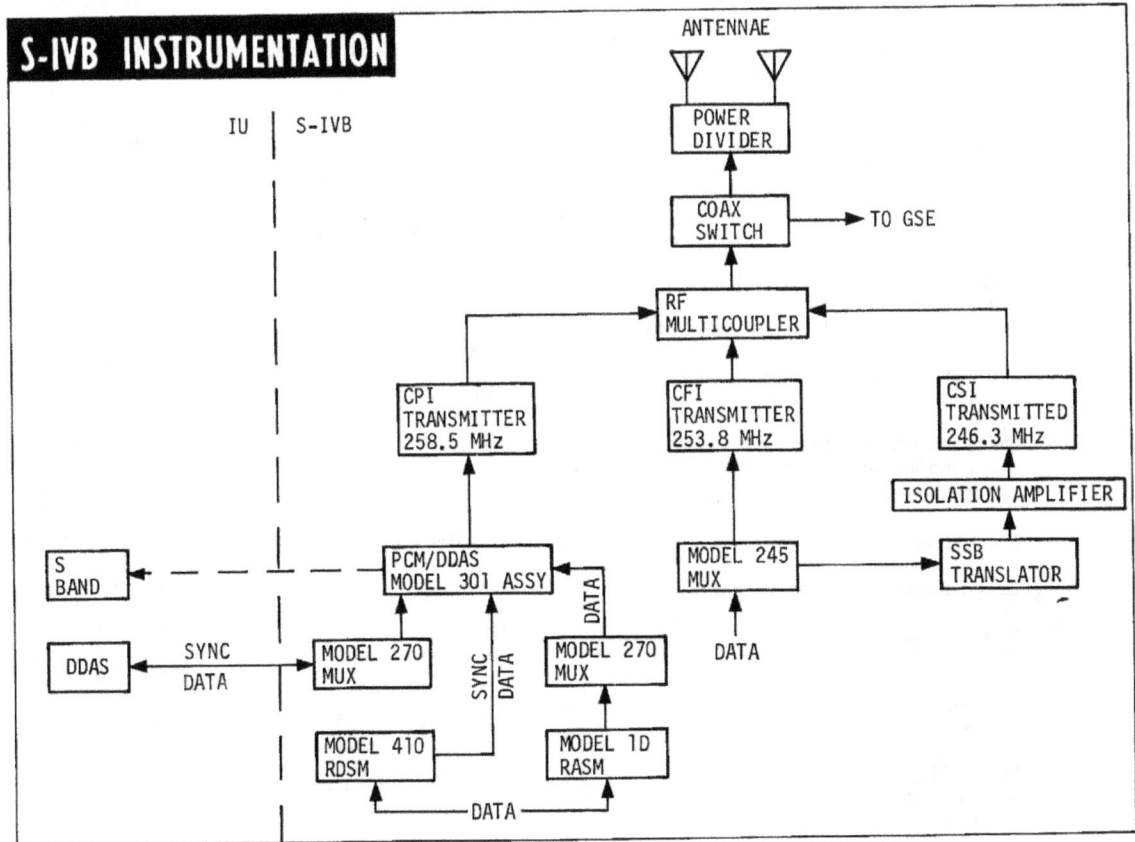

Figure 6-23

| STAGE MEASUREMENTS | | | |
|---|---|---|---|
| TYPE | QTY | TYPE | QTY |
| ACCELERATION | 5 | MISCELLANEOUS | 7 |
| ACOUSTIC | 4 | STRAIN | 38 |
| TEMPERATURE | 146 | RPM | 2 |
| PRESSURE | 77 | TOTAL | 436 |
| VIBRATION | 32 | | |
| FLOWRATE | 4 | | |
| POSITION | 2 | | |
| DISCRETE SIGNALS | 79 | | |
| LIQUID LEVEL | 7 | | |
| VOLTAGE, CURRENT, FREQUENCY | 33 | | |

Figure 6-24

(pot) type (5000 ohms) and the strain gauge type. The pot type transducers are used on 12 samples per second (S/S) measurements, and where the temperature environment is no colder than -100 degrees. The strain gauge transducers are used on all 120 S/S measurements, and where it is established that temperature at the transducer will be colder than -100 degrees F. All strain gauge types have a high and low simulation calibration capability.

3.  Flowmeters. Conventional turbine type flowmeters with frequency to direct current (dc) (0 to 5 volts) converters are supplied for flow measurements.

4.  Positions of actuators and metering valves are measured by conventional potentiometers.

5.  Events are points in time where a switch opens or closes and are not conditioned but fed directly to multiplexers.

6.  Liquid level sensors use a capacitive bridge type control unit to signal and condition the measurement for each data channel.

7.  Voltages, currents, and frequencies are conditioned to 0 to 5 volt analog signals.

8.  Miscellaneous measurements are conditioned to 0 to 5 volts.

9.  Speed. Magnetic pickups with frequency to dc converters transform the lox and $LH_2$ turbopump speeds to 0 to 5 volt analog signals.

An S-IVB stage measurement summary is given in figure 6-24.

TELEMETRY SYSTEM

The telemetry system consists of a pulse-code-modulator (PCM) digital data acquisition system (DDAS) for prelaunch checkout, a PCM frequency modulated (PCM/FM) system, a FM/FM system, and a single sideband (SS/FM) system for launch information.

The PCM/FM system consists of two model 270 multiplexers, one model 1D remote analog submultiplexer (RASM), one model 410 remote digital submultiplexer (RDSM), and a model 301 PCM assembly. The model 301 output is routed to the IU for redundant transmission at S-band frequencies. The FM/FM system consists of a proportional band width subcarrier oscillator assembly (SCO) and a model 345 multiplexer. The SS/FM system consists of a model 245 multiplexer, a single sideband translator and an isolation amplifier. (For a more detailed description of the telemetry components, see Section VII.)

One model 270 multiplexer will receive synchronization from, and input data to, the S-IVB stage DDAS. The second model 270 multiplexer will receive synchronization from, and input data to, the IU DDAS. The model 270 multiplexer receiving synchronization from the IU DDAS will have up to six of its prime channels allocated to a remote analog submultiplexer.

Analog parameters designated as flight control data (FCD) are fed in parallel to both model 270 multiplexers. FCD bi-level, or discrete functions are placed on that model 270 multiplexer receiving synchronization from the IU DDAS (via summing network) and also input to the remote digital submultiplexer or DDAS directly. Analog functions may be placed upon either model 270 multiplexer. Low level measurements located aft are carried by the remote analog submultiplexer while those forward require individual amplifiers. Bi-level measurements will normally use either the RDSM aft or the DDAS forward.

RADIO-FREQUENCY SYSTEM

The RF subsystem consists of a PCM RF assembly, bi-directional coupler, RF detectors, dc amplifiers, coaxial switch, dummy load, RF power divider, and associated cabling (figure 6-25).

Omnidirectional antenna pattern coverage is provided by the folded-sleeve dipoles. One is located 11 degrees from fin plane toward fin plane 1; the other is located 11 degrees from fin plane 4 toward fin plane 3. The electrical phasing between antennae is 180 degrees, thereby reducing nulls off the nose and tail. The effective radiating power of the system is 20 watts nominal (+43 dbm) and 16 watts minimum.

The antennae are designed to operate with a 50-ohm impedance and a voltage standing wave ratio (VSWR) no greater than 1.5:1 over the 225 to 260 MHz band. Transmitted and reflected power measurements are routed through the DDAS for ground checkout, by use of the bi-directional coupler to provide RF outputs without interferences. The coaxial switch provides a means of connecting a dummy load to the RF assembly for ground testing purpose.

The Model II PCM RF assembly uses the PCM pulse train to modulate the VHF carrier (258.5 MHz) in the transmitter.

# RF SUBSYSTEMS

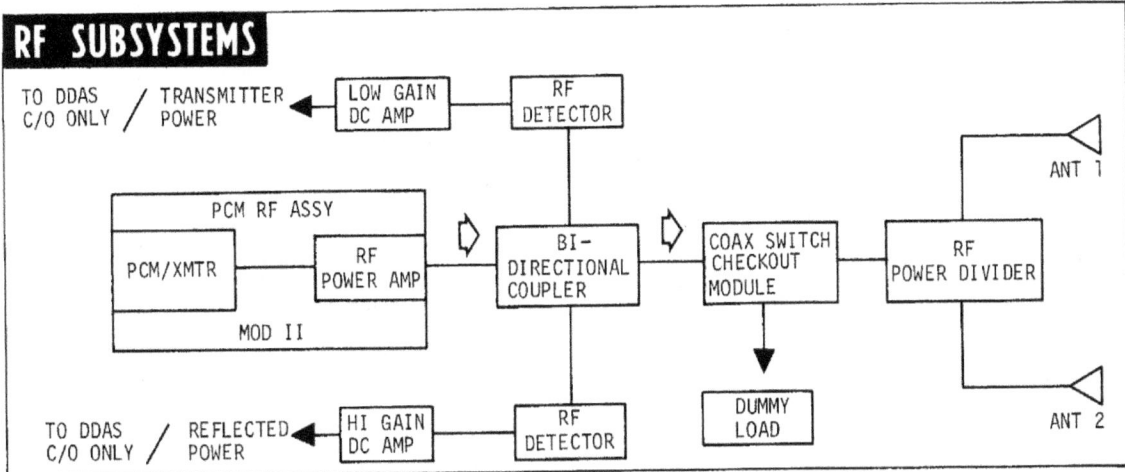

Figure 6-25

The transmitter output is amplified in the RF power amplifier. The radiation pattern coverage of the two antennae is designed to provide coverage to at least two ground stations simultaneously during flight.

## ORDNANCE

The S-IVB ordnance systems include the separation, ullage rocket, ullage rocket jettison, and propellant dispersion (flight termination) systems.

## SEPARATION SYSTEM

The third plane separation system for S-II/S-IVB is located at the top of the S-II/S-IVB interstage (figure 5-20). The separation plane is at station 2746.5. Ordnance for the third plane separation consists of two exploding bridgewire (EBW) firing units, two EBW detonators, one detonator block assembly and a detonating fuse assembly (figure 6-26). The EBW firing units are on the S-II/S-IVB interstage slightly below the third separation plane. The leads of the EBW firing units are attached to the EBW detonators which are installed in the detonator block assembly. The detonator block assembly is mounted just inside the skin of the vehicle, and the ends of the detonating fuse assembly are installed within the detonator block assembly. The detonating fuse assembly is mounted around the periphery of the vehicle beneath the tension strap.

The two EBW firing units for third plane separation provide redundant signal paths for initiation of the detonating fuse assembly. The function of the ordnance train is similar to that described in the separation system discussion in Section V. Detonation of the detonating fuse assembly severs the tension strap attaching the S-II/S-IVB interstage at station 2746.5 (figure 5-20). A sequence of events for S-IC/S-II/S-IVB separations and a block diagram of the separation systems is contained in figure 5-21.

At the time of separation, four retrorocket motors mounted on the interstage structure below the separation plane fire to decelerate the S-II stage. For information on the S-II retrorocket system, refer to Section V.

## ULLAGE ROCKET SYSTEM

To provide propellant settling and thus ensure stable flow of lox and LH$_2$ during J-2 engine start, the S-IVB stage requires a small acceleration. This acceleration is provided by two ullage rockets.

The S-IVB ullage rocket system (figure 6-26) consists of two EBW firing units, two EBW detonators, two confined detonating fuse (CDF) manifolds, nine CDF assemblies, two separation blocks, four CDF initiators, and two ullage rockets. The EBW firing units, EBW detonators, and CDF manifolds are mounted on the S-IVB aft skirt. The CDF assemblies connect the manifolds to the separation blocks and then to the CDF initiators. The rockets are within fairings mounted diametrically opposite each other on the S-IVB aft skirt. The rockets are canted outward from the vehicle to reduce effects of exhaust impingement, and to reduce the resulting moment if one rocket fails.

A separation block is used between the stage and each ullage rocket to allow jettison and maintain CDF continuity. The separation block, an inert item, is located on the skin of the S-IVB aft skirt under the ullage rocket fairing. Each block consists of two machined pieces of aluminum. The upper piece holds the ends of the CDF assemblies to the initiators, while the lower piece holds the CDF assemblies from the manifolds. The separation block forms a housing or connector that holds the CDF assembly ends together to ensure propagation and to contain the detonation of the connection. At jettison, the block slips apart with the lower portion remaining on the stage and the upper portion falling away with the rocket and fairing.

Each ullage rocket has a single grain, five point star configuration, internal burning, polymerized solid propellant that is case bonded in a 4135 steel case. The propellant weighs approximately 58.8 pounds, and burns for 3.8 seconds, developing a thrust of 3390 pounds (175,000 feet, 70 degrees F).

The firing sequence begins with the arming of the EBW firing units by charging the storage capacitors to 2300 volts. At S-II engine shutdown, the EBW units receive a trigger signal which discharges the storage capacitors, releasing high energy pulses to the EBW detonators, and thereby exploding the bridgewires. The resulting detonations propagate through the CDF manifolds, CDF assemblies, separation blocks and to the CDF initiators which cause the ullage rockets to ignite. A crossover CDF assembly between CDF manifolds provides redundancy and added system reliability.

## ULLAGE ROCKET JETTISON SYSTEM

To reduce weight, the ullage rockets and their fairings are jettisoned after J-2 engine start. The system, located on the S-IVB aft skirt, uses two EBW firing units, two EBW detonators, one detonator block, two CDF assemblies, four frangible nuts, and two spring-loaded jettison assemblies (figure 6-26).

The EBW firing units are armed by charging their storage capacitors to 2300 volts about five seconds after the S-IVB ullage rockets have stopped firing. A trigger signal releases the high voltage pulse to explode the bridgewire in the EBW detonator. Either detonator will detonate both CDF

assemblies (figure 6-26) through the detonator block. The detonation propagates through the CDF assemblies to detonate and fracture the frangible nuts. This frees the bolts that secure the ullage rocket and fairing assemblies to the aft skirt. The spring loaded jettison assemblies propel the spent rocket and fairing assemblies away from the vehicle.

## PROPELLANT DISPERSION SYSTEM

The S-IVB propellant dispersion system (PDS) provides for termination of vehicle flight during the S-IVB first engine firing boost burn period if the vehicle flight path varies beyond its prescribed limits or if continuation of vehicle flight creates a safety hazard. The S-IVB PDS may be safed after the launch escape tower is jettisoned. The system is installed in compliance with Air Force Eastern Test Range (AFETR) Regulation 127-9 and AFETR Safety Manual 127-1.

The S-IVB PDS is a dual channel, parallel redundant system composed of two segments (figure 6-26). The radio frequency segment receives, decodes and controls the propellant dispersion commands. The ordnance train segment

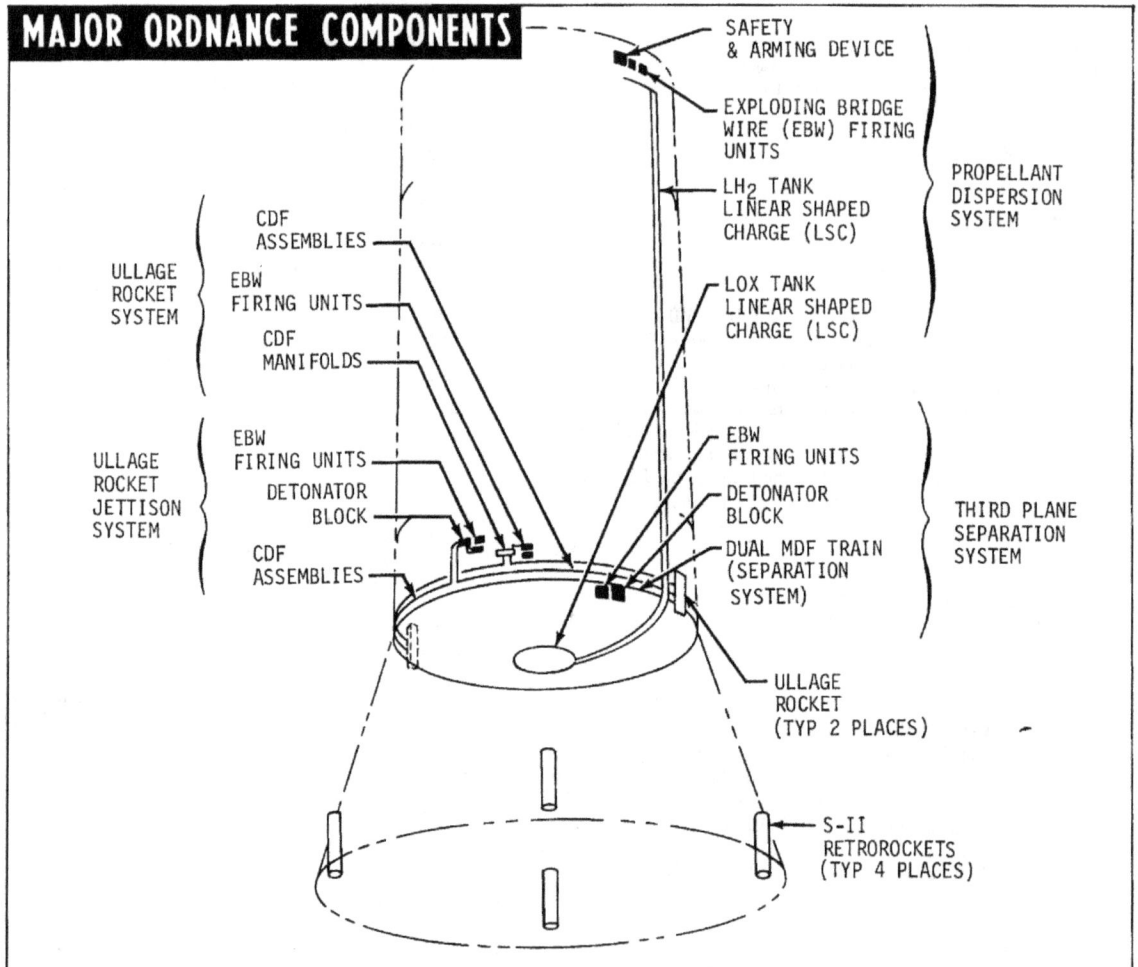

Figure 6-26 (Sheet 1 of 2)

# MAJOR ORDNANCE COMPONENTS

ULLAGE ROCKET SYSTEM

ULLAGE ROCKET JETTISON SYSTEM

PROPELLANT DISPERSION SYSTEM

THIRD PLANE SEPARATION SYSTEM

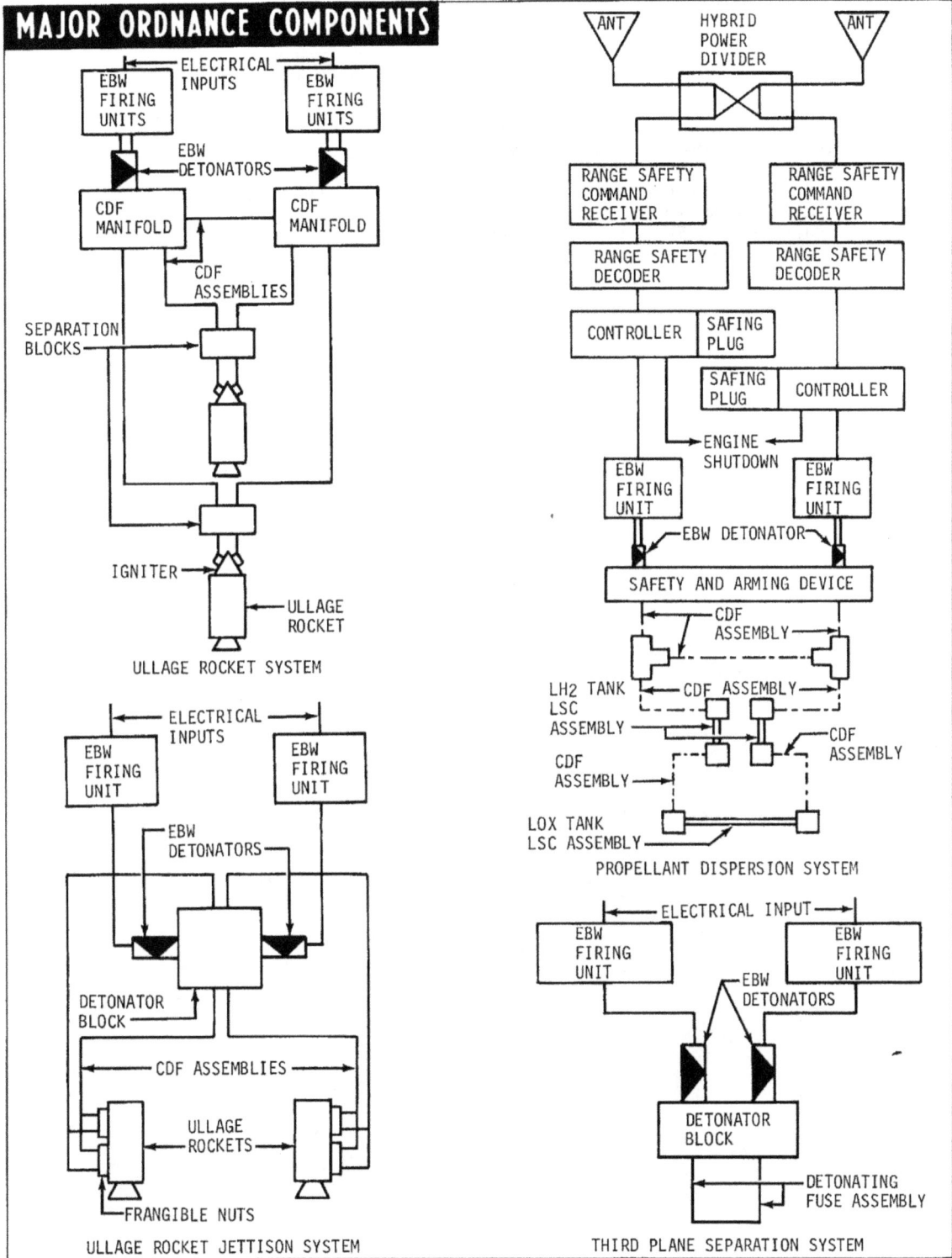

Figure 6-26 (Sheet 2 of 2)

consists of two EBW firing units, two EBW detonators, one safety and arming (S&A) device (shared by both channels), seven CDF assemblies, two CDF tees, and three linear shaped charge (LSC) assemblies.

Should emergency termination become necessary, two coded messages are transmitted to the launch vehicle by the range safety officer. The first command arms the EBW firing units and initiates S-IVB stage engine cutoff. The second command, which is delayed to permit charging of the EBW firing units, discharges the storage capacitors across the exploding bridgewires in the EBW detonators mounted on the S&A device. The resulting explosive wave propagates through the S&A device inserts and through the remainder of the ordnance train to sever the $LH_2$ and lox tanks.

A description of the S&A device is included in the PDS discussion in Section IV.

The linear shaped charges for the $LH_2$ and lox tanks are RDX loaded at 150 grains per foot. Two assemblies are used to cut two 20.2-foot long parallel openings in the side of the $LH_2$ tank. One assembly is used to cut a 47-inch diameter hole in the bottom of the lox tank.

Following S-IVB engine cutoff at orbit insertion, the PDS is electrically safed by ground command.

# SECTION VII

# INSTRUMENT UNIT

## TABLE OF CONTENTS

## INTRODUCTION

The Instrument Unit (IU) is a cylindrical structure installed on top of the S-IVB stage (see figure 7-1). The IU contains the guidance, navigation, and control equipment which will guide the vehicle through its earth orbits and subsequently into its mission trajectory. In addition, it contains telemetry, communications, tracking, and crew safety systems, along with their supporting electrical power and environmental control systems.

This section of the Flight Manual contains a description of the physical characteristics and functional operation for the equipment installed in the IU.

VII

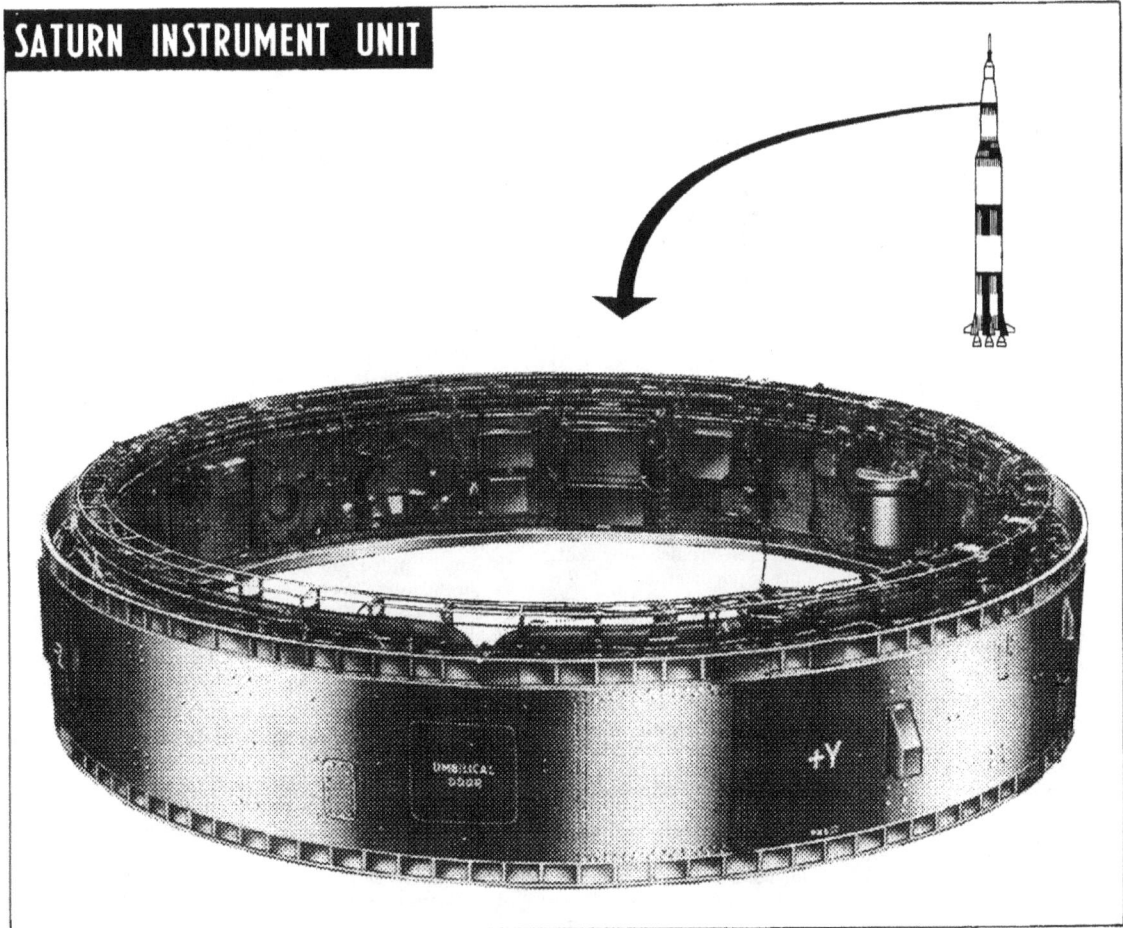

## SATURN INSTRUMENT UNIT

Figure 7-1

## STRUCTURE

The basic IU structure is a short cylinder fabricated of an aluminum alloy honeycomb sandwich material (see figure 7-2). The structure is fabricated from three honeycomb sandwich segments of equal length. The top and bottom edges are made from extruded aluminum channels bonded to the honeycomb sandwich. This type of construction was selected for its high strength-to- weight ratio, acoustical insulation,and thermal conductivity properties. The cylinder is manufactured in three 120 degree segments (figure 7-4), which are joined by splice plates into an integral structure. The three segments are the access door segment, the flight control computer segment, and the ST-124-M3 segment. The access door segment has an umbilical door, as well as an equipment/personnel access door. The access door has the requirement to carry flight loads, and still be removable at any time prior to flight.

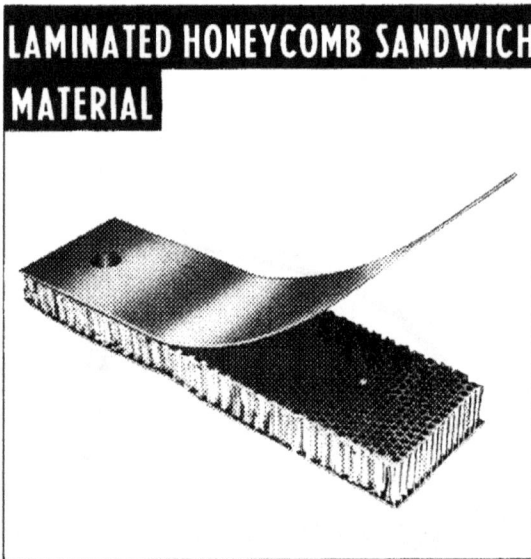

Figure 7-2

Attached to the inner surface of the cylinder are cold plates which serve both as mounting structure and thermal conditioning units for the electrical/electronic equipment. Mounting the electrical/electronic equipment around the inner circumference of the IU leaves the center of the unit open to accommodate the convex upper tank bulkhead of the S-IVB stage and the landing gear of the Lunar Excursion Module (LEM).

Cross section "A" of figure 7-3 shows equipment mounting pads bolted and bonded to the honeycomb structure. This method is used when equipment is not mounted on thermal conditioning cold plates. The bolts are inserted through the honeycomb core, and the bolt ends and nuts protrude through the outside surface. Cross section"B" shows a thermal conditioning cold plate mounting panel bolted to brackets which, in turn, are bolted on the honeycomb structure. The bolts go through the honeycomb core with the bolt heads protruding through the outer surface. Cross section "C" shows the cable tray supports bolted to inserts, which are potted in the honeycomb core at the upper and lower edges of the structure.

Figure 7-4 shows the relative locations of all equipment installed in the IU.

## ENVIRONMENTAL CONTROL SYSTEM

The environmental control system (ECS) maintains an acceptable operating environment for the IU equipment during preflight and flight operations. The ECS is composed of the following:

1. The thermal conditioning system (TCS) which maintains a circulating coolant temperature to the electronic equipment of 59 ($\pm$ 1) degrees F.

2. The preflight purging system which maintains a supply of temperature and pressure regulated air/GN$_2$ in the IU/S-IVB equipment area.

3. The gas-bearing supply system which furnishes GN$_2$ to the ST-124-M3 inertial platform gas bearings.

4. The hazardous gas detection sampling equipment which monitors the IU/S-IVB forward interstage area for the presence of hazardous vapors.

## THERMAL CONDITIONING SYSTEM

Thermal conditioning panels, also called cold plates, are located in both the IU and S-IVB stage; there are up to sixteen in each. Each cold plate contains tapped bolt holes in a grid pattern which provides flexibility of component mounting.

The cooling fluid circulated through the TCS is a mixture of 60% methanol and 40% demineralized water, by weight. Each cold plate is capable of dissipating at least 420 watts.

A functional flow diagram is shown in figure 7-5. The main coolant loop is the methanol-water mixture. Two heat exchangers are employed in the system. One is used during the preflight mode and employs a GSE supplied, circulating, methanol-water solution as the heat exchanging medium. The other is the flight mode unit, and it employs demineralized water and the principle of sublimation to effect the cooling.

The manifold, plumbing, and both accumulators are manually filled during the prelaunch preparations. The accumulators serve as dampers to absorb pressure fluctuations and thermal expansion, and also as reservoirs to replace lost or expended fluids. There are flexible diaphragms in each accumulator, backed by low pressure GN$_2$.

During operation of the TCS, the methanol-water coolant is circulated through a closed loop by electrically driven, redundant pumps. The flow is from the heat exchanger past the accumulator, through the pumps, through a temperature sensor to an orifice assembly. The orifice assembly diverts part of the coolant to the cold plates in the S-IVB stage and part to the cold plates, gas bearing heat exchanger, inertial platform, LVDC/LVDA and flight control computer in the IU. Return flow is through a modulating flow control assembly which regulates the amount of coolant flowing into the heat exchangers or around them. Operation of this valve is based on fluid temperature.

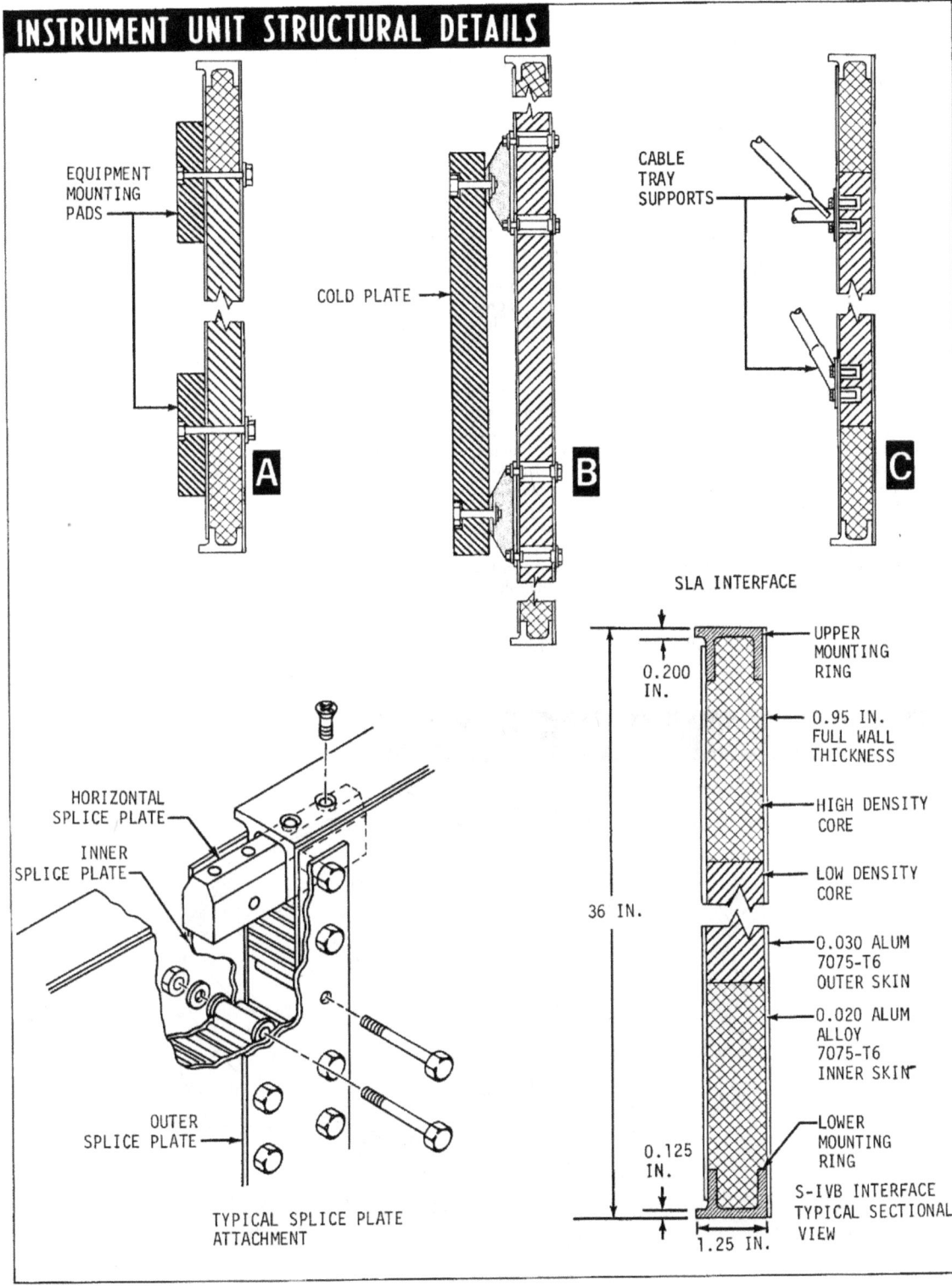

# INSTRUMENT UNIT STRUCTURAL DETAILS

EQUIPMENT
MOUNTING
PADS

COLD PLATE

CABLE
TRAY
SUPPORTS

A

B

C

HORIZONTAL
SPLICE PLATE

INNER
SPLICE PLATE

OUTER
SPLICE PLATE

TYPICAL SPLICE PLATE
ATTACHMENT

SLA INTERFACE

UPPER
MOUNTING
RING

0.200
IN.

0.95 IN.
FULL WALL
THICKNESS

HIGH DENSITY
CORE

LOW DENSITY
CORE

36 IN.

0.030 ALUM
7075-T6
OUTER SKIN

0.020 ALUM
ALLOY
7075-T6
INNER SKIN

LOWER
MOUNTING
RING

0.125
IN.

S-IVB INTERFACE
TYPICAL SECTIONAL
VIEW

1.25 IN.

Figure 7-3

During the preflight mode, the sublimator heat exchanger is inactive due to the high ambient pressure (one atmosphere), and a solenoid valve which blocks the water flow. The preflight heat exchanger is operated from the GSE, and cools the closed loop fluid.

Approximately 180 seconds after liftoff, the water solenoid valve is opened and the sublimator heat exchanger becomes active. During the period between GSE disconnect and sublimator activation, the residual cooling in the system is sufficient to preclude equipment overheating.

The sublimator element is a porous plate. Since the sublimator is not activated until approximately 180 seconds after launch, the ambient temperature and pressure outside the porous plates are quite low. Water flows readily into the porous plates and attempts to flow through the pores. However, the water freezes when it meets the low temperature of the space environment, and the resulting ice blocks the pores (see figure 7-6).

As heat is generated by equipment, the temperature in the methanol-water solution rises. This heat is transferred within the sublimator to the demineralized water. As the water temperature rises, it causes the ice in the pores to sublime. The vapor is vented overboard. As the heat flow decreases, ice plugs are formed in the pores, decreasing the water flow. Thus, the sublimator is a self-regulating system. $GN_2$ for the methanol-water and water accumulators is stored in a 165 cubic inch sphere in the IU, at a pressure of 3,000 psig. The sphere is filled by applying high pressure $GN_2$ through the umbilical. A solenoid valve controls the flow into the sphere, and a pressure transducer indicates to the GSE when the sphere is pressurized. The output of the sphere is applied to the accumulators through a filter and a pressure regulator,

which reduces the 3,000 psig to 15 psia. An orifice regulator further reduces the pressure at the accumulator to 5 psia, the differential of 10 psia being vented into the IU.

## PREFLIGHT AIR/GN$_2$ PURGE SYSTEM

The preflight air/GN$_2$ purge system (see figure 7-7) consists primarily of flexible ducting located above the IU payload interface. The system distributes ground supplied, temperature and pressure regulated, filtered air or GN$_2$ through openings in the ducting. During preflight phases, ventilating air is furnished. During fueling, inert GN$_2$ is furnished to prevent accumulation of a hazardous and corrosive atmosphere.

## GAS BEARING SUPPLY

Gaseous nitrogen, for the ST-124-M3 stable platform, is stored in a two cubic foot sphere in the IU, at a pressure of 3,000 psig (see figure 7-5). The sphere is filled by applying high pressure GN$_2$ through the umbilical, under control of the IU pneumatic console. A low pressure switch monitors the sphere and, if the pressure falls below 1,000 psig, the ST-124-M3 stable platform is shut down to preclude damage to the gas bearing.

Output of the sphere is through a filter and a pressure regulator. The regulator reduces the sphere pressure to a level suitable for gas bearing lubrication. Pressure at the gas bearing is sampled and applied, as a control pressure to the regulator. This provides for a constant pressure across the gas bearing. From the main regulator, the gas flows through a heat exchanger where its temperature is stabilized, then through another filter and on to the gas bearing. Spent gas is then vented into the IU.

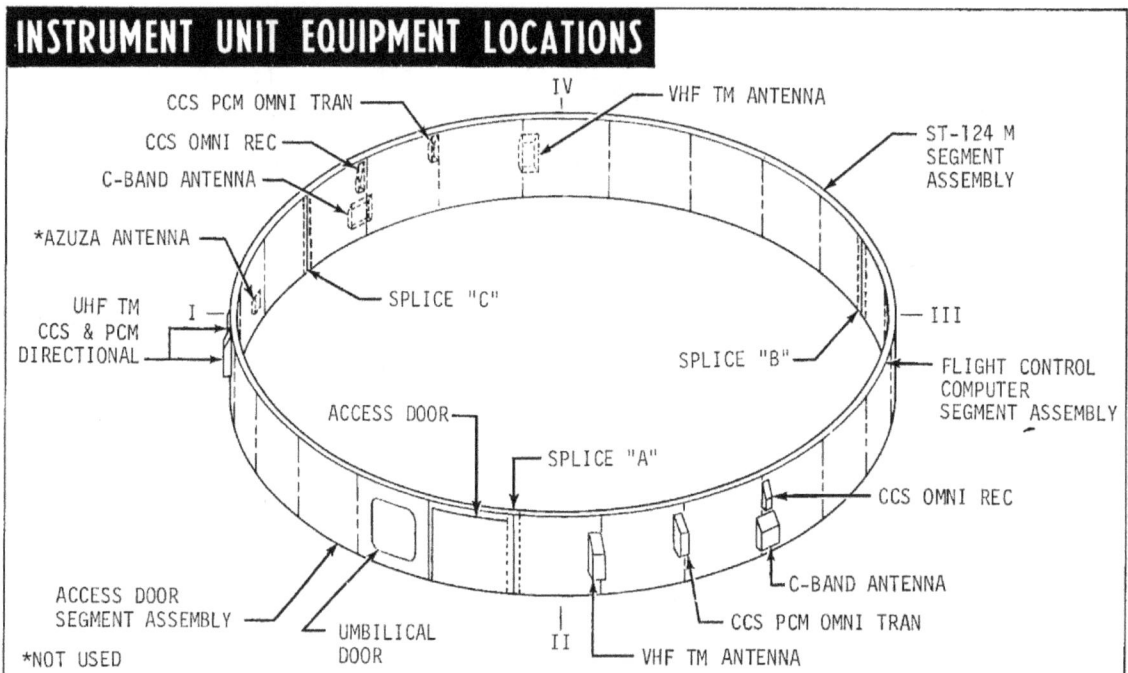

**INSTRUMENT UNIT EQUIPMENT LOCATIONS**

Figure 7-4 (Sheet 1 of 5)

# INSTRUMENT UNIT EQUIPMENT LOCATIONS

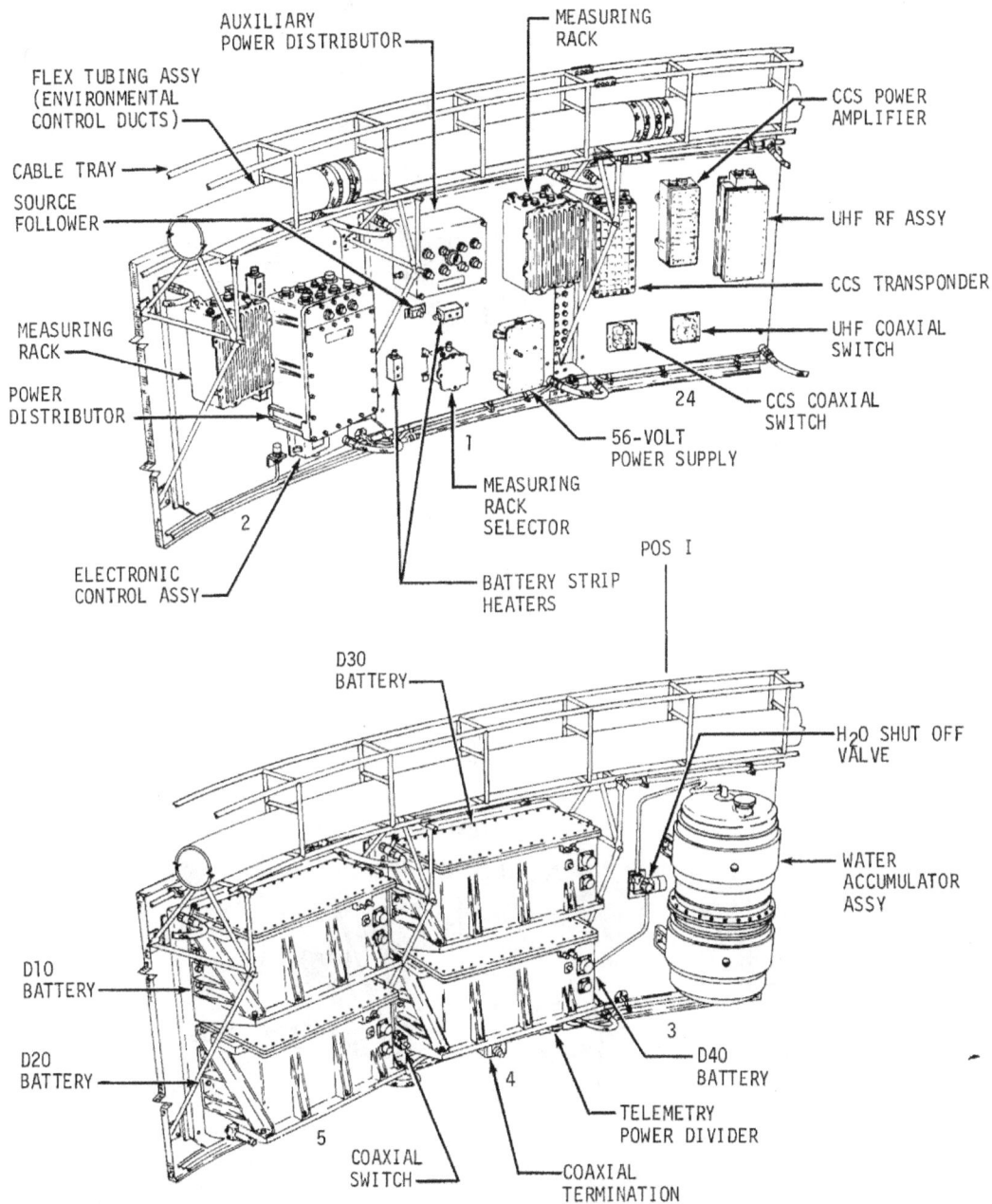

Figure 7-4 (Sheet 2 of 5)

# INSTRUMENT UNIT EQUIPMENT LOCATIONS

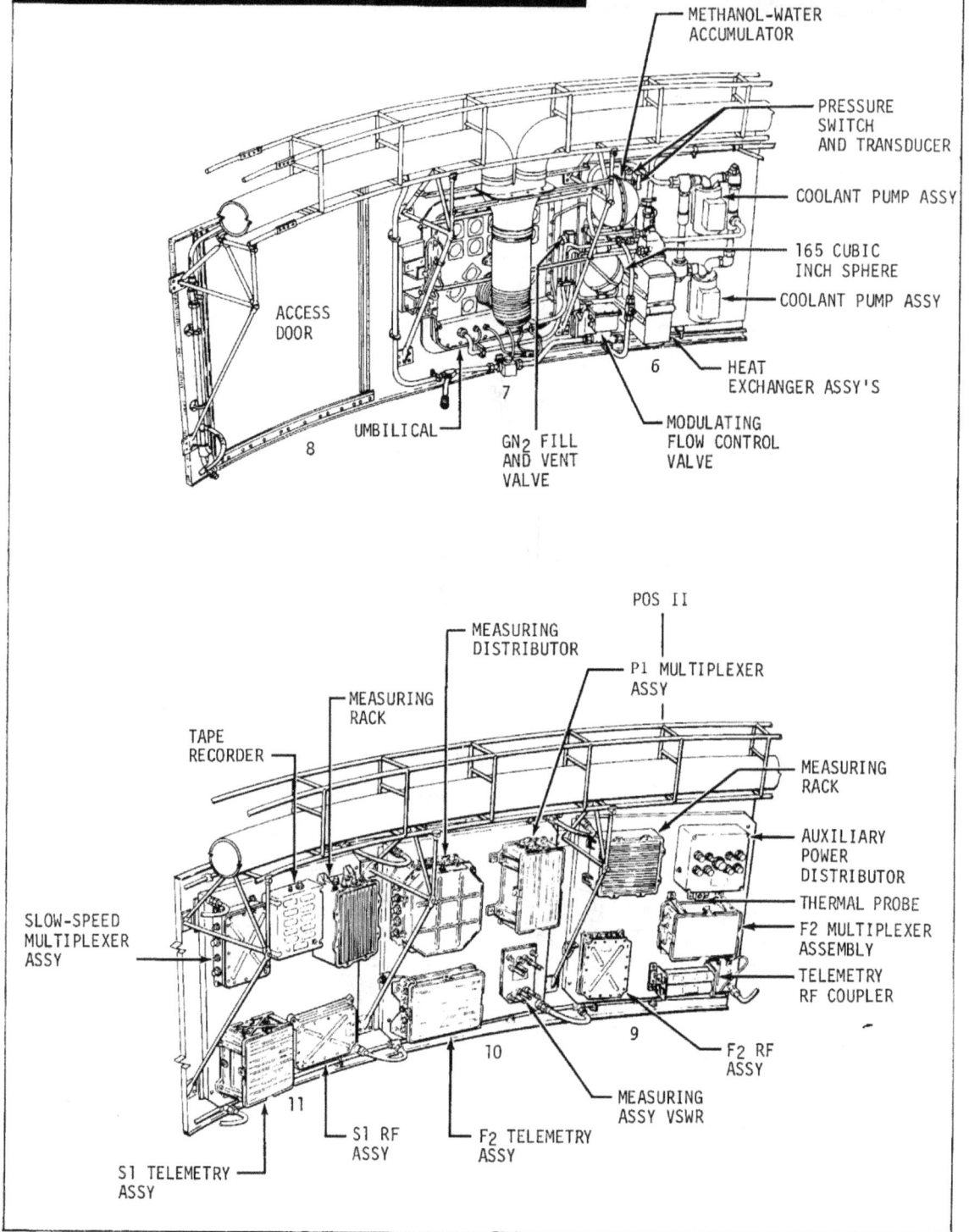

METHANOL-WATER
ACCUMULATOR

PRESSURE
SWITCH
AND TRANSDUCER

COOLANT PUMP ASSY

165 CUBIC
INCH SPHERE

COOLANT PUMP ASSY

HEAT
EXCHANGER ASSY'S

ACCESS
DOOR

7

6

MODULATING
FLOW CONTROL
VALVE

8

UMBILICAL

GN$_2$ FILL
AND VENT
VALVE

POS II

MEASURING
DISTRIBUTOR

P1 MULTIPLEXER
ASSY

MEASURING
RACK

MEASURING
RACK

TAPE
RECORDER

AUXILIARY
POWER
DISTRIBUTOR

THERMAL PROBE

F2 MULTIPLEXER
ASSEMBLY

TELEMETRY
RF COUPLER

SLOW-SPEED
MULTIPLEXER
ASSY

9

F2 RF
ASSY

10

MEASURING
ASSY VSWR

11

S1 RF
ASSY

F2 TELEMETRY
ASSY

S1 TELEMETRY
ASSY

Figure 7-4 (Sheet 3 of 5)

# INSTRUMENT UNIT EQUIPMENT LOCATIONS

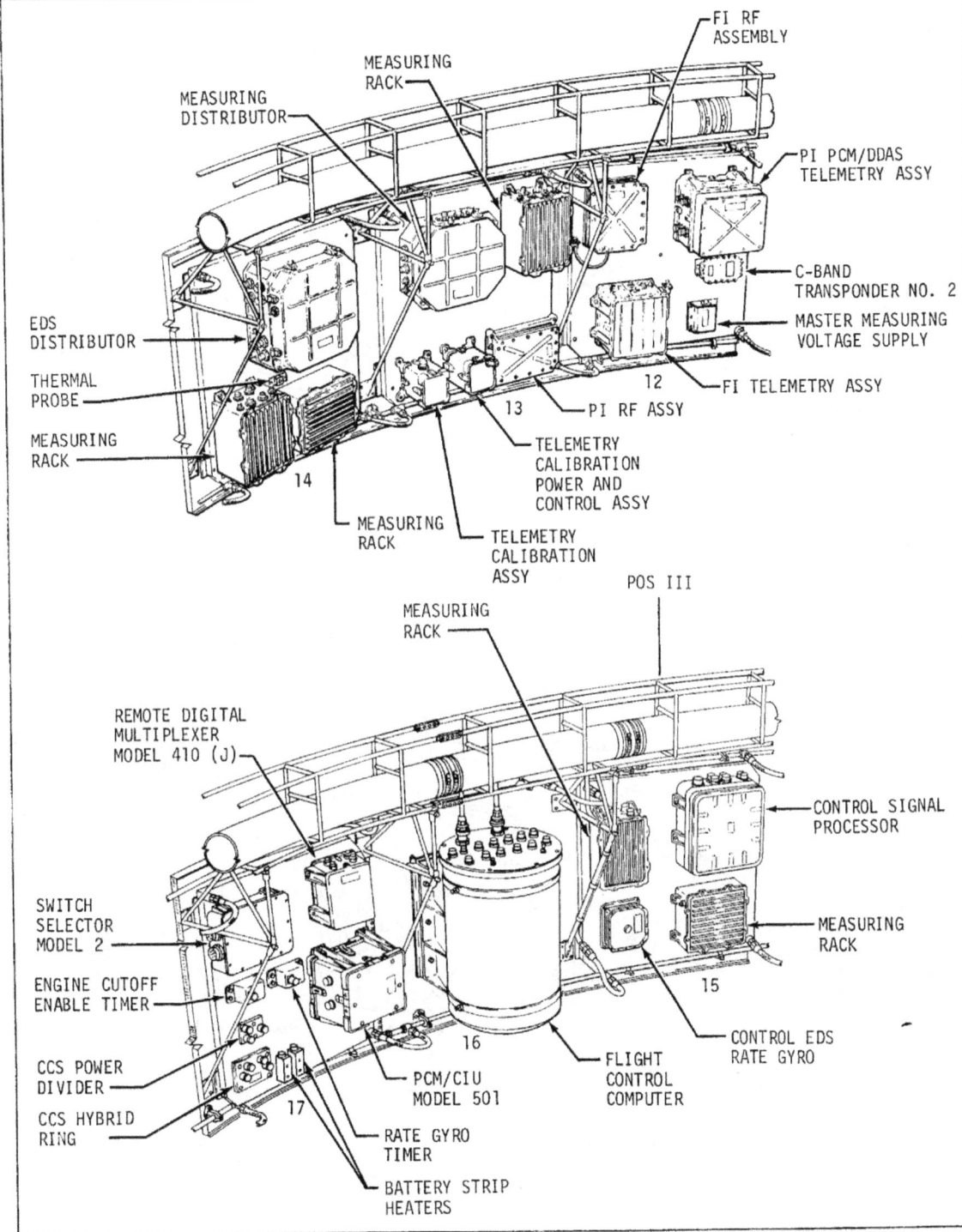

FI RF ASSEMBLY

MEASURING RACK

MEASURING DISTRIBUTOR

PI PCM/DDAS TELEMETRY ASSY

C-BAND TRANSPONDER NO. 2

MASTER MEASURING VOLTAGE SUPPLY

EDS DISTRIBUTOR

THERMAL PROBE

MEASURING RACK

12

FI TELEMETRY ASSY

13

PI RF ASSY

14

TELEMETRY CALIBRATION POWER AND CONTROL ASSY

MEASURING RACK

TELEMETRY CALIBRATION ASSY

POS III

MEASURING RACK

REMOTE DIGITAL MULTIPLEXER MODEL 410 (J)

CONTROL SIGNAL PROCESSOR

SWITCH SELECTOR MODEL 2

MEASURING RACK

ENGINE CUTOFF ENABLE TIMER

15

CCS POWER DIVIDER

CONTROL EDS RATE GYRO

CCS HYBRID RING

16

FLIGHT CONTROL COMPUTER

17

PCM/CIU MODEL 501

RATE GYRO TIMER

BATTERY STRIP HEATERS

Figure 7-4 (Sheet 4 of 5)

# INSTRUMENT UNIT EQUIPMENT LOCATIONS

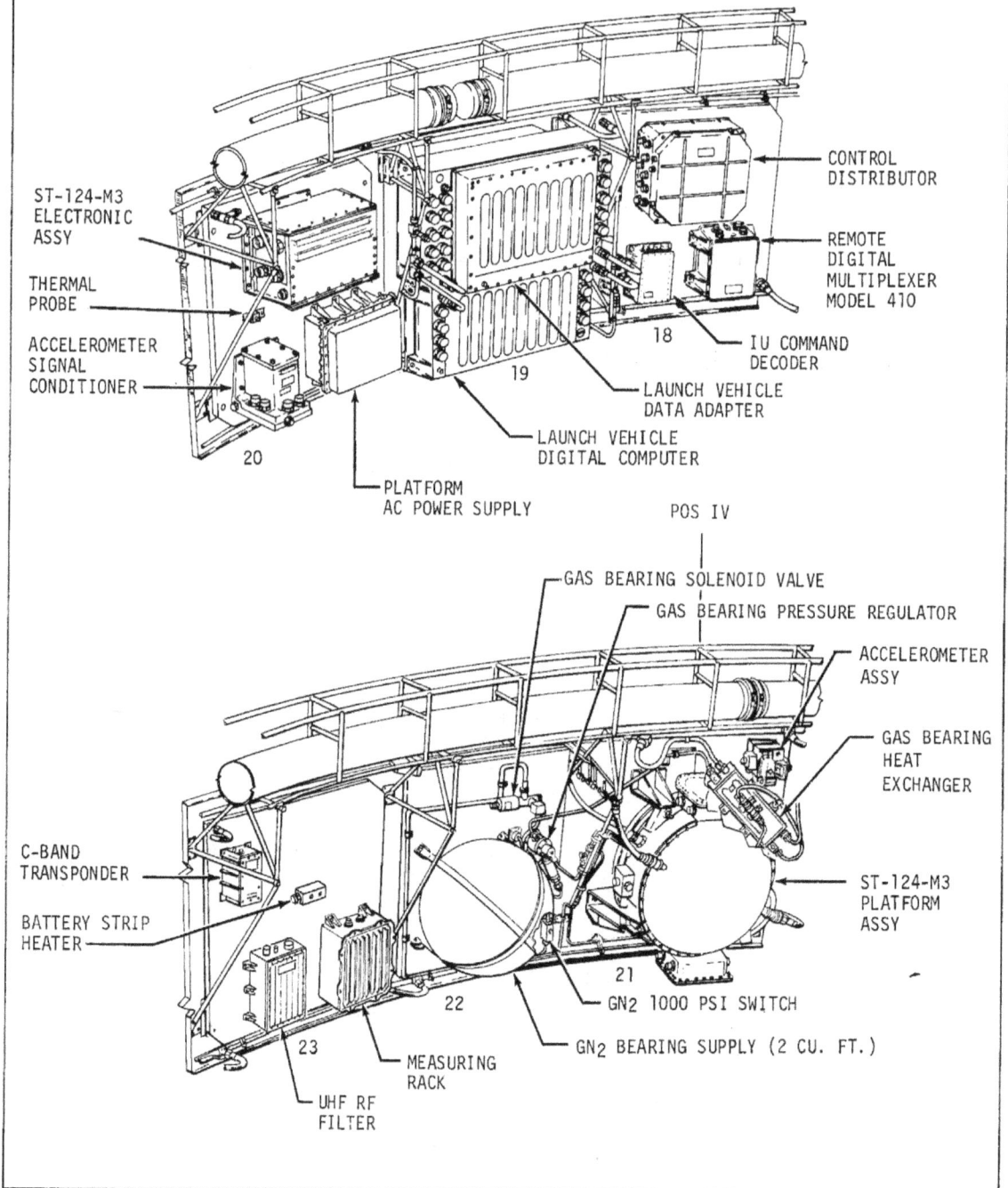

ST-124-M3
ELECTRONIC
ASSY

THERMAL
PROBE

ACCELEROMETER
SIGNAL
CONDITIONER

20

PLATFORM
AC POWER SUPPLY

CONTROL
DISTRIBUTOR

REMOTE
DIGITAL
MULTIPLEXER
MODEL 410

18

IU COMMAND
DECODER

LAUNCH VEHICLE
DATA ADAPTER

19

LAUNCH VEHICLE
DIGITAL COMPUTER

POS IV

GAS BEARING SOLENOID VALVE

GAS BEARING PRESSURE REGULATOR

ACCELEROMETER
ASSY

GAS BEARING
HEAT
EXCHANGER

ST-124-M3
PLATFORM
ASSY

C-BAND
TRANSPONDER

BATTERY STRIP
HEATER

23

UHF RF
FILTER

22

MEASURING
RACK

21

GN2 1000 PSI SWITCH

GN2 BEARING SUPPLY (2 CU. FT.)

Figure 7-4 (Sheet 5 of 5)

# THERMAL CONDITIONING SYSTEM FLOW DIAGRAM

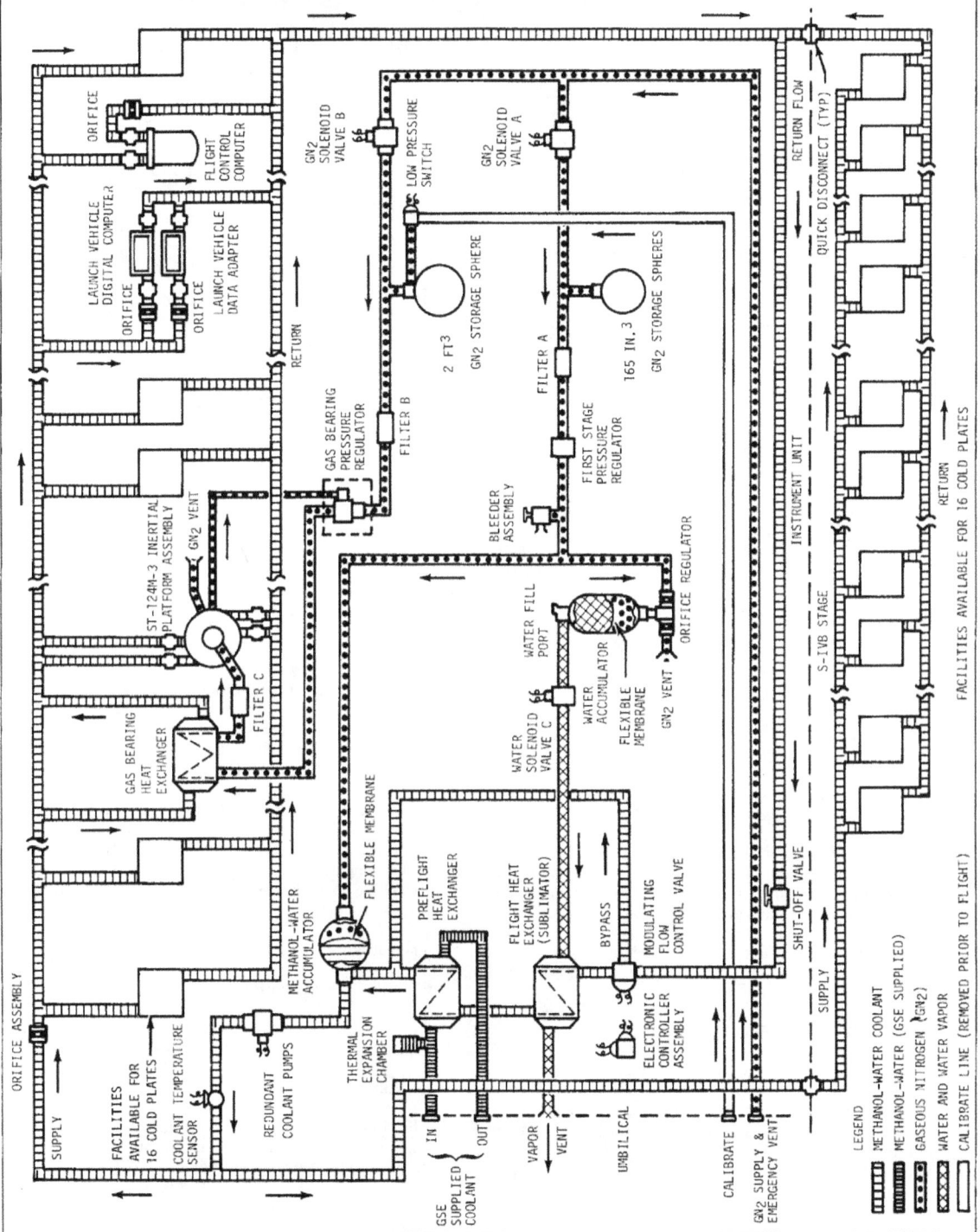

Figure 7-5

## SUBLIMATOR DETAILS

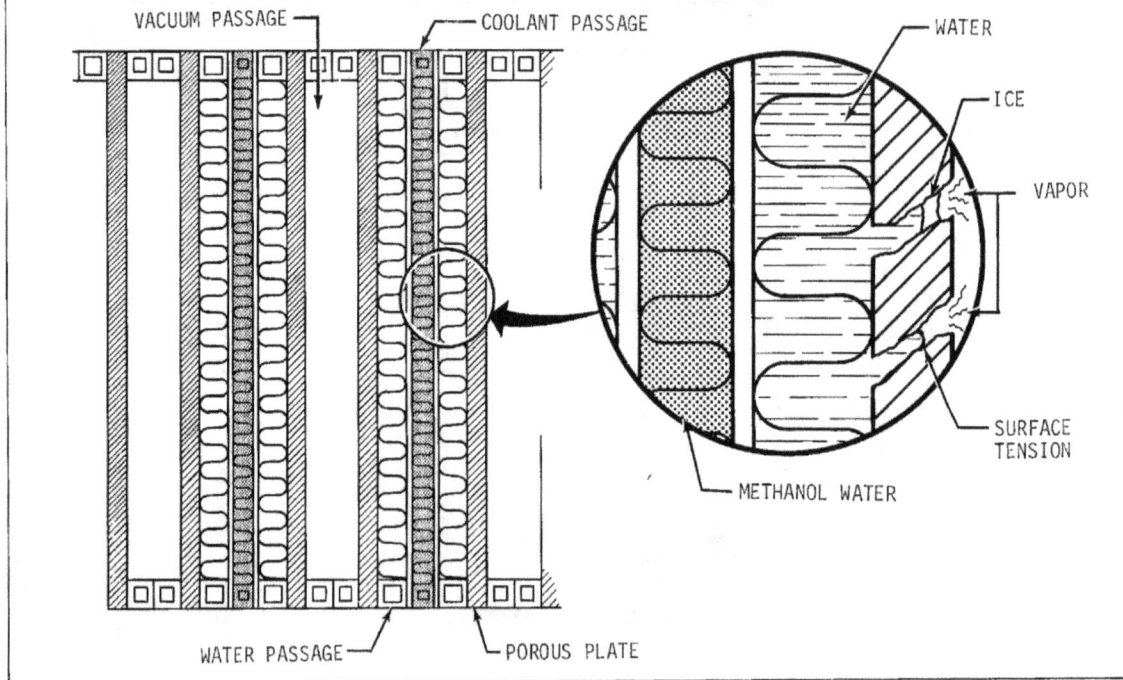

VACUUM PASSAGE — | — COOLANT PASSAGE

WATER

ICE

VAPOR

SURFACE TENSION

METHANOL WATER

WATER PASSAGE — | — POROUS PLATE

Figure 7-6

### Hazardous Gas Detection

The hazardous gas detection system is used to monitor for the presence of hazardous gases in the IU and S-IVB stage forward compartments during vehicle fueling. The monitoring operation is continuous from the start of vehicle fueling to umbilical disconnect at liftoff.

The hazardous gas detection sampling equipment consists of four tubes which are open ended between panels 1 and 2, 7 and 8, 13 and 14, and 19 and 20. The tubes are connected to a quick disconect coupling on a single tube (see figure 7-8).

The hazardous gas detection equipment (GSE) extracts samples through the four tubes, and monitors the samples for the presence of hazardous gases.

### ELECTRICAL POWER SYSTEMS

Primary flight power for the IU equipment is supplied by silver zinc batteries at a nominal voltage level of 28 (± 2) vdc. During prelaunch operations, primary power is supplied by the GSE. Where ac power is required within the IU, it is developed by solid state dc to ac inverters. Power distribution within the IU is accomplished through power distributors which are, essentially, junction boxes and switching circuits.

### BATTERIES

Silver-zinc primary flight batteries are installed in the IU

during prelaunch preparations. These batteries are physically and electrically identical, but each is connected to a separate bus in a power distributor. Flight components are connected to these busses in such a manner as to distribute the electrical load evenly between the batteries.

An attractive feature of the silver-zinc batteries is their high efficiency. Their ampere-hour rating is about four times as great as that of a lead-acid or nickle-cadmium battery of the same weight. The low temperature performance of the silver-zinc batteries is also substantially better than the others.

The battery characteristics are listed in figure 7-9.

### POWER CONVERTERS

The IU electrical power systems contain a 56-volt power supply and a 5-volt measuring voltage supply.

### 56-Volt Power Supply

The 56-volt power supply furnishes the power required by the ST-124-M3 platform electronic assembly and the accelerometer signal conditioner. It is basically a dc to dc converter that uses a magnetic amplifier as a control unit. It converts the unregulated 28 vdc from the batteries to a regulated 56 vdc. The 56-volt power supply is connected to the platform electronic assembly through the power and control distributors.

# PREFLIGHT AIR/GN₂ PURGE SYSTEM

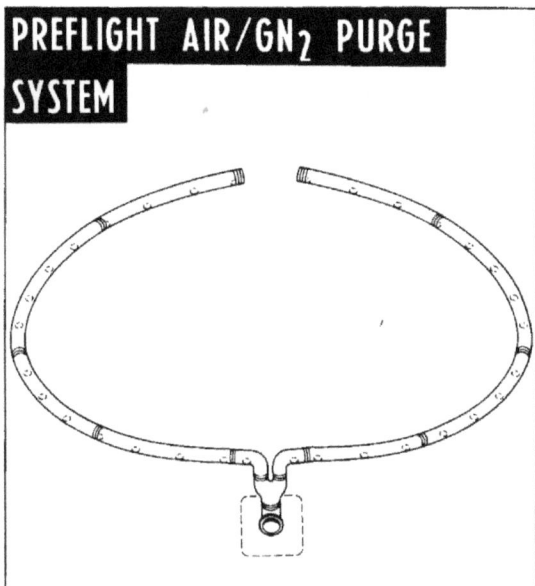

Figure 7-7

# HAZARDOUS GAS DETECTION SAMPLING EQUIPMENT LAYOUT

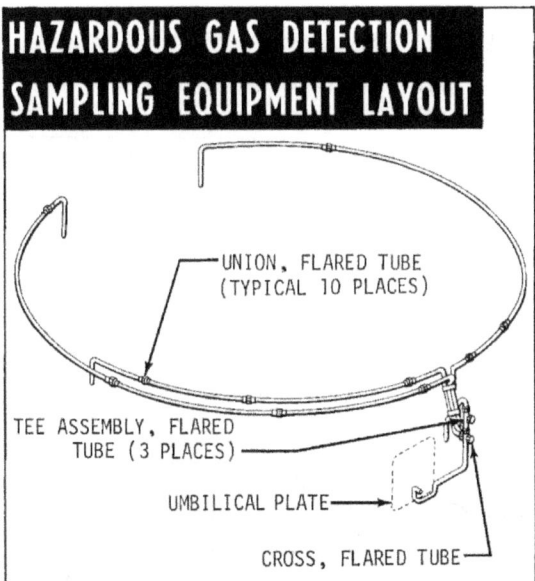

UNION, FLARED TUBE (TYPICAL 10 PLACES)

TEE ASSEMBLY, FLARED TUBE (3 PLACES)

UMBILICAL PLATE

CROSS, FLARED TUBE

Figure 7-8

### 5-Volt Measuring Voltage Supply

The 5-volt measuring voltage supply converts unregulated 28 vdc to a closely regulated 5 ($\pm$ .005) vdc for use throughout the IU measuring system. This regulated voltage is used primarily as excitation for measurement sensors (transducers), and as a reference voltage for inflight calibration of certain telemetry channels. Like the 56-volt supply, it is basically a dc to dc converter.

| IU BATTERY CHARACTERISTICS | |
|---|---|
| Type | Map 4240-Dry charge |
| Material | Alkaline silver-zinc |
| Cells | 20 (with taps for selecting 18 or 19 cells if required to reduce high voltage) |
| Nominal Voltage | 1.5 per cell |
| Electrolyte | Potassium hydroxide (KOH) in demineralized water |
| Output Voltage | +28 $\pm$2 vdc |
| Output Current | 35 amperes for a 10 hour load period (if used within 72 hours of activation) |
| Gross Weight | 165 pounds each |

Figure 7-9

### DISTRIBUTORS

The distribution system within the IU is comprised of the following:

 2 Measuring distributors
 1 Control distributor
 1 Emergency Detection System (EDS) distributor
 1 Power distributor
 2 Auxiliary power distributors.

#### Measuring Distributors

The primary function of the measuring distributors is to collect all measurements that are transmitted by the IU telemetry system, and to direct them to their proper telemetry channels. These measurements are obtained from instrumentation transducers, functional components, and various signal and control lines. The measuring distributors also distribute the output of the 5-volt measuring voltage supply throughout the measuring system.

Through switching capabilities, the measuring distributors can change the selection of measurements monitored by the telemetry system. The switching function transfers certain measurements to channels which had been allotted to expended functions. If it were not for this switching, these channels would be wasted for the remainder of the flight.

#### Control Distributor

The control distributor serves as an auxiliary power distributor for IU segment 603. It provides distribution of 28-volt power to small current loads, and distributes 56 vdc from the 56-volt power supply to the ST-124-M3 inertial platform assembly. The control distributor provides power and signal switching during prelaunch checkout for testing various guidance, control, and EDS functions, requested by the launch vehicle data adapter through the switch selector.

## Emergency Detection System Distributor

The EDS distributor provides the only electrical link between the spacecraft and the LV. All EDS signals from the LV are routed to the logic circuits in the EDS distributor. EDS output signals from these logic circuits are then fed to the spacecraft and to the IU telemetry. Also, EDS signals from the spacecraft are routed back through the IU EDS logic circuits before being sent to the S-IVB, S-II, and S-IC stages.

## Power Distributor

The power distributor provides primary distribution for all 28-volt power required by IU components. Inflight 28-volt battery power, or prelaunch ESE-supplied 28-volt power, is distributed by the power distributor.

The power distributor also provides paths for command and measurement signals between the ESE and IU components. The power distributor connects the IU component power return and signal return lines to the IU single point ground and to the umbilical supply return bus. These return lines are connected to the common bus in the power distributor, directly or indirectly, through one of the other distributors.

Silver-zinc batteries supply the inflight 28-vdc power. See figure 7-10 for a typical power distribution block diagram. Each battery is connected to a separate bus in the distributor. The flight components are connected to the busses in such a manner as to distribute the load evenly across the batteries.

## Auxiliary Power Distributors

Two auxiliary power distributors supply 28 vdc power to small current loads. Both auxiliary power distributors receive 28 vdc from each of the battery busses in the power distributor, so that current loads on each of the batteries may be evenly distributed. Relays in the auxiliary power distributors provide power ON/OFF control for IU components both during the prelaunch checkout and while in flight. These relays are controlled by the ESE and by the switch selector.

## IU GROUNDING

All IU grounding is referenced to the outer skin of the LV. The power system is grounded by means of hardwires routed from the power distributor COM bus to a grounding stud attached to the LV skin. All COM busses in the various other distributors are wired back to the COM bus in the power distributor. This provides for a single point ground.

Equipment boxes are grounded by direct metal-to-metal contact with cold plates or other mounting surfaces which are common to the LV skin. Most cabling shields are grounded to a COM bus in one of the distributors. However, where shielded cables run between equipment boxes, and not through a distributor, only one end of the shield would be grounded.

During prelaunch operations, the IU and GSE COM busses are referenced to earth ground. To ensure the earth ground reference until after all umbilicals are ejected, two, single wire grounding cables are connected to the IU below the

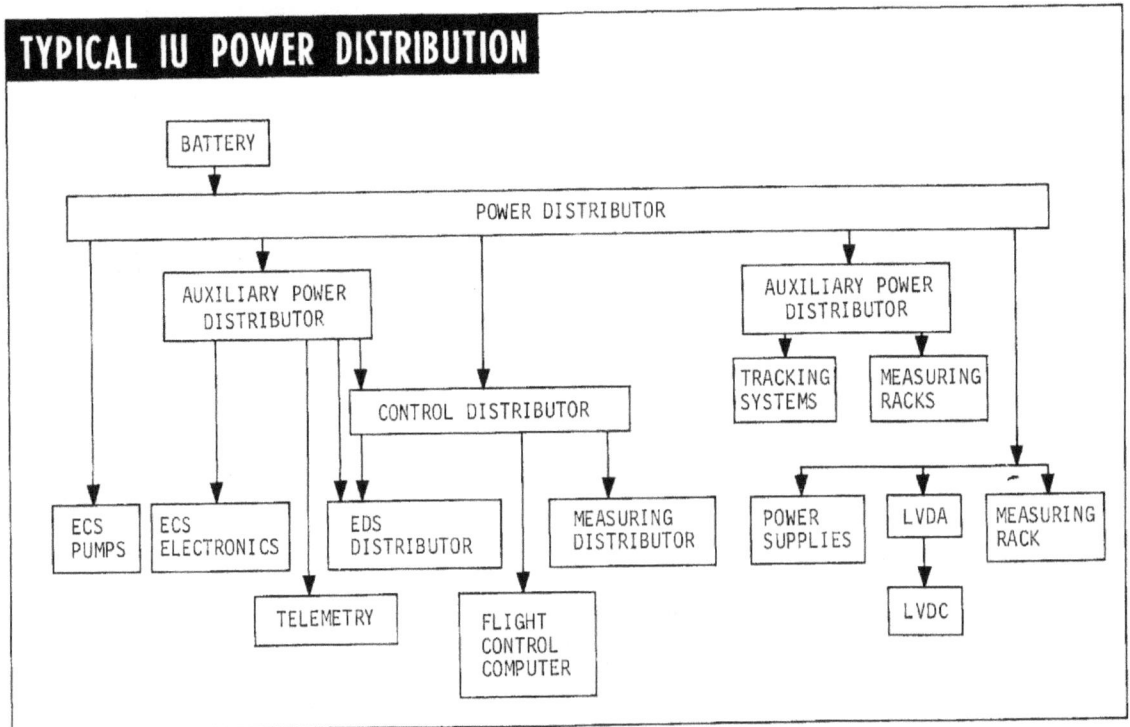

Figure 7-10

umbilical plates. These are the final conductors to be disconnected from the IU.

## EMERGENCY DETECTION SYSTEM

The EDS is one element of several crew safety systems. EDS design is a coordinated effort of crew safety personnel from several NASA centers.

The EDS senses initial development of conditions which could cause vehicle failure. The EDS reacts to these emergency situations in either of two ways. If breakup of the vehicle is imminent, an automatic abort sequence is initiated. If, however, the emergency condition is developing slowly enough, or is of such a nature that the flight crew can evaluate it and take action, only visual indications are provided to the flight crew. Once an abort sequence has been initiated, either automatically or manually, it is irrevocable and runs to completion.

The EDS is comprised of sensing elements, such as signal processing and switching circuitry, relay and diode logic circuitry, electronic timers and display equipment, all located in various places on the flight vehicle. Only that part of the EDS equipment located in the IU will be discussed here.

There are nine EDS rate gyros installed in the IU. Three gyros monitor each of the three axes (pitch, roll, and yaw) thus providing triple redundancy.

The control signal processor provides power to the nine EDS rate gyros, as well as receiving inputs from them. These inputs are processed and sent to the EDS distributor and to the flight control computer.

The EDS distributor serves as a junction box and switching device to furnish the spacecraft display panels with emergency signals if emergency conditions exist. It also contains relay and diode logic for the automatic abort sequence.

There is an electronic timer which is activated at liftoff, and which produces an output 30 seconds later. This output energizes relays in the EDS distributor which allows multiple engine shutdown. This function is inhibited during the first 30 seconds of launch.

Inhibiting of automatic abort circuitry is also provided by the LV flight sequencing circuits through the IU switch selector. This inhibiting is required prior to normal S-IC engine cutoff and other normal LV sequencing. While the automatic abort capability is inhibited, the flight crew must initiate a manual abort, if an angular-overrate or two-engine-out condition arises.

See Section III for a more complete discussion of the overall EDS. Section III includes abort limits, displays, controls, diagrams, and a description of the voting logic.

## NAVIGATION, GUIDANCE AND CONTROL

The Saturn V launch vehicle is guided from its launch pad into earth orbit by navigation, guidance and control equipment located in the IU. An all inertial system, using a space stabilized platform for acceleration and attitude measurements, is utilized. A launch vehicle digital computer (LVDC) is used to solve guidance equation's and a flight control computer (analog) is used for the flight control functions.

In the following discussions, the terms "navigation," "guidance," and "control" are used according to these definitions:

Navigation is the determination of the flight vehicle's present position and velocity from measurements made on board the vehicle.

Guidance is the computation of maneuvers necessary to achieve the desired flight path.

Control is the execution of the guidance maneuver by controlling the proper hardware.

Consider the block diagram of the overall Saturn V guidance and control subsystem shown in figure 7-11. The three-gimbal stabilized platform (ST-124-M3) provides a space-fixed coordinate reference frame for attitude control and for navigation (acceleration) measurements. Three integrating accelerometers, mounted on the gyro-stabilized inner gimbal of the platform, measure the three components of velocity resulting from vehicle propulsion. The accelerometer measurements are sent through the launch vehicle data adapter (LVDA) to the LVDC. In the LVDC, the accelerometer measurements are combined with the computed gravitational acceleration to obtain velocity and position of the vehicle.

The LVDA is the input/output device for the LVDC. It performs the necessary processing of signals to make these signals acceptable to the LVDC.

According to the guidance scheme (programmed into the computer), the maneuvers required to achieve the desired end conditions are determined by the LVDC. The instantaneous position and velocity of the vehicle are used as inputs. The result is the required thrust direction (guidance command) and the time of engine cutoff.

Control of the launch vehicle can be divided into attitude control and discrete control functions. For attitude control, the instantaneous attitude of the vehicle is compared with the desired vehicle attitude (computed according to the guidance scheme). This comparison is performed in the LVDC. Attitude correction signals are derived from the difference between the existing attitude angles (platform gimbal angles) and the desired attitude angles. In the flight control computer, these attitude correction signals are combined with signals from control sensors to generate the control commands for the engine actuators. The required thrust direction is obtained by gimbaling the engines in the propelling stage to change the thrust direction of the vehicle. In the S-IC and S-II, the four outboard engines are gimbaled to control roll, pitch, and yaw. Since the S-IVB stage has only one engine, an auxiliary propulsion system (APS) is used for roll control during powered flight. The APS provides complete attitude control during coast flight of the S-IVB/IU/spacecraft.

Guidance information stored in the LVDC (e.g., position, velocity) can be updated through the IU command system by data transmission from ground stations. The IU command system provides the general capability of changing or inserting information into the LVDC.

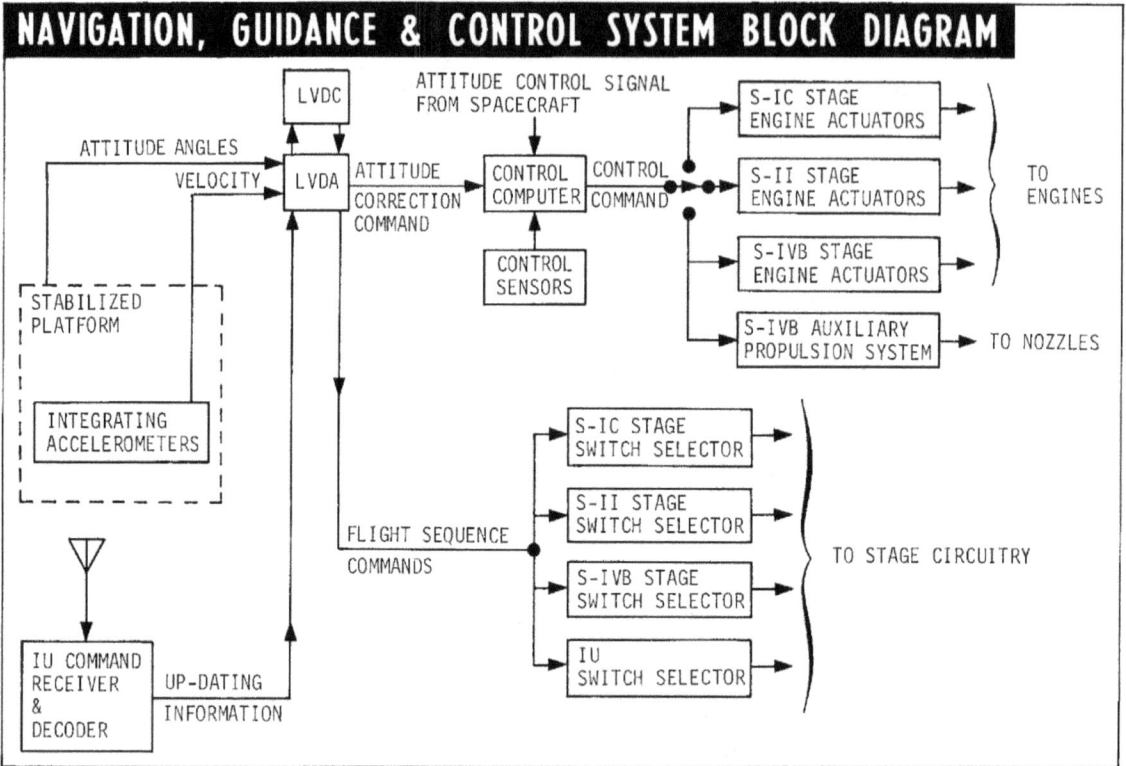

# NAVIGATION, GUIDANCE & CONTROL SYSTEM BLOCK DIAGRAM

Figure 7-11

## NAVIGATION SCHEME

### Powered Flight

The basic navigation scheme is shown in figure 7-12. Gimbal resolvers supply platform position in analog form to the LVDA. An analog-to-digital converter in the LVDA converts the signal to the digital format required by the LVDC.

Platform integrating accelerometers sense acceleration components, and mechanically integrate them into velocity. The LVDA provides signal conditioning. Within the LVDC, initial velocity imparted by the spinning earth, gravitational velocity, and the platform velocities are algebraically summed together. This vehicle velocity is integrated by the LVDC to determine vehicle position.

Acceleration can be defined as the rate-of-change (derivative) of velocity. Velocity is the rate-of-change of position (distance). Velocity is the integral of acceleration, and position is the integral of velocity. Therefore, position is obtained by integrating acceleration twice.

### Orbital Flight

During orbital coast flight, the navigational program continually computes the vehicle position, velocity, and acceleration from equations of motion which are based on vehicle conditions at the time of orbital insertion. In orbit, navigation and guidance information in the LVDC can be updated by digital data transmission through the command

and communications system.

Additional navigational computations are used in maintaining vehicle attitude during orbit. These computations establish a local vertical which is used as a reference for attitude control. The attitude of the vehicle roll axis will be maintained at 90 degrees with respect to the local vertical.

## GUIDANCE COMPUTATIONS

The guidance function of the launch vehicle is accomplished by computing the necessary flight maneuvers to meet the desired end conditions of the flight (e.g., inserting the spacecraft into the desired trajectory). Guidance computations are performed within the LVDC by programmed guidance equations, which use navigation data and mission requirements as their inputs. These computations are actually a logical progression of computed formulas which have a guidance command as their solution. After the desired attitude has been determined by the "best path" program, the guidance computations might be reduced into the following equation: $\chi - \phi = \psi$ (See figure 7-13.)

where:

$\chi$ is the desired attitude

$\phi$ is the vehicle attitude

$\psi$ is the attitude error command

## NAVIGATION SCHEME

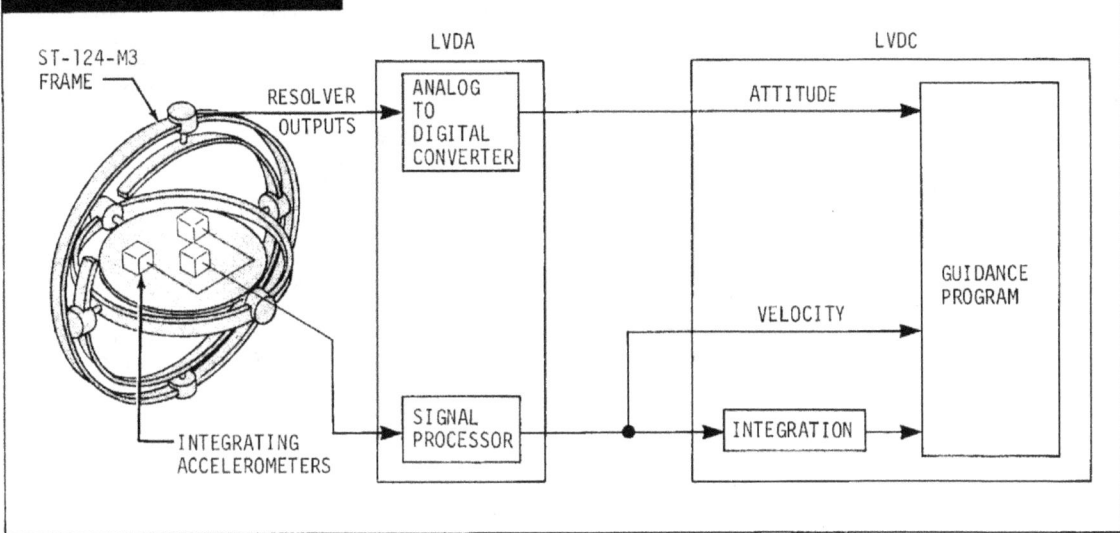

Figure 7-12

## GUIDANCE EQUATION

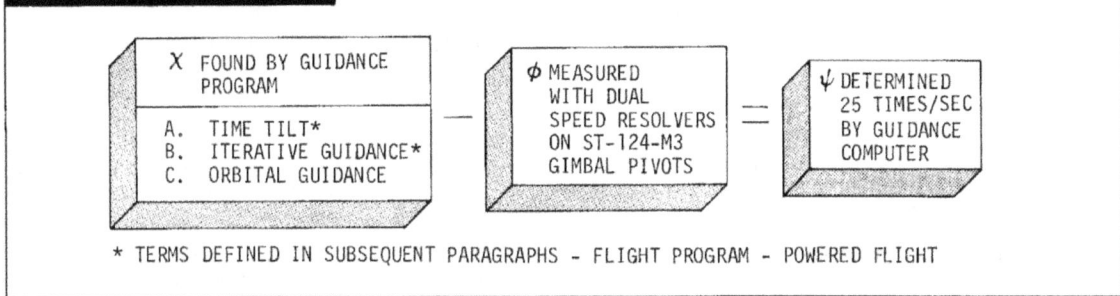

Figure 7-13

## CONTROL SUBSYSTEM

The control subsystem (figure 7-14) is designed to control and maintain vehicle attitude by forming the steering commands used to control the engines of the active stage.

Vehicle attitude is achieved by gimbaling the four outboard engines of the S-IC stage, the four outboard engines of the S-II stage, or the single engine of the S-IVB stage. These engines are gimbaled by hydraulic actuators. Roll attitude control on the S-IVB stage cannot, of course, be controlled with a single engine. Therefore, roll control of the S-IVB stage is accomplished by the APS (see figure 7-15). During the coast period of the mission, the S-IVB APS will be used to control the vehicle attitude in all three axes.

The control system accepts guidance computations from the LVDC/LVDA guidance system. These guidance commands, which are actually attitude error signals, are then combined with measured data from the various control sensors. The resultant output is the command signal to the various engine actuators and APS nozzles.

The final computations (analog) are performed within the flight control computer. This computer is also the central switching point for command signals. From this point, the signals are routed to their associated active stages, and to the appropriate attitude control devices.

## CONTROL SYSTEM COMPONENTS

### Control Signal Processor

The control signal processor demodulates the ac signals from the control-EDS rate gyros into dc analog signals, required by the flight control computer. The control signal processor compares the output signals from the triple redundant gyros, and selects one each of the pitch, yaw, and roll signals for the flight control computer. The control signal processor supplies the control-EDS rate gyro package with the necessary control and reference voltages. EDS and DDAS rate gyro monitoring signals also originate within the control signal processor, thus accounting for the EDS portion of the control-EDS rate gyro name.

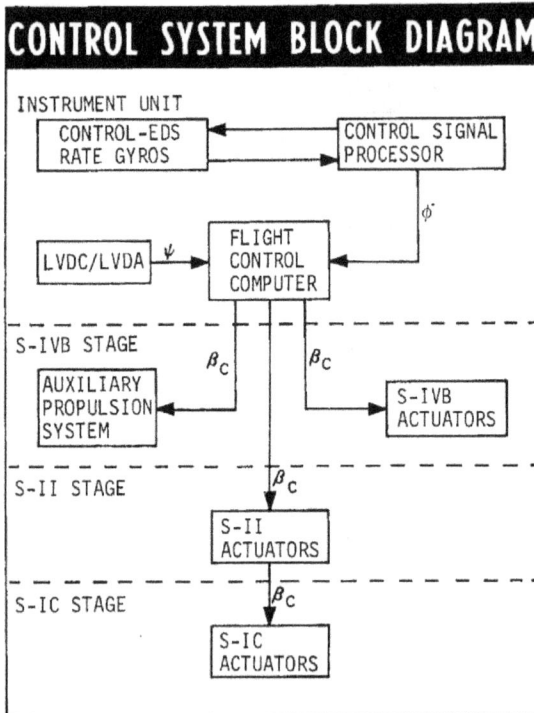

# CONTROL SYSTEM BLOCK DIAGRAM

INSTRUMENT UNIT

```
CONTROL-EDS          CONTROL SIGNAL
RATE GYROS           PROCESSOR

                                    φ̇
         ψ    FLIGHT
LVDC/LVDA     CONTROL
              COMPUTER

S-IVB STAGE    β_C      β_C

AUXILIARY                      S-IVB
PROPULSION                     ACTUATORS
SYSTEM

S-II STAGE         β_C

              S-II
              ACTUATORS

S-IC STAGE         β_C

              S-IC
              ACTUATORS
```

Figure 7-14

## Flight Control Computer

The flight control computer is an analog computer which converts attitude correction commands ($\psi$) and angular change rates ($\phi$) into APS thruster nozzle and/or engine actuator positioning commands.

Input signals to the flight control computer include:

1. Attitude correction commands ($\psi$) from the LVDC/LVDA or spacecraft

2. Angular change rates ($\phi$) from the control-EDS rate gyro package, via the control signal processor.

Since one of the inputs to the flight control computer is from the control-EDS rate gyros, an excessive attitude correction commands from the LVDC is limited within the flight control computer to protect the vehicle structure.

Output signals from the flight control computer include:

1. Command signals to the engine actuators ($\beta$c)

2. Command signals to the APS thruster nozzles

3. Telemetry outputs which monitor internal operations and functions.

## ST-124-M3 Inertial Platform Assembly

The gimbal configuration of the ST-124-M3 offers unlimited freedom about the X & Y axes, but is limited to $\pm45$ degrees about its Z axis (vehicle yaw at launch). See figure 7-16.

The gimbal system allows the inertial gimbal rotational freedom. Three single-degree-of-freedom gyroscopes have their input axes aligned along an orthogonal inertial coordinates system: $X_1$, $Y_1$, and $Z_1$ of the inertial gimbal. The signal generator, which is fixed to the output axis of each gyro, generates electrical signals proportional to torque disturbances. These signals are transmitted through the servo electronics which terminate in the gimbal pivot servo-torque motors. The servo loops maintain the inner gimbal rotationally fixed in inertial space.

The inner gimbal has three, pendulous, integrating, gyroscopic accelerometers, oriented along the inertial coordinates $X_1$, $Y_1$, and $Z_1$. Each accelerometer measuring head contains a pendulous, single-degree-of-freedom gyro. The rotation of the measuring head is a measure of acceleration along the input axis of the accelerometer. Since acceleration causes the accelerometer shaft to be displaced as a function of time, the shaft position (with respect to a zero reference) is proportional to velocity, and the accelerometer is referred to as an integrating accelerometer.

Vehicle attitude is measured with respect to the inertial platform, using dual speed (32:1) resolvers located at the gimbal pivot points. The outputs of these angle encoders are converted into a digital count in the LVDA.

During prelaunch, the ST-124-M3 platform is held aligned to the local vertical by a set of gas bearing leveling pendulums. The pendulum output is amplified in the platform, and then transmitted to the ground equipment alignment amplifier. The alignment amplifier provides a signal to the torque drive amplifier, and then to the platform gyro torque generator. The vertical alignment system levels the platform to an accuracy of $\pm2.5$ arc seconds.

The azimuth alignment is accomplished by means of a theodolite on the ground and two prisms on the platform - one fixed and one servo driven. The theodolite maintains the azimuth orientation of the movable prism, and the computer computes a mission azimuth, and programs the inner gimbal to that azimuth. The laying system has an accuracy of $\pm5$ arc seconds.

At approximately liftoff minus seventeen seconds, the platform is released to maintain an inertial reference initiated at the launch point. At this time, the LVDC begins navigation, using velocity accumulations derived from the ST-124-M3 inertial platform.

## Platform Electronic Assembly

The ST-124-M3 platform electronic assembly (PEA) contains the electronics, other than those located in the platform, required for the inertial gimbal and the accelerometer stabilization. Switching electronics for controlling platform system power and checkout functions are also located in the ST-124-M3 platform electronic assembly.

# SATURN V ENGINES, ACTUATORS AND NOZZLE ARRANGEMENT

NOTES:

1. ALL SIGNAL ARROWS INDICATE POSITIVE VEHICLE MOVEMENTS.
2. VEHICLE PITCHES AROUND THE "Y" AXIS
3. ENGINE ACTUATOR LAYOUTS SHOWN AS VIEWED FROM AFT END OF VEHICLE.
4. DIRECTIONS AND POLARITIES SHOWN ARE TYPICAL FOR ALL STAGES.
5. $+\beta$ INDICATES ENGINE DEFLECTION REQUIRED TO CORRECT FOR POSITIVE VEHICLE MOVEMENT.
6. CG = CENTER OF GRAVITY
   F = NOZZLES ON
   EXT = ACTUATOR EXTENDED
   RET = ACTUATOR RETRACTED
   $\beta$ = THRUST VECTOR ANGULAR DEFLECTION

S-IC & S-II ACTUATOR LAYOUTS

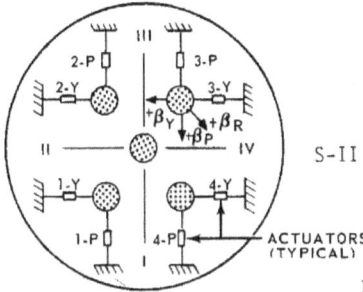

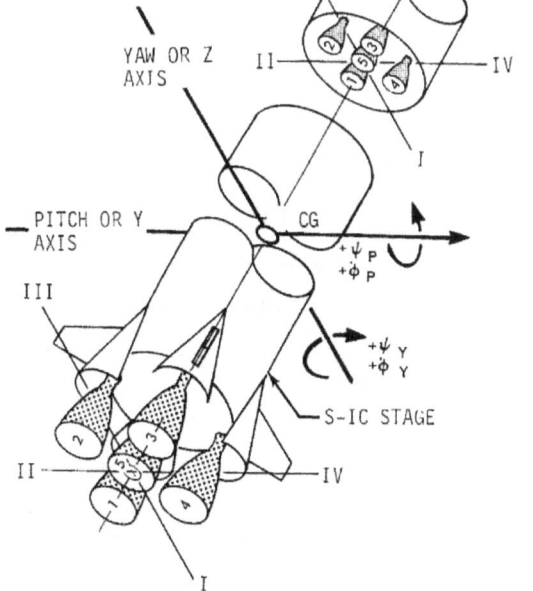

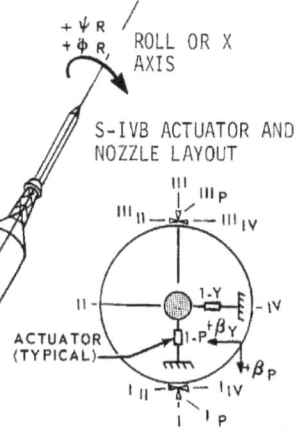

S-IVB ACTUATOR AND NOZZLE LAYOUT

### S-IC & S-II POLARITY TABLE

| ACTUATOR NO. | ACTUATOR MOVEMENT | | |
|---|---|---|---|
| | $+\phi_R$ | $+\phi_Y$ | $+\phi_P$ |
| 1-Y | RET | RET | |
| 1-P | EXT | | RET |
| 2-Y | EXT | RET | |
| 2-P | RET | | EXT |
| 3-Y | RET | EXT | |
| 3-P | EXT | | EXT |
| 4-P | EXT | EXT | |
| 4-Y | RET | | RET |

### S-IVB POLARITY TABLE

| ACTUATOR NO. | SIGNAL & ACTION | | | |
|---|---|---|---|---|
| | $+\psi_R$ | $-\psi_R$ | $+\psi_Y$ | $+\psi_P$ |
| I-Y | | | EXT | |
| I-P | | | | RET |
| ENGINE NO. | | | | |
| I IV | | F | | |
| I P | | | | |
| I II | F | | | |
| III II | | F | | |
| III P | | | | |
| III IV | F | | | |

| | CONDITIONS DURING COAST | | | | | |
|---|---|---|---|---|---|---|
| | $+\psi_R$ | $-\psi_R$ | $+\psi_Y$ | $-\psi_Y$ | $+\psi_P$ | $-\psi_P$ |
| I IV | | F | | F | | |
| I P | | | | | F | |
| I II | F | | F | | | |
| III II | | F | F | | | |
| III P | | | | | | F |
| III IV | F | | | F | | |

Figure 7-15

# PLATFORM GIMBAL CONFIGURATION

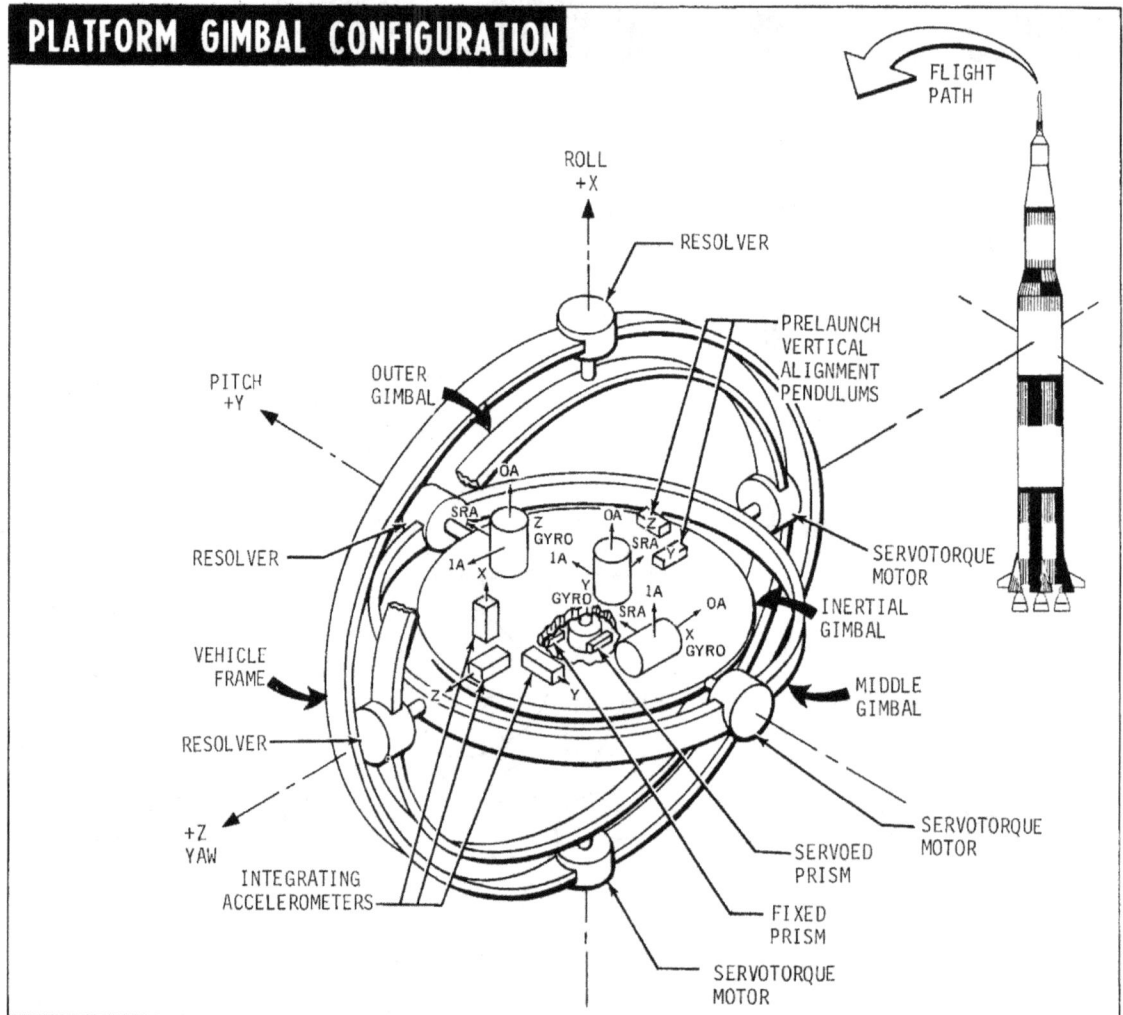

Figure 7-16

The PEA includes the following circuitry:

1.  Amplifiers, modulators, and stabilization networks for the platform gimbal and accelerometer servo loops

2.  Relay logic for signal and power control

3.  Amplifiers for the gyro and accelerometer pick-off coil excitation

4.  Automatic checkout selection and test circuitry for servo loops

5.  Control circuitry for the heaters and gas supply.

## ST-124-M3 AC Power Supply

The ST-124-M3 platform ac power supply furnishes the power required to run the gyro rotors, and provides excitation for the platform gimbal synchros. It is also the frequency source for the resolver chain references and for gyro and accelerometer servo systems carrier.

The supply produces a three-phase, sine wave output which is fixed at 26 volts (rms) line-to-line at a frequency of 400 (± 0.01) Hertz. Three single-phase, 20-volt reference outputs (square wave) of 4.8 kHz, 1.02 kHz, and 1.6 kHz are also provided. With a normal input voltage of 28 vdc, the supply is capable of producing a continuous 250 va output.

## Accelerometer Signal Conditioner

The accelerometer signal conditioner accepts the velocity signals from the accelerometer optical encoders and shapes them before they are passed on to the LVDA/LVDC.

Each accelerometer requires four shapers: a sine shaper and cosine shaper for the active channel, and a sine shaper and cosine shaper for the redundant channels. Also included are four buffer amplifiers for each accelerometer, one for each sine and cosine output.

Accelerometer outputs are provided for telemetry and ground checkout in addition to the outputs to the LVDA.

### LV Digital Computer and LV Data Adapter

The LVDC and LVDA comprise a modern electronic digital computer system. The LVDC is a relatively high-speed computer with the LVDA serving as its input/output device. Any signal to or from the computer is routed through the LVDA. The LVDA serves as central equipment for interconnection and signal flow between the various systems in the IU. See figure 7-17 and 7-18 for LVDC and LVDA characteristics.

The LVDA and LVDC are involved in four main operations:

1.  Prelaunch checkout

2.  Navigation and guidance computations

3.  Vehicle sequencing

4.  Orbital checkout.

The LVDC is a general purpose computer which processes data under control of a stored program. Data is processed serially in two arithmetic sections which can, if so programmed, operate concurrently. Addition, subtraction, and logical extractions are performed in one arithmetic section, while multiplication and division are performed in the other.

The principal storage device is a random access, ferrite-core memory with separate controls for data and instruction addressing. The memory can be operated in either a simplex or duplex mode. In duplex operation, memory modules are operated in pairs with the same data being stored in each module. Readout errors in one module are corrected by using data from its mate to restore the defective location. In simplex operation, each module contains different data, which doubles the capacity of the memory. However, simplex operation decreases the reliability of the LVDC, because the ability to correct readout errors is sacrificed. The memory operation mode is program controlled. Temporary storage is provided by static registers, composed of latches, and by shift registers, composed of delay lines and latches.

Computer reliability is increased within the logic sections by the use of triple modular redundancy. Within this redundancy scheme, three separate logic paths are voted upon to correct any errors which develop.

| LAUNCH VEHICLE DIGITAL COMPUTER CHARACTERISTICS | |
|---|---|
| ITEM | DESCRIPTION |
| Type | General Purpose, Digital, Stored Program |
| Memory | Random Access, Ferrite (Torodial) Core, with a Capacity of 32,768 words of 28 Bits each. |
| Speed | Serial Processing at 512,000 Bits Per Seconds |
| Word Make-Up | Memory = 28 Bits<br>Data = 26 Bits Plus 2 Parity Bits<br>Instruction = 13 Bits Plus 1 Parity Bit |
| Programming | 18 Instruction Codes<br>10 Arithmetic<br>6 Program Control<br>1 Input/Output<br>1 Store |
| Timing | Computer Cycle = 82.03 $\mu$ sec.<br>Bit Time = 1.95 $\mu$ sec.<br>Clock Time = 0.49 $\mu$ sec. |
| Input/Output | External, Program Controlled |

Figure 7-17

| LAUNCH, VEHICLE DATA ADAPTER CHARACTERISTICS | |
|---|---|
| ITEM | DESCRIPTION |
| Input/Output Rate | Serial Processing at 512,000 Bits Per Second |
| Switch Selector | 8 Bit Input<br>15 Bit Output |
| Telemetry Command Receiver | 14 Bits for Input Data |
| Data Transmitter | 38 Data and Identification Bits Plus Validity Bit and Parity Bit |
| Computer Interface Unit | 15 Bits Address Plus 1 Data Request Bit<br>10 Bits for Input Data Plus 1 Bit for Data Ready Interrupt |
| Delay Lines | 3 Four-Channel Delay Lines for Normal Operation<br>1 Four-Channel Delay Line for Telemetry Operations |
| Output to Launch Computer | 41 Data and Identification Bits Plus Discrete Outputs |
| Input From RCA-110 GCC | 14 Bits for Data Plus Interrupt |

Figure 7-18

# FLIGHT PROGRAM

A flight program is defined as a set of instructions which controls the LVDC operation from seconds before liftoff until the end of the launch vehicle mission. These instructions are stored within the LVDC memory.

The flight program performs many functions during the launch vehicle mission. These functions include navigation, guidance, attitude control, event sequencing, data management, ground command processing, and hardware evaluation.

For purposes of discussion, the flight program is divided into five subelements: the powered flight major loop, the orbital flight program, the minor loop, interrupts, and telemetry.

The powered flight major loop contains guidance and navigation calculations, timekeeping, and all repetitive functions which do not occur on an interrupt basis. The orbital flight program consists of an executive routine concerned with IU equipment evaluation during orbit, and a telemetry time-sharing routine, to be employed while the vehicle is over receiving stations. In addition, in the orbital flight program, all navigation, guidance, and timekeeping computations are carried out on an interrupt basis, keyed to the minor loop. The minor loop contains the platform gimbal angle and accelerometer sampling routines and control system computations. Since the minor loop is involved with vehicle control, minor loop computations are executed at the rate of 25 times per second during the powered phase of flight. However, in earth orbit, a rate of only ten executions per second is required for satisfactory vehicle control.

## PRELAUNCH AND INITIALIZATION

Until just minutes before launch, the LVDC is under control of the ground control computer (GCC). At T-8 minutes, the GCC issues a prepare-to-launch (PTL) command to the LVDC. The PTL routine performs the following functions:

1. Executes an LVDC/LVDA self-test program and telemeters the results

2. Monitors accelerometer inputs, calculates the platform-off-level indicators, and telemeters accelerometer outputs and time

3. Performs reasonableness checks on particular discrete inputs and alerts

4. Interrogates the LVDC error monitor register

5. Keeps all flight control system ladder outputs zeroed, which keeps the engines in a neutral position for launch

6. Processes the GRR interrupt and transfers LVDC control to the flight program

7. Samples platform gimbal angles.

At T-22 seconds, the launch sequencer issues a GRR alert signal to the LVDC and GCC. At T-17 seconds, a GRR interrupt signal is sent to the LVDC and GCC. With the receipt of this signal, the PTL routine transfers control of the LVDC to the flight program.

When the GRR interrupt is received by the LVDC, the following events take place:

1. The LVDC sets time base zero ($T_0$).

2. Gimbal angles and accelerometer values are sampled and stored for use by flight program routines.

3. Time and accelerometer readings are telemetered.

4. All flight variables are initialized.

5. The GCC is signaled that the LVDC is under control of the flight program.

During the time period between GRR and liftoff, the LVDC begins to perform navigational calculations, and processes the minor loops. At T-8.9 seconds, engine ignition command is issued. At T-0, liftoff occurs, and a new time base ($T_1$) is initiated.

## POWERED FLIGHT MAJOR LOOP

The major loop contains the navigation and guidance calculations, timekeeping, and other repetitive operations of the flight program. Its various routines are subdivided by function. Depending upon mode of operation and time of flight, the program will follow the appropriate sequence of routines.

The accelerometer processing routine accomplishes two main objectives: it accumulates velocities as measured by the platform, and detects velocity measurement errors through "reasonableness" tests.

The boost navigation routine combines gravitational acceleration with measured platform data to compute position and velocity.

The "pre-iterative" guidance mode, or "time-tilt" guidance program, is that part of the flight program which performs from GRR until the end of the S-IC burn. The guidance commands issued during the time-tilt phase are functions of time only. This phase of the program is referred to as open loop guidance, since vehicle dynamics do not affect or influence the guidance commands. When the launch vehicle has cleared the mobile launcher, the time-tilt program first initiates a roll maneuver to align the vehicle with the proper azimuth. After this command, roll and yaw commands remain at zero, and the vehicle is gradually pitched about the vehicle's Y axis to its predetermined boost heading. Rate limiting of the output commands prevents the flight control system from maneuvering the LV at rates which exceed safe limits.

The iterative guidance mode (IGM) routine, or "path adaptive" guidance, commences after second/stage ignition, and continues until the end of S-IVB first burn. Cutoff occurs when the velocity required for earth orbit has been reached. IGM is used again during S-IVB second burn. IGM is based on optimizing techniques, using the calculus of variations to determine a minimum propellant flight path which satisfies mission requirements. Since the IGM considers vehicle dynamics, it is referred to as closed loop guidance.

## INTERRUPTS

An interrupt routine permits interruption of the normal program operation to free the LVDC for priority work, and may occur at any time within the program. When an interrupt occurs, the interrupt transfers LVDC control to a special subroutine which identifies the interrupt source, performs the necessary subroutines, and then returns to the point in the program where the interrupt occurred. Figure 7-19 is a list of interrupts in the order of decreasing priority.

| INTERRUPTS | |
|---|---|
| Decreasing Priority | Function |
| 1 | Minor Loop Interrupt |
| 2 | Switch Selector Interrupt |
| 3 | Computer Interface Unit Interrupt |
| 4 | Temporary Loss of Control |
| 5 | Command Receiver Interrupt |
| 6 | Guidance Reference Release |
| 7 | S-II Propellant Depletion/Engine Cutoff |
| 8 | S-IC Propellant Depletion/Engine Cutoff "A" |
| 9 | S-IVB Engine Out "B" |
| 10 | Program Re-Cycle (RCA-110A Interrupt) |
| 11 | S-IC Inboard Engine Out "A" |
| 12 | Command LVDA/RCA-110A Interrupt |

Figure 7-19

## TELEMETRY ROUTINE

A programmed telemetry feature is also provided as a method of monitoring LVDC and LVDA operations. The telemetry routine transmits specified information and data to the ground via IU telemetry equipment. In orbit, telemetry data must be stored at times when the vehicle is not within range of a ground receiving station. This operation is referred to as data compression. The stored data is transmitted on a time-shared basis with real-time telemetry when range conditions are favorable.

## DISCRETE BACKUPS

Certain discrete events are particularly important to the flight program, since they periodically reset the computer time base which is the reference for all sequential events. These significant time base events are:

$T_1$    Liftoff (LO)
$T_2$    S-IC center engine cutoff (CECO)
$T_3$    S-IC outboard engine cutoff (OECO)
$T_4$    S-II cutoff
$T_5$    S-IVB cutoff (boost phase)
$T_6$    S-IVB restart
$T_7$    S-IVB cutoff (orbital phase)

Since switch selector outputs are a function of time (relative to one of the time bases), no switch selector output could be generated if one of the discrete signals were missed. A backup routine is provided to circumvent such a failure. The discrete backup routine simulates these critical signals if they do not occur when expected.

In the cases of the backup routines for LO and CECO, special routines are established as a double safety check. In both cases, motion as well as time are confirmed before a backup discrete is used. For LO, the backup routine is entered 17.5 seconds after GRR. If the vertical acceleration exceeds 6.544 ft/sec$^2$ for four computation cycles, the vehicle is assumed to be airborne and the liftoff discrete is issued. For CECO, an assurance is made than an on-the-pad firing of the S-II stage cannot occur if $T_1$ is accidentally set. Before $T_2$ can be initiated, velocity along the downrange axis is tested for a minimum of 500 m/sec.

Refer to MODE AND SEQUENCE CONTROL, Section VII, for a discussion of the discrete time bases $T_1$ through $T_7$.

The execution time for any given major loop, complete with minor loop computations and interrupts, is not fixed. The average execution time for any given major loop in powered flight, complete with minor loop computation and interrupt processing, is called the normal computation cycle for that mode. The computation cycle is not fixed for two reasons. First, the various flight modes of the program have different computation cycle lengths. Second, even in a given flight mode, the uncertainties of discrete and interrupt processing, and the variety of possible paths in the loop preclude a fixed computation cycle length.

## MODE AND SEQUENCE CONTROL

Mode and sequence control involves most of the electrical/electronic systems in the launch vehicle. However, in this section the discussion will deal mainly with the switch selector and associated circuitry.

The LVDC memory contains a predetermined number of sets of instructions which, when initiated, induce portions of the launch vehicle electrical/electronic systems to operate in a particular mode. Each mode consists of a predetermined sequence of events. The LVDC also generates appropriate discrete signals such as engine ignition, engine cutoff, and stage separation.

Mode selection and initiation can be accomplished by an automatic LVDC internal command, an external command from ground checkout equipment or IU command system, or by the flight crew in the spacecraft.

The flexibility of the mode and sequence control scheme is such that no hardware modification is required for mode and flight sequence changes. The changes are accomplished by changing the instructions and programs in the LVDC memory.

## SWITCH SELECTOR

Many of the sequential operations in the launch vehicle that are controlled by the LVDC are performed through a switch selector located in each stage. The switch selector decodes digital flight sequence commands from the LVDA/LVDC, and activates the proper stage circuits to execute the commands. The outputs of the switch selector drive relays either in the units affected or in the stage sequencer.

Each switch selector can activate, one at a time, 112 different circuits in its stage. The selection of a particular stage switch selector is accomplished through the command code. Coding of flight sequence commands and decoding by the stage switch selectors reduces the number of interface lines between stages, and increases the flexibility of the system with respect to timing and sequence. In the launch vehicle, which contains four switch selectors, 448 different functions can be controlled, using only 28 lines from the LVDA. Flight sequence commands may be issued at time intervals of 100 milliseconds.

To maintain power isolation between vehicle stages, the switch selectors are divided into sections. The input sections (relay circuits) of each switch selector receive their power from the IU. The output sections (decoding circuitry and drivers) receive their power from the stage in which the switch selector is located. The inputs and outputs are coupled together through a diode matrix. This matrix decodes the 8-bit input code, and activates a transistorized output driver, thus producing a switch selector output.

The output signals of the LVDA switch selector register, with the exception of the 8-bit command, are sampled at the control distributor in the IU and sent to IU PCM telemetry. Each switch selector also provides three outputs to the telemetry system within its stage.

The switch selector is designed to execute flight sequence commands given by the 8-bit code or by its complement. This feature increases reliability and permits operation of the system, despite certain failures in the LVDA switch selector register, line drivers, interface cabling, or switch selector relays.

The flight sequence commands are stored in the LVDC memory, and are issued according to the flight program. When a programmed input/output instruction is given, the LVDC loads the 15-bit switch selector register with the computer data.

The switch selector register, bits 1 through 8, represents the flight sequence command. Bits 9 through 13 select the switch selector to be activated. Bit 14 resets all the relays in the switch selectors in the event data transfer is incorrect, as indicated by faulty verification information received by the LVDA. Bit 15 activates the addressed switch selector for execution of the command. The switch selector register is loaded in two passes by the LVDC: bits 1 through 13 on the first pass, and either bit 14 or bit 15 on the second pass, depending on the feedback code. The LVDA/LVDC receives the complement of the code after the flight sequence command (bits 1 through 8) has been picked up by the input relays of the switch selector. The feedback (verification information) is returned to the LVDA, and compared with the original code in the LVDC. If the feedback agrees, the LVDC/LVDA sends a read command to the switch selector. If the verification is not correct, a reset command is given (forced reset), and the LVDC/LVDA reissues the 8-bit command in complement form.

Figure 7-20 illustrates the Saturn V switch selector functional configuration. All switch selector control lines are connected through the control distributor in the IU to the LVDC and the electrical support equipment.

The LVDC switch selector interconnection diagram is shown in figure 7-21. All connections between the LVDA and the switch selectors, with the exception of the stage select inputs, are connected in parallel.

## OPERATION SEQUENCE

The Saturn V operation sequence starts during the prelaunch phase at approximately T-24 hours, when the electrical power from the ground support equipment is applied to all stages of the launch vehicle. During this time, the sequencing is controlled from the launch control center/mobile launcher complex, utilizing both manual and automatic control to check out the functions of the entire launch vehicle. After the umbilicals are disconnected, the sequencing is primarily controlled by the flight program within the LVDC.

Since flight sequencing is time phased, the sequencing operation is divided into seven primary time bases. Each time base is related to a particular flight event. Four alternate time bases have also been provided for. These time bases are defined as follows:

### Time Base No. 1 ($T_1$)

$T_1$ is initiated by a liftoff signal, provided by deactuation of the liftoff relay in the IU at umbilical disconnect. However, as a safety measure, the LVDC will not recognize the liftoff signal and start $T_1$ prior to receiving guidance reference release (GRR) plus 16.0 seconds.

A backup method for starting $T_1$ is provided should the LVDC fail to receive or recognize the liftoff signal. If $T_1$ is not initiated within 17.5 seconds after GRR, the LVDC will monitor the vertical accelerometer. If a significant positive acceleration (in excess of 1 g) exists, the LVDC assumes liftoff has occurred and begins $T_1$. A compensating time adjustment is made by the LVDC.

No "negative backup" (i.e., provisions for the LVDC to return to prelaunch conditions) is provided because, in the event $T_1$ began by error, the launch vehicle could safety complete $T_1$ on the pad without catastrophic results.

### Time Base No. 2 ($T_2$)

The S-IC center engine is cut off by the LVDC, through the S-IC switch selector, at a predetermined time. At this time, the LVDC monitors the downrange accelerometer. If sufficient downrange velocity exists, the LVDC will start $T_2$.

Use of the downrange velocity reading provides a safeguard against starting $T_2$ on the pad, should $T_1$ be started without liftoff. Furthermore, if $T_2$ is not established, no subsequent time bases can be started. This ensures a safe vehicle, requiring at least one additional failure to render the vehicle unsafe on the pad.

### Time Base No. 3 ($T_3$)

$T_3$ is initiated at S-IC outboard engine cutoff by either of two redundant outboard engines cutoff signals. However, the LVDC must arm outboard engines propellant depletion cutoff prior to starting $T_3$. Outboard engines propellant depletion cutoff is armed several seconds prior to calculated outboard engines cutoff.

# SWITCH SELECTOR FUNCTIONAL CONFIGURATION

NOTES:   SIGNAL RETURN LINES FROM THE SWITCH SELECTORS, THROUGH THE CONTROL
         DISTRIBUTOR, TO THE LVDA ARE NOT SHOWN IN THIS FIGURE.
         THE LETTERS USED TO LABEL INTERSTAGE CONNECTIONS BETWEEN UNITS ARE
         NOT ACTUAL PIN OR CABLE CONNECTORS.  THE LETTER CODE IS DENOTED
         BELOW:

         a = 8-DIGIT COMMAND (8 LINES)
         b = FORCE RESET (REGISTER) (1 LINE + 1 REDUNDANT LINE)
         c = REGISTER VERIFICATION (8 LINES)
         d = READ COMMAND (1 LINE + 1 REDUNDANT LINE)
         e =
         f =        STAGE SELECT LINES
         g =      (1 LINE + 1 REDUNDANT LINE)
         h =
         j = b, c, d, e, f, g, AND h TO IU TELEMETRY
         k = REGISTER TEST
             ZERO INDICATE     TO STAGE TELEMETRY
             SW SEL OUTPUT      (1 LINE EACH)
         m = +28 VDC FROM THE INSTRUMENT UNIT

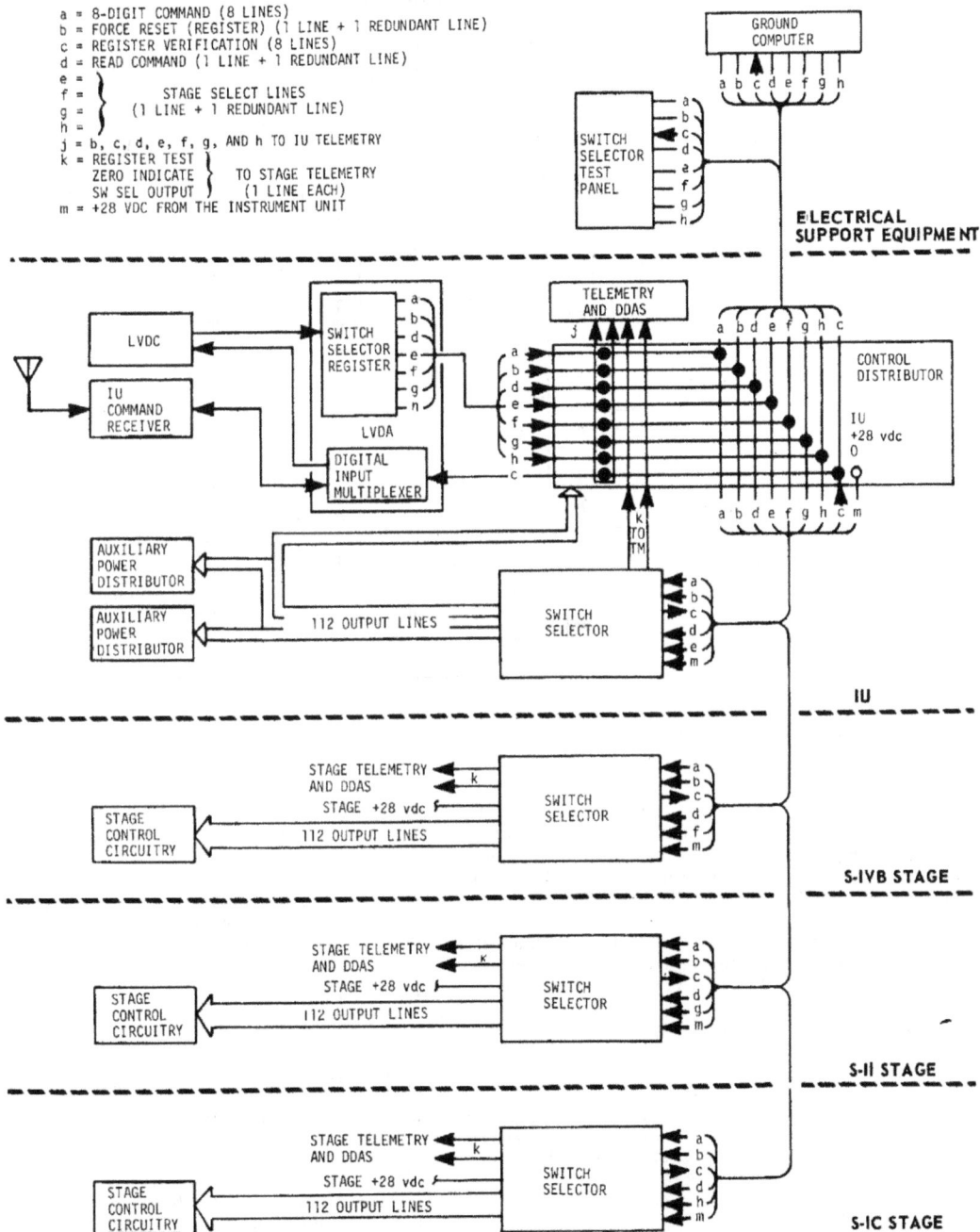

Figure 7-20

# LVDC-SWITCH SELECTOR INTERCONNECTION DIAGRAM

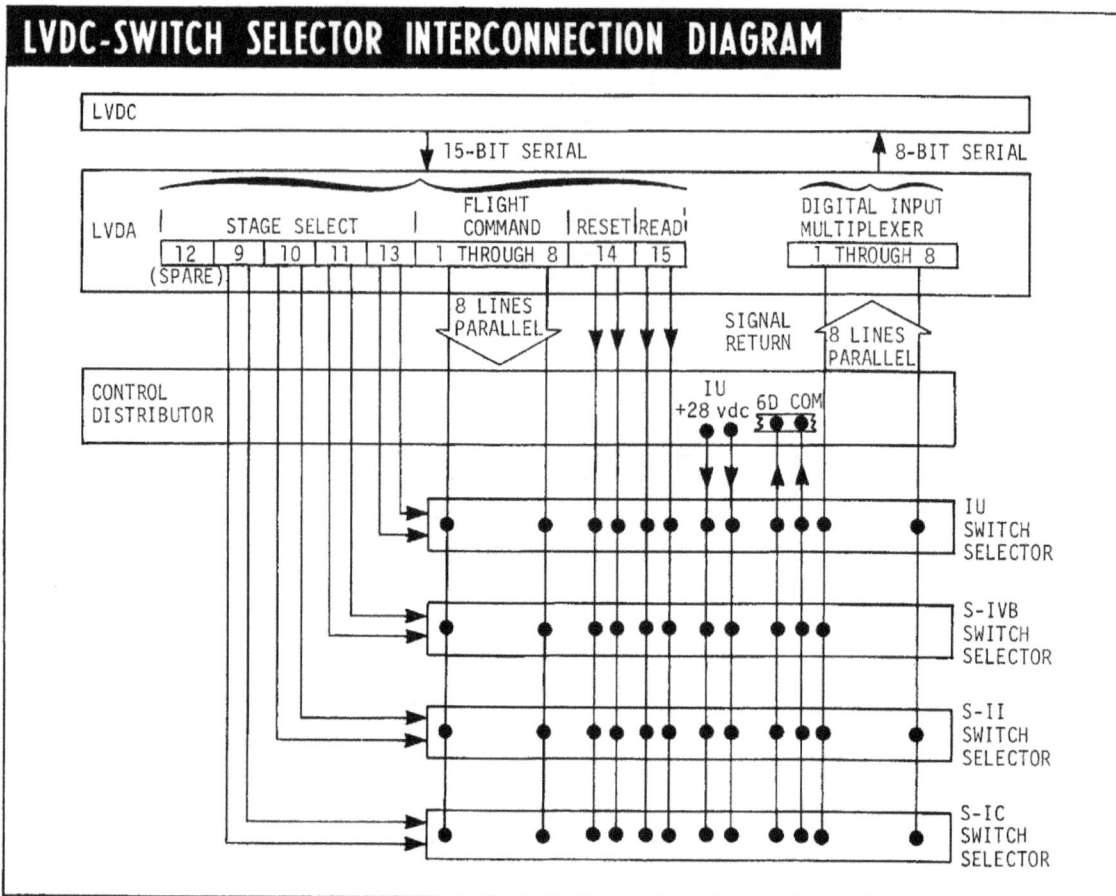

Figure 7-21

## Time Base No. 4 (T4)

After arming S-II lox depletion cutoff, the LVDC initiates $T_4$ upon receiving either of two signals: S-II engines cutoff, or S-II engines out. The S-II engines cutoff signal is the primary signal for starting $T_4$. The S-II engines out signal from the thrust OK circuitry is a backup.

As a safeguard against trying to separate the S-II stage with the thrust of the engines present, a redundant S-II engines cutoff command is issued by the LVDC at the start of $T_4$.

## Alternate Time Base No. 4a (T4a)

$T_{4a}$ is programmed for use in early staging of the S-IVB stage. This time base is initiated by the LVDC upon receiving either of two signals: spacecraft initiation of S-II/S-IVB separation "A", or spacecraft initiation of S-II/S-IVB separation "B". Starting of $T_{4a}$ is inhibited until $T_3$ + 1.4 seconds.

## Time Base No. 5 (T5)

$T_5$ is initiated by the deactuation of the S-IVB thrust OK pressure switches at S-IVB cutoff. The LVDC starts $T_5$ after

receiving any two of four functions monitored by the LVDC. The functions are: S-IVB engine out "A"; S-IVB engine out "B"; S-IVB velocity cutoff, which is issued by the LVDC; and/or loss of thrust, determined by LVDC using accelerometer readings.

A redundant S-IVB cutoff command is issued at the start of $T_5$ as a safeguard against having started time base 5 with the thrust of the S-IVB engine present.

## Time Base No. 6 (T6)

$T_6$ is initiated by the LVDC upon solving the restart equation with the prerequisite of approximately two hours 30 minutes in time base 5. However, the start of $T_6$ can be inhibited by the TRANSLUNAR INJECTION INHIBIT signal from the spacecraft.

With spacecraft control of the launch vehicle (guidance switchover), $T_6$ is initiated by the LVDC upon solving the restart equation, or it can be initiated by the spacecraft S-IVB IGNITION SEQUENCE START signal. Start of $T_6$, in this case, can also be inhibited by the spacecraft TRANSLUNAR INJECTION INHIBIT signal.

Alternate Time Base No. 6a (T$_{6a}$)

T$_{6a}$ is programmed for use should the oxygen-hydrogen burner malfunction between the times T$_6$ + 48 seconds and T$_6$ + 5 minutes and 41.3 seconds. This alternate time base is initiated by the LVDC upon receiving an oxygen-hydrogen burner malfunction signal from the S-IVB stage. Upon completion of T$_{6a}$, the LVDC returns to T$_6$.

Alternate Time Base No. 6b (T$_{6b}$)

T$_{6a}$ is programmed for use should the oxygen-hydrogen burner malfunction between the times T$_6$ + 48 seconds and T$_6$ + 5 minutes and 41.3 seconds. This alternate time base is initiated by the LVDC upon receiving an oxygen-hydrogen burner malfunction signal from the S-IVB stage. Upon completion of T$_{6a}$, the LVDC returns to T$_6$.

Alternate Time Base No. 6c (T$_{6c}$)

T$_{6c}$ is programmed for use should a failure occur which would require a delay in the S-IVB restart attempt. The TRANSLUNAR INJECTION INHIBIT signal from the spacecraft is required by the LVDC before this alternate time base is initiated. Upon completion of T$_{6c}$, the LVDC returns to T$_5$, updated by time elapsed in T$_6$ and T$_{6c}$.

Time Base No. 7 (T$_7$)

After a predetermined time, sufficient to allow the S-IVB engine to establish thrust OK, the LVDC starts T$_7$ after receiving any two of four functions monitored by the LVDC. The functions are: S-IVB engine out "A"; S-IVB engine out "B"; S-IVB engine cutoff, which is issued by the LVDC; and/or loss of thrust, determined by LVDC using accelerometer readings.

As a safeguard against starting T$_7$ with the thrust of the S-IVB engine present, a redundant S-IVB engine cutoff command is issued at the start of T$_7$.

## MEASUREMENTS AND TELEMETRY

The instrumentation within the IU consists of a measuring subsystem, a telemetry subsystem, and an antenna subsystem. This instrumentation is for the purpose of monitoring certain conditions and events which take place within the IU, and for transmitting monitored signals to ground receiving stations. Telemetry data is used on the ground for the following purposes:

1. Prior to launch, to assist in the checkout of the launch vehicle

2. During vehicle flight, for immediate determination of vehicle condition, and for verification of commands received by the IU command system

3. Postflight scientific analysis of the mission.

## MEASUREMENTS

The requirement for measurements of great variety and quantity has dictated the use of transducers of many types and at many locations. However, a discussion of the transducers used requires a level of detail beyond the scope of this manual.

Figure 7-22 provides a summary of the measurements made in the IU.

| IU MEASUREMENT SUMMARY | |
|---|---|
| TYPE | QTY |
| Acceleration | 4 |
| Acoustics | 1 |
| Angular Velocity | 33 |
| Control | 21 |
| Flow Rate | 11 |
| Guidance | 33 |
| Position | 21 |
| Pressure | 14 |
| RF and Telemetry | 26 |
| Signals | 71 |
| Strain | 16 |
| Temperature | 57 |
| Vibration | 28 |
| Voltage, Current, Frequency | 20 |
| TOTAL | 356 |

Figure 7-22

Signal conditioning is accomplished by amplifiers or converters located in measuring racks. There are ten measuring racks in the IU, and 20 signal conditioning modules in each. Each signal conditioning module contains, in addition to its conditioning circuitry, two relays and circuitry to simulate its transducers at both their high range and low range extremities. These relays, and transducer simulation circuitry are used for prelaunch calibration of the signal conditioners.

Conditioned signals are routed to their assigned telemetry channel by the measuring distributors. Switching functions connect different sets of measurements to the same telemetry channels during different flight periods. These switching functions, controlled from the ground through the umbilical, connect measurements not required during flight to digital data acquisition system channels for ground checkout, and return the channels to flight measurements after checkout.

## TELEMETRY

The function of the telemetry system is to format and transmit measurement signals received from the measuring distributor.

The approximately 350 measurements made on the IU are transmitted via four telemetry links. The three modulation techniques used are:

1. Pulse Code Modulation/Frequency Modulation (PCM/FM)

2. Frequency Modulation/Frequency Modulation (FM/FM)

3. Single Sideband/Frequency Modulation (SS/FM)

The PCM/FM data is transmitted over a VHF band, a UHF band, and the CCS transponder. The SS/FM and the two FM/FM links are transmitted over separate VHF bands.

Multiplexing

In order for the four IU telemetry links to handle in excess of 350 separate measurements, these links must be "shared". By proper multiplexing, it is possible to transmit several different signals simultaneously from one telemetry system. Both frequency sharing and time sharing multiplexing techniques are used to accomplish this. Refer to figure 7-23 for a block diagram of the IU telemetry system.

Two Model 270 multiplexers (MUX-270) are used in the IU telemetry system. The MUX-270 is a time sharing

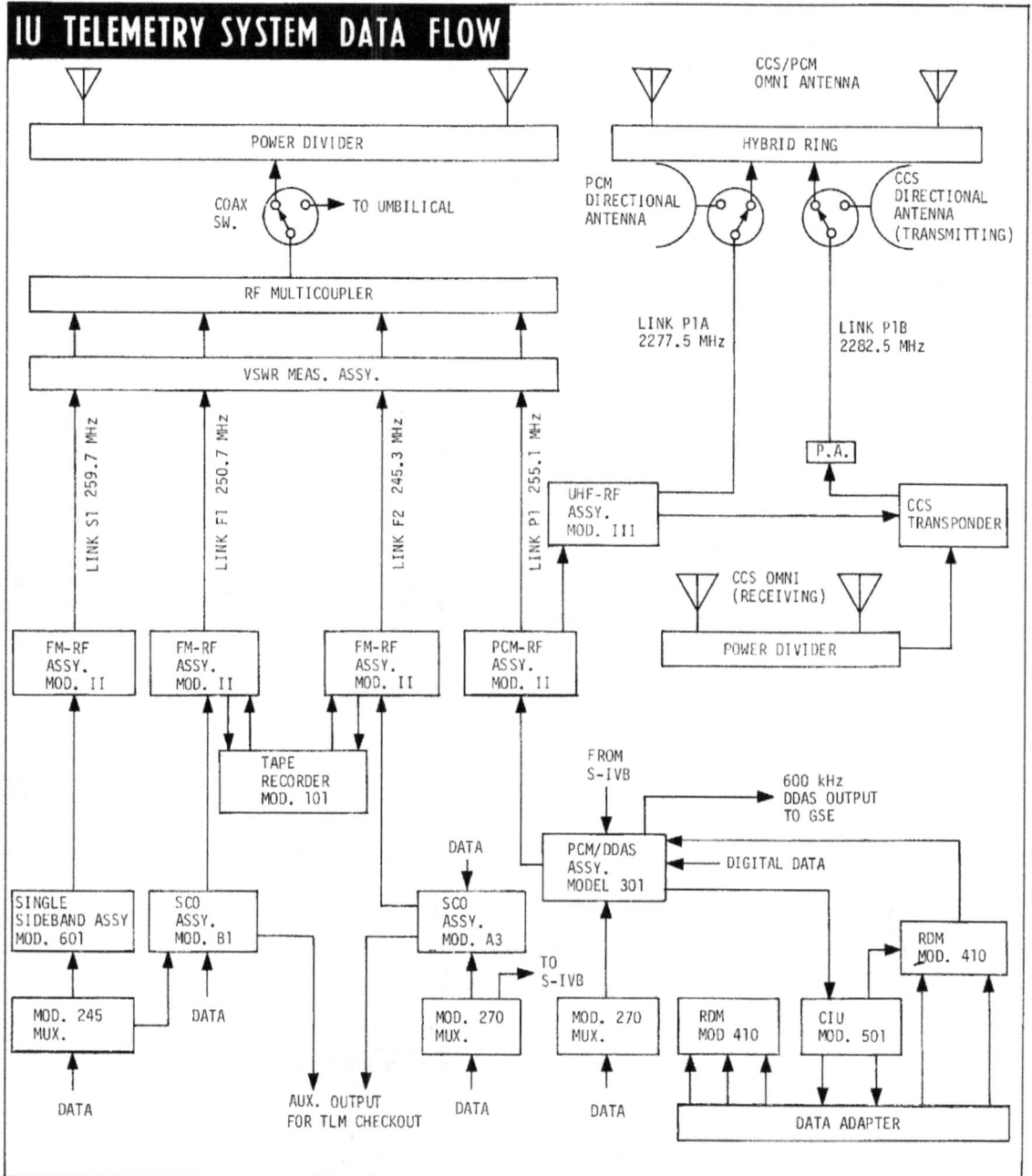

Figure 7-23

multiplexer. Each one operates as a 30 x 120 (30 primary channels, each sampled 120 times per second) multiplexer with provisions for submultiplexing individual primary channels to form ten subchannels, each sampled at 12 times per second. Twenty-seven of the 30 primary channels are used for analog data, while the remaining three are used for references. Ten-channel submultiplexer modules, which plug into the MUX-270, can be use; to submultiplex any primary data channel, providing a 10 to 1 increase in the quantity of channels, in exchange for a 10 to 1 decrease in sampling rates. Any proportion of the 23 data channels can be submultiplexed or sampled at the 120 per second rate.

The MUX-270 also has an integral calibration generator for inflight calibration capability. Upon command, the calibration generator seeks the next available master frame, and applies a sequence of five calibration voltages to all data channels. Each level is sustained for one master frame, and approximately 400 milliseconds are required for the full sequence.

The Model 245 multiplexer (MUX-245) is also a time sharing type. The MUX-245 has provisions for up to 80 analog data inputs. Outputs are read out on 16 channels. Each of these 16 channels can provide a continuous single data signal or a time shared output of two, four, or five data signals on a twelve second repetition cycle. The number of time shared outputs is dependent upon plug-in modules, of which there are four types: the dummy module connects an input directly to an output channel; the two-channel, four-channel, and five-channel modules provide time sharing, as indicated by their numbers.

The Model 410 remote digital multiplexer (RDM-410) accepts ten 10-bit parallel words, and transfers this data to the Model 301 PCM/DDAS according to a programmed format.

During launch vehicle staging, retrorockets create patches of ionized atmosphere which interfere with reliable transmission of telemetry data. To preclude loss of data, or transmission of incorrect data, telemetry transmission is stopped at these times, and the data is put on magnetic recording tape for transmission at a more favorable time. The tape recorder which accomplishes this is the Model 101, a two-channel unit. Tape speed is 60 inches per second, with a recording time of 180 seconds.

Low level conditioned analog signals are fed to subcarrier oscillators (SCO). The Model B1 SCO has a capacity of 28 continuous data channel inputs. Each input is applied to a separate channel within the SCO, and each channel produces a different output frequency. These output frequencies are combined within the SCO assembly, and the composite signal is used to frequency modulate an FM-RF assembly producing the FM/FM telemetry signals.

The SS/FM telemetry link is made up of an FM-RF unit (amplitude) modulated by SS subcarriers. The Model 601 single sideband unit (SS-601) has a capacity of 15 channels. A MUX-245 furnishes inputs to the SS-601 which, in turn, modulates the FM-RF unit to produce the SS/FM telemetry link. The MUX-245 sequentially samples 75 signals, and provides time sharing for the 15 channel SS-601 unit.

The PCM/FM system performs a dual function. During flight, it serves as a telemetry link, and during prelaunch checkout, it serves as an IU interface with the digital GSE. PCM techniques provide the high degree of accuracy required for telemetering certain signal types. The PCM-301 unit accepts analog inputs from MUX-270 or RDM-410 units, or direct inputs in digital form. All inputs are digitized and encoded. Output of the PCM-301 unit is a serial train of digital data which modulates the PCM-RF transmitter.

All of the RF assemblies are essentially the same. All use combinations of solid state and vacuum tube electronics. Frequency outputs of each unit are, of course, different, and are applied to the antenna subsystem.

## ANTENNA SUBSYSTEM

The antenna subsystem includes that equipment from the output of the RF units through the radiating or receiving elements.

### VSWR Monitor

Voltage standing wave ratio (VSWR) between the antenna and RF unit is checked and telemetered to the ground. The VSWR is a ratio of RF power being reflected, and is an indication of the efficiency of the RF unit, transmission coupling, and antenna elements.

### Multicoupler

The multicoupler simultaneously couples two, three, or four RF signals into a common antenna without mutual interference and with maximum efficiency.

### Coaxial Switch

The coaxial switch connects a coaxial transmission line to either of two other coaxial transmission lines.

### Power Divider

The power divider splits RF power from one transmission line equally into two other transmission lines.

## RADIO COMMAND SYSTEM

### COMMAND COMMUNICATIONS SYSTEM (CCS)

The CCS provides for digital data transmission from ground stations to the LVDC. This communications link is used to update guidance information, or to command certain other functions through the LVDC. Command data originates in the Mission Control Center, and is sent to remote stations of the MSFN for transmission to the launch vehicle.

At the time of spacecraft separation from the IU/S-IVB, the IU CCS transmitter will be commanded off for a short period of time, to preclude interference with the spacecraft S-band transponder. After adequate separation of the spacecraft and IU/S-IVB, the IU/S-IVB, the IU CCS transmitter will be commanded on again, to provide for psuedo-random noise (PRN) turnaround ranging, to facilitate IU/S-IVB tracking.

The CCS equipment located in the IU consists of:

1. Antenna systems
    a. transmitting and receiving
    b. directional and omni

2. Antenna switching elements
    a. coaxial switches
    b. hybrid rings
    c. power divider

3. S-band transponder:
    a phase coherent receiver-transmitter

4. Command decoder, which precludes unauthorized command data entry.

Figure 7-24 is a block diagram of the overall CCS. Command messages are transmitted from the unified S-band ground stations on a carrier frequency of 2101.8 MHz, modulated by a subcarrier of 70 kHz, which is modulated by a digital message. The transmitted message is received by the airborne transponder, where demodulation is accomplished. The resulting digital message is passed on to the command decoder, where it is checked for authenticity before being passed to the LVDC. The IU PCM telemetry system verifies receipt of the message.

The command decoder and the LVDC are programmed for the acceptance of seven different message types. These seven message types are:

1. Update LVDC
2. Execute update
3. Enter switch selector mode
4. Enter closed-loop test
5. Execute subroutine command (e.g., telemeter flight control measurements)
6. Memory sector dump
7. Telemeter single memory address.

## SATURN TRACKING INSTRUMENTATION

The purpose of radio tracking is the determination of the vehicle's trajectory. Tracking data is used for mission control, range safety, and post-flight evaluation of vehicle performance.

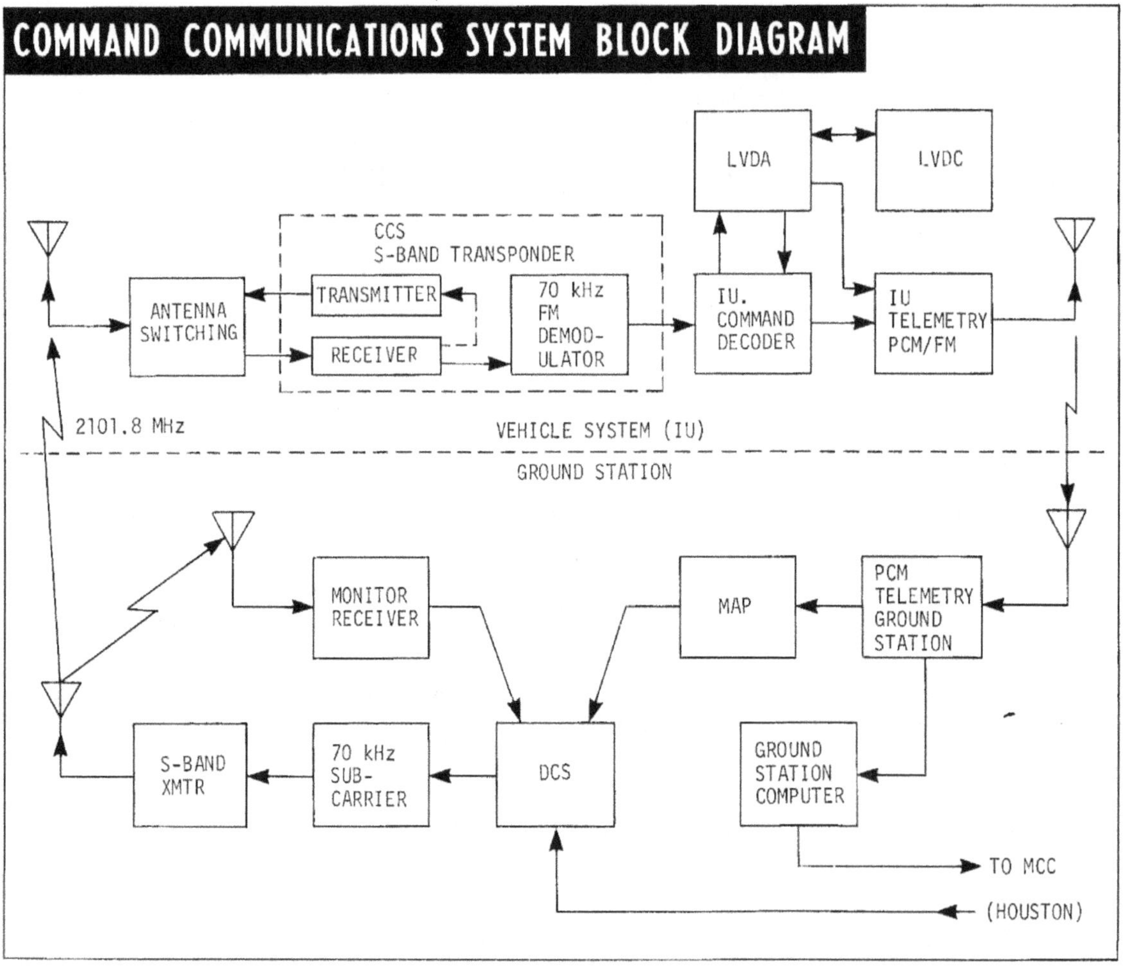

Figure 7-24

The Saturn V IU carries two C-band radar transponders. The tracking of the launch vehicle may be divided into four phases: powered flight into earth orbit, orbital flight, injection into mission trajectory, and coast flight after injection.

Continuous tracking is required during powered flight into earth orbit. Because of the long burning time (approximately 11 minutes and 17 seconds) of the three-stage launch vehicle, the end of the powered flight phase cannot be covered sufficiently from land-based tracking stations. Therefore, tracking ships will be located in the Atlantic to obtain the tracking data, during insertion, which is required for orbit determination. The number of stations which can "see" the vehicle depends on the launch azimuth.

In addition, the launch vehicle will be tracked from S-band stations at Cape Kennedy and on the Atlantic tracking ships. These stations have dual tracking capability: i.e., they can simultaneously track the two S-band transponders on the vehicle, one in the IU and the other in the Apollo spacecraft. The S-band station on Bermuda has only a single capability, and will track the Apollo spacecraft transponder. Refer to Radio Command Systems for additional information on the S-band equipment.

During orbital flight, tracking is accomplished by S-band stations of the MSFN and by C-band radar stations. The S-band stations, including the Deep Space Instrumentation Facility, can track the Apollo spacecraft to the moon, and will also be involved in tracking after injection. Tracking information collected during orbital flight may be used to update the Saturn guidance before injection into mission trajectory.

## C-BAND RADAR

The function of the C-band radar transponder is to increase the range and accuracy of the radar ground stations equipped with AN/FPS-16, and AN/FPQ-6 radar systems. C-band radar stations at the Kennedy Space Center, along the Atlantic Missile Range, and at many other locations around the world, provide global tracking capabilities. Two C-band radar transponders are carried in the IU to provide radar tracking capabilities independent of the vehicle attitude. This arrangement is more reliable than the antenna switching circuits necessary if only one transponder were used.

The transponder consists of a single, compact package. Major elements include an integrated RF head, an IF amplifier, a decoder, overinterrogation protection circuitry, a fast recovery solid-state modulator, a magnetron, a secondary power supply, and transducers for telemetry channels. The complete unit weighs 5.5 pounds, and has a volume of only 100 cubic inches.

The transponder receives coded or single-pulse interrogation from ground stations, and transmits a single-pulse reply in the same frequency band.

Six conditioned telemetry outputs are provided: input signal level, input PRF, temperature, incident power, reflected power, and reply PRF.

The characteristics of the C-band radar transponder are given in figure 7-25.

## GROUND SUPPORT EQUIPMENT

The IU, because of its complex nature, requires the services of many types of GSE (mechanical, pneudraulic, electrical, electronic) and personnel. This section of the manual is limited to a very brief description of the IU GSE.

There are three primary interfaces between the IU and its GSE. One is the IU access door, used during prelaunch preparations for battery installation, ordnance servicing, servicing IU equipment, S-IVB forward dome and LEM servicing. The second interface is the umbilical, through which the IU is furnished with ground power, purging air/GN$_2$ methanol-water for environmental control, and hardwire links with electrical/electronic checkout equipment. The third interface is the optical window, through which the guidance system ST-124-M3 stable platform is aligned.

## IU ACCESS DOOR

The structure of the IU consists of three 120-degree segments of aluminum honeycomb sandwich, joined to form a cylindrical ring. After assembly of the IU, a door assembly provides access to the electronic equipment inside the structure. This access door has been designed to act as a load supporting part of the structure in flight.

Work platforms, lights, and air-conditioning are used inside the IU to facilitate servicing operations. When the spacecraft is being fueled through the IU access door, a special protective cover is installed inside the IU to protect components from any possible volatile fuel spillage.

Approximately 20 hours before launch, the IU flight batteries, each weighing 165 pounds, are activated in the battery shop and installed in the IU through the access door.

At approximately T-6 hours, the service equipment is removed and the access door is secured.

## IU UMBILICAL

The physical link between the IU and the GSE is through the umbilical connection, located adjacent to the access door. The umbilical is made up of numerous electrical connectors, two pneudraulic couplings and an air conditioning duct. The electrical connectors provide ground power and the electrical/electronic signals necessary for prelaunch checkout of the IU equipment. The pneudraulic couplings provide for circulation of GSE supplied methanol-water coolant fluid for the IU/S-IVB ECS. The air conditioning duct provides for compartment cooling air or purging GN$_2$.

The umbilical is retracted at liftoff, and a spring loaded door on the IU closes to cover the connectors.

## OPTICAL ALIGNMENT

The IU contains a window through which the ST-124-M3 stable platform has its alignment checked and corrected by a theodolite located in a hut on the ground and a computer feedback loop. By means of this loop, the launch azimuth can be monitored, updated and verified to a high degree of accuracy.

| C-BAND TRANSPONDER CHARACTERISTICS | |
|---|---|
| Receiver Characteristics | |
| Frequency (Tunable externally) | 5400 to 5900 MHz (set to 5690 $\pm$8 MHz) |
| Frequency stability | $\pm$2.0 MHz |
| Bandwidth (3 db) | 10 MHz |
| Off-frequency rejection | 50 db image; 80 db minimum, 0.15 to 10,000 MHz |
| Sensitivity (99% reply) | -65 dbm over entire frequency range and all environments |
| Maximum input signal | -20 dbm |
| Interrogation code | Single or double pulse |
| Pulse width | 0.2 to 5.0 $\mu$sec (single phase), 0.2 to 1.0 $\mu$sec (double pulse) |
| Pulse spacing | Continuously settable between 5 and 12 $\mu$sec (set to 8 $\pm$0.05 $\mu$sec) |
| Decoder limits | $\pm$0.25 $\mu$sec accept, $\pm$ 0.85 $\mu$sec reject (5 to 12 $\mu$sec) |
| Transmitter Characteristics | |
| Frequency (Tunable externally) | 5400 to 5900 MHz (set to 5765 $\pm$2 MHz) |
| Peak power output | 400 watts minimum, 700 watts nominal |
| Pulse width | 1.0 $\pm$ 0.1 $\mu$sec |
| Pulse jitter | 0.020 $\mu$sec maximum for signals above -55 dbm |
| Pulse rise time (10% to 90%) | 0.1 $\mu$sec maximum |
| Duty cycle | 0.002 maximum |
| VSWR of load | 1.5:1 maximum |
| Pulse repetition rate | 10 to 2000 pps; overinterrogation protection allows interrogation at much higher rates with count-down; replies during overinterrogation meet all requirements |
| Transponder Characteristics | |
| Recover time | 50 $\mu$sec single pulse, 62 $\mu$sec double pulse maximum for input signal levels differing by up to 65 db (recovers to full sensitivity with no change in transmitter reply power or frequency with multiple radars interrogating simultaneously) |
| Fixed delay | Settable 2 $\pm$ 0.1 and 3.0 to 0.01 $\mu$sec (set to 3.0 $\pm$ 0.01 $\mu$sec) |
| Delay variation with signal level | 50 nanoseconds maximum from -65 dbm to 0 dbm |
| Power requirements | 24 to 30 volts |
| Primary current drain | 0.7 ampere standby; 0.9 ampere at 1000 pps |
| Weight | 5.5 lbs |

Figure 7-25

## IU/SLA INTERFACE

### MECHANICAL INTERFACE

The IU and spacecraft-LM adapter (SLA) are mechanically aligned with three guide pins and brackets as shown in figure 7-26. These pins facilitate the alignment of the close tolerance interface bolt holes, as the two units are joined during vehicle assembly. Six bolts are installed around the circumference of the interface and sequentially torqued, using a special MSFC designed wrench assembly. These six bolts secure the IU/SLA mechanical interface. (See figure 7-27.)

### ELECTRICAL INTERFACE

The electrical interface between the IU and spacecraft consists of three 61 pin connectors. (see figure 7-28.) The

definition and function of each connector is presented in the following paragraphs.

#### IU/Spacecraft Interface Connector J-1

This connector provides lines for power, control, indication circuitry and EDS circuitry.

#### IU/Spacecraft Interface Connector J-2

This connector provides lines for power, control and indications for the Q-ball circuitry and the EDS circuitry.

#### IU/Spacecraft Interface Connector J-3

This connector provides lines for power, control, and indication circuitry and EDS circuitry.

# IU/SLA ALIGNMENT

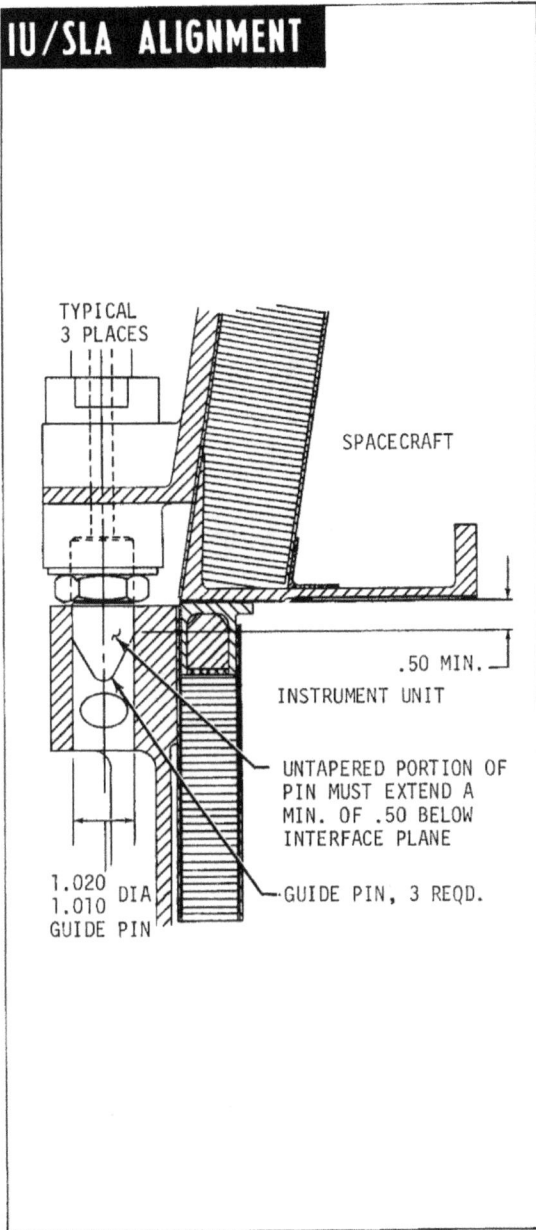

TYPICAL
3 PLACES

SPACECRAFT

.50 MIN.

INSTRUMENT UNIT

UNTAPERED PORTION OF
PIN MUST EXTEND A
MIN. OF .50 BELOW
INTERFACE PLANE

1.020
1.010 DIA
GUIDE PIN

GUIDE PIN, 3 REQD.

Figure 7-26

# IU/SLA MECHANICAL ALIGNMENT

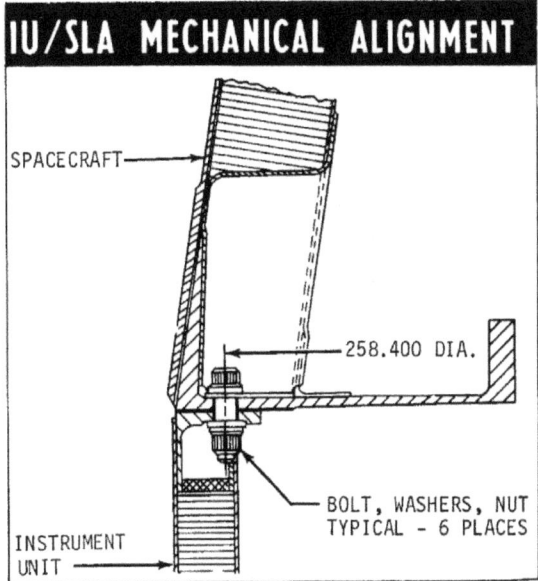

SPACECRAFT

258.400 DIA.

BOLT, WASHERS, NUT
TYPICAL - 6 PLACES

INSTRUMENT
UNIT

Figure 7-27

# IU/SLA INTERFACE ELECTRICAL CONNECTORS

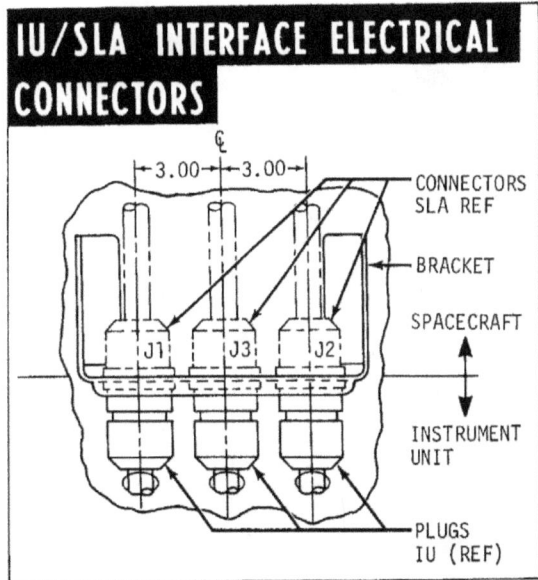

3.00    3.00

CONNECTORS
SLA REF

BRACKET

SPACECRAFT

J1    J3    J2

INSTRUMENT
UNIT

PLUGS
IU (REF)

Figure 7-28

# SECTION VIII
# GROUND SUPPORT INTERFACE

## TABLE OF CONTENTS

## LAUNCH COMPLEX 39

Launch Complex 39 (LC-39), Kennedy Space Center, Florida, provides all the facilities necessary to the assembly, checkout, and launch of the Apollo/Saturn space vehicle. The vehicle assembly building (VAB) provides a controlled environment in which the vehicle is assembled and checked out on a mobile launcher (ML). The space vehicle and the launch structure are then moved as a unit by the crawler-transporter to the launch site, where vehicle launch is accomplished after propellant loading and final checkout. The major elements of the launch complex shown in figure 8-1, are the vehicle assembly building (VAB), the launch control center (LCC), the mobile launcher (ML), the crawler-transporter (C-T), the crawlerway, the mobile service structure (MSS), and the launch pad.

## LC-39 FACILITIES AND EQUIPMENT

### VEHICLE ASSEMBLY BUILDING

The VAB is located adjacent to Kennedy Parkway, about five miles north of the KSC industrial area. Its purpose is to provide a protected environment for receipt and checkout of the propulsion stages and instrument unit (IU), erection of the vehicle stages and spacecraft in a vertical position on the ML, and integrated checkout of the assembled space vehicle.

The VAB, as shown in figure 8-2 is a totally enclosed structure covering eight acres of ground. It is a structural steel building approximately 525 feet high, 518 feet wide, and 716 feet long. The siding is insulated aluminum except where translucent fiberglass sandwich panels are used in part of the north and south walls.

The principal operational elements of the VAB are the low bay area and high bay area. A 92-foot wide transfer aisle extends through the length of the VAB and divides the low and high bay areas into equal segments (See figure 8-3).

### Low Bay Area

The low bay area provides the facilities for receiving, uncrating, checkout, and preparation of the S-II, and S-IVB stages, and the IU. The low bay area, located in the southern section of the VAB, is approximately 210 feet high, 442 feet wide, and 274 feet long. There are eight stage preparation and checkout cells, four of which are equipped with systems to simulate interface operations between the stages and the IU.

Work platforms, made up of fixed and folded sections, fit about the various sections as required. The platforms are bolted, to permit vertical repositioning, to the low bay structure. Access from fixed floor levels to the work platforms is provided by stairs.

### High Bay Area

The high bay area provides the facilities for erection and checkout of the S-IC stage; mating and erection operations of the S-II stage, S-IVB stage, IU, and spacecraft; and integrated checkout of the assembled space vehicle. The high bay area, which is located in the northern section of the building, is approximately 525 feet high, 518 feet wide, and 442 feet long. It contains four checkout bays, each capable of accommodating a fully assembled, Saturn V space vehicle.

Access to the vehicle at various levels is provided from air conditioned work platforms that extend from either side of the bay to completely surround the launch vehicle. Each platform is composed of two biparting sections which can be positioned in the vertical plane. The floor and roof of each section conform to and surround the vehicle. Hollow seals on the floor and roof of the section provide an environmental seal between the vehicle and the platform.

Each pair of opposite checkout bays is served by a 250-ton bridge crane with a hook height of 462 feet. The wall framing between the bays and the transfer aisle is open above the 190-foot elevation to permit movement of components from the transfer aisle to their assembly position in the checkout bay.

The high bay doors provide an inverted T-shaped opening 456 feet in height. The lower portion of the opening is closed by doors which move horizontally on tracks. The upper portion of the opening is closed by seven vertically moving doors.

### Utility Annex

The utility annex, located on the west side of the VAB, supports the VAB, LCC and other facilities in the VAB area. It provides air conditioning, hot water, compressed air, water for fire protection, and emergency electrical power.

### Helium/Nitrogen Storage-VAB Area

The gas storage facility at the VAB provides high-pressure gaseous helium and nitrogen. It is located east of the VAB and south of the crawlerway. The roof deck of the building is removable to permit installation and removal of pressure vessels through the roof. This facility is serviced from the converter/compressor facility by a 6,000 psig gaseous helium line and a 6,000 psig gaseous nitrogen line.

### LAUNCH CONTROL CENTER

The LCC (figure 8-4) serves as the focal point for overall direction, control, and surveillance of space vehicle checkout and launch. The LCC is located adjacent to the VAB and at a sufficient distance from the launch pad (three miles) to permit the safe viewing of liftoff without requiring site hardening. An enclosed personnel and cabling bridge connects the VAB and LCC at the third floor level.

**VIII**

# LAUNCH COMPLEX 39

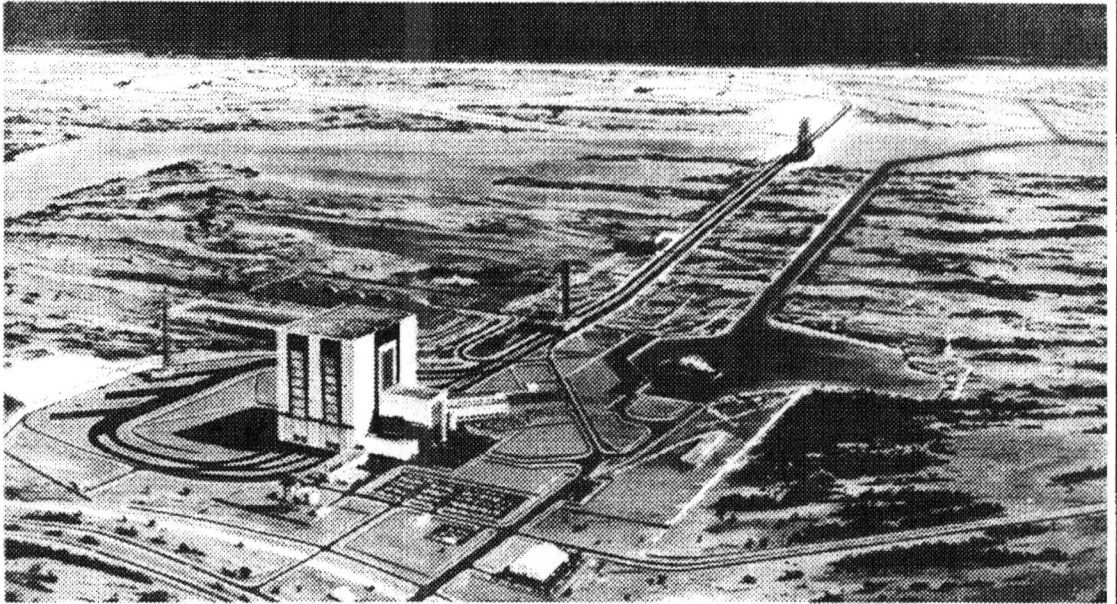

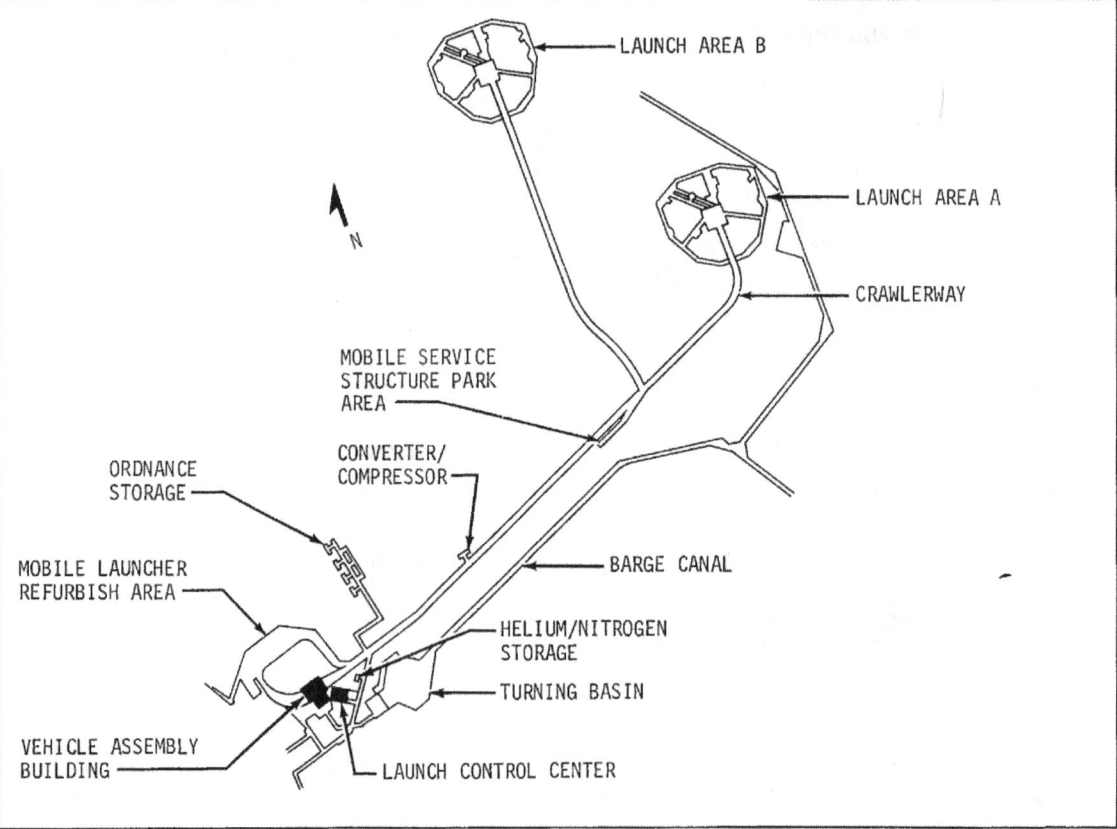

Figure 8-1

# VEHICLE ASSEMBLY BUILDING

VERTICAL ASSEMBLY BUILDING

LAUNCH CONTROL CENTER

Figure 8-2

# VAB INTERIOR

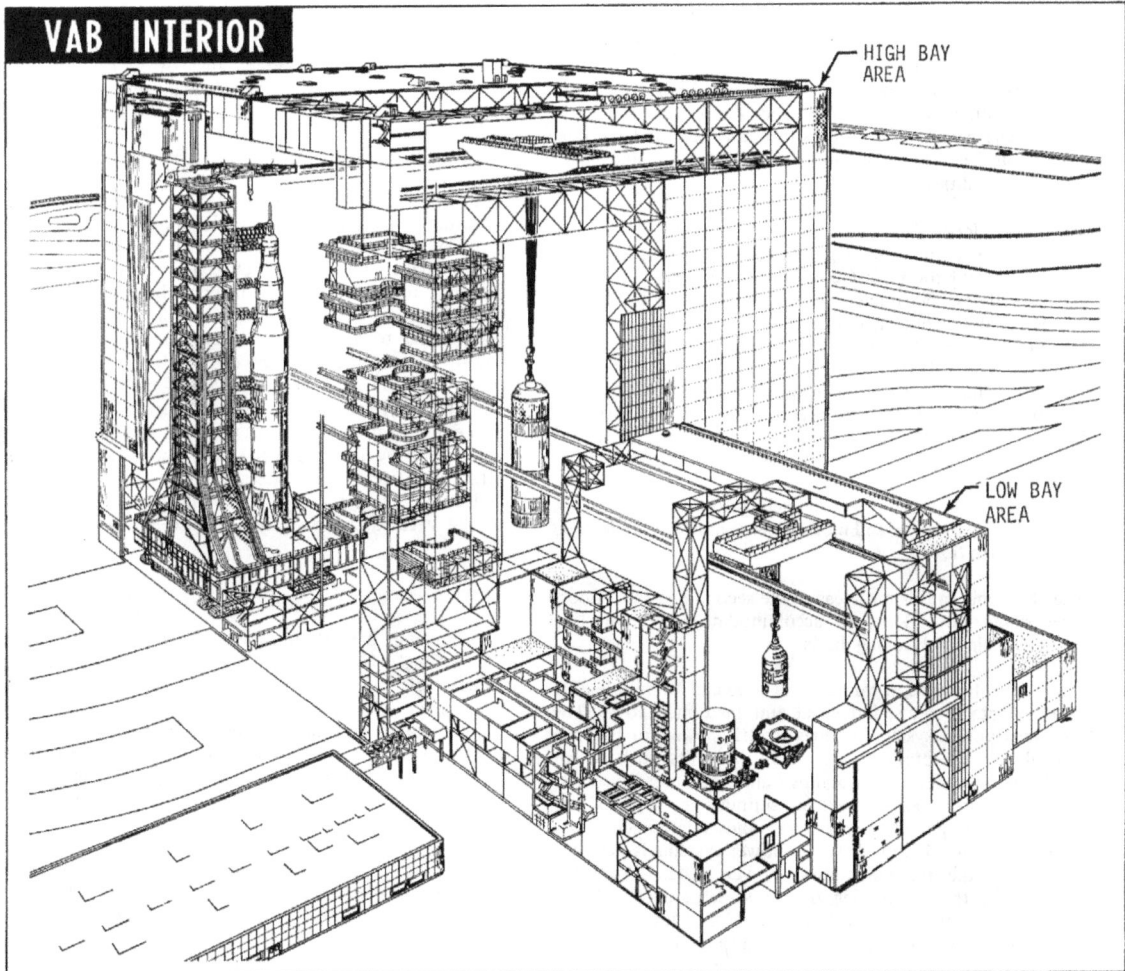

HIGH BAY AREA

LOW BAY AREA

Figure 8-3

## LCC EXTERIOR

Figure 8-4

The LCC is a four-story structure approximately 380 by 180 feet. The ground floor is devoted to service and support functions such as cafeteria, offices, shops, laboratories, the communications control room, and the complex control center. The second floor houses telemetry, RF and tracking equipment, in addition to instrumentation and data reduction facilities.

The third floor is divided into four separate but similar control areas, each containing a firing room, computer room, mission control room, test conductor platform area, visitor gallery, offices and frame rooms. Three of the four firing rooms, contain control, monitoring and display equipment for automatic vehicle checkout and launch.

Direct viewing of the firing rooms and the launch area is possible from the mezzanine level through specially designed, laminated, and tinted glass windows. Electrically controlled sun louvers are positioned outside the windows.

The display rooms, offices, launch information exchange facility (LIEF) rooms, and mechanical equipment are located on the fourth floor.

The electronic equipment areas of the second and third floors have raised false floors to accommodate interconnecting cables and air conditioning ducts.

The power demands in this area are large and are supplied by two separate systems, industrial and instrumentation. The industrial power system supplies electric power for lighting, general use receptacles, and industrial units such as air conditioning, elevators, pumps and compressors. The instrumentation power system supplies power to the electronic equipment, computers, and related checkout equipment. This division between power systems is designed to protect the instrumentation power system from the adverse effects of switching transients, large cycling loads, and intermittent motor starting loads. Communication and signal cable provisions have been incorporated into the design of the facility. Cable troughs extend from the LCC via the enclosed bridge to each ML location in the VAB high bay area. The LCC is also connected by buried cableways to the

ML refurbishing area and to the pad terminal connection room (PTCR) at the launch pad. Antennas on the roof provide an RF link to the launch pads and other facilities at KSC.

MOBILE LAUNCHER

The mobile launcher (figure 8-5) is a transportable steel structure which, with the crawler-transporter, provides the capability to move the erected vehicle to the launch pad. The ML is divided into two functional areas, the launcher base and the umbilical tower. The launcher base is the platform on which a Saturn V vehicle is assembled in the vertical position, transported to a launch site, and launched. The umbilical tower, permanently erected on the base, is the means of ready access to all important levels of the vehicle during the assembly, checkout, and servicing periods prior to launch. The equipment used in the servicing, checkout, and launch is installed throughout both the base and tower sections of the ML. The intricate vehicle-to-ground interfaces are established and debugged in the convenient and protected environment of the VAB, and moved undisturbed aboard the ML to the pad.

Launcher Base

The launcher base (figure 8-6) is a two story steel structure 25 feet high, 160 feet long, and 135 feet wide. Each of the three levels provides approximately 12,000 square feet of floor space. The upper deck, designated level 0, contains, in addition to the umbilical tower, the four holddown arms and the three tail service masts. Level A, the upper of the two internal levels, contains 21 compartments and level B has 22 compartments. There is a 45-foot square opening through the ML base for first stage exhaust.

Access to the base interior is provided by personnel/equipment access doors opening into levels A and B and equipment access hatches located on levels O and A.

The base has provisions for attachment to the crawler-transporter, six launcher-to-ground mount mechanisms, and four extensible support columns.

# MOBILE LAUNCHER

GSCU FLOW CONTROL VALVE BOX

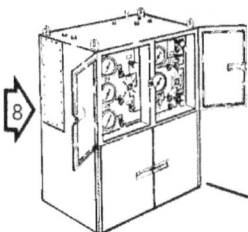

IU PNEUMATIC CONSOLE

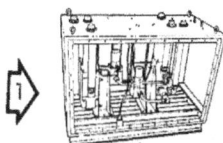

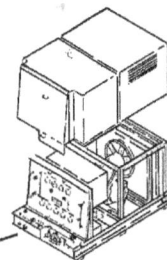

IU GROUND SUPPORT
COOLING UNIT (2 UNITS)

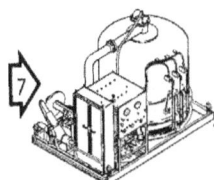

S-IVB GAS HEAT EXCHANGER

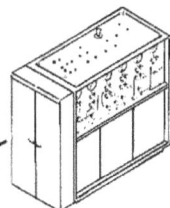

S-IVB PNEUMATIC CONSOLE "A"

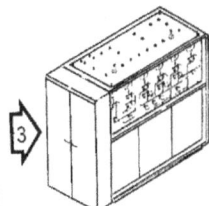

S-IVB PNEUMATIC CONSOLE "B"

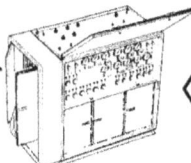

S-IVB APS PNEUMATIC CONSOLE

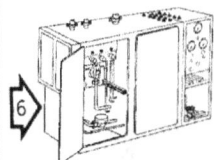

S-II PNEUMATIC CONSOLE S7-41A

S-II LH$_2$ HEAT EXCHANGER A7-71

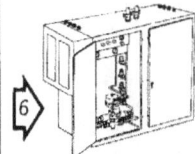

S-II PNEUMATIC CONSOLE S7-41B

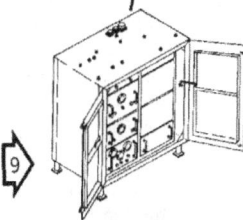

S-IC FWD UMBILICAL
SERVICE CONSOLE

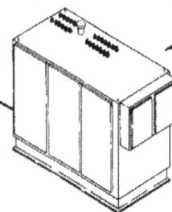

S-II PNEUMATIC CONSOLE S7-41C

Figure 8-5 (Sheet 1 of 2)

## MOBILE LAUNCHER

1. GSCU Flow Control Valve Box
   Selects either GSCU for operation of one unit while the other recirculates.

2. Ground Support Cooling Unit
   Supplies water-methanol to the heat exchanger in the IU thermal conditioning system to absorb heat in the IU generated by electronic equipment.

3. S-IVB Pneumatic Console A&B
   Regulates and controls helium and nitrogen gases for leak testing, functional checkout, propellant loading, purge, and propellant unloading.

4. S-IVB APS Pneumatic Console
   Regulate and distribute helium and nitrogen gases during checkout and propellant loading.

5. S-II LH$_2$ Heat Exchanger A7-71
   Provides gases to the S-IC stage for the following:
   1. Fuel tank pressurization
   2. LOX tank pre-pressurization
   3. Thrust Chamber jacket chilldown

6. S-II Pneumatic Consoles S7-41A, B, & C
   Regulate, control, and monitor gases for S-II stage during standby, prelaunch, and launch.

7. S-IVB Gas Heat Exchanger
   Supplies cold helium or hydrogen for the following:
   1. Lox and Fuel Tank Pre-Pressurization
   2. Thrust chamber jacket chilldown
   3. Pressurize engine turbine start bottle

8. IU Pneumatic Console
   Regulates, monitors, and controls pneumatic pressure to pressurize, checkout, and test the air bearing spheres and related pneumatic and electro-mechanical circuitry.

9. S-IC Forward Umbilical Service Console
   Supplies nitrogen from three regulation modules to S-IC stage pneumatic systems (camera lens purge, camera eject, and lox and fuel tank preservation) through the forward umbilical plate.

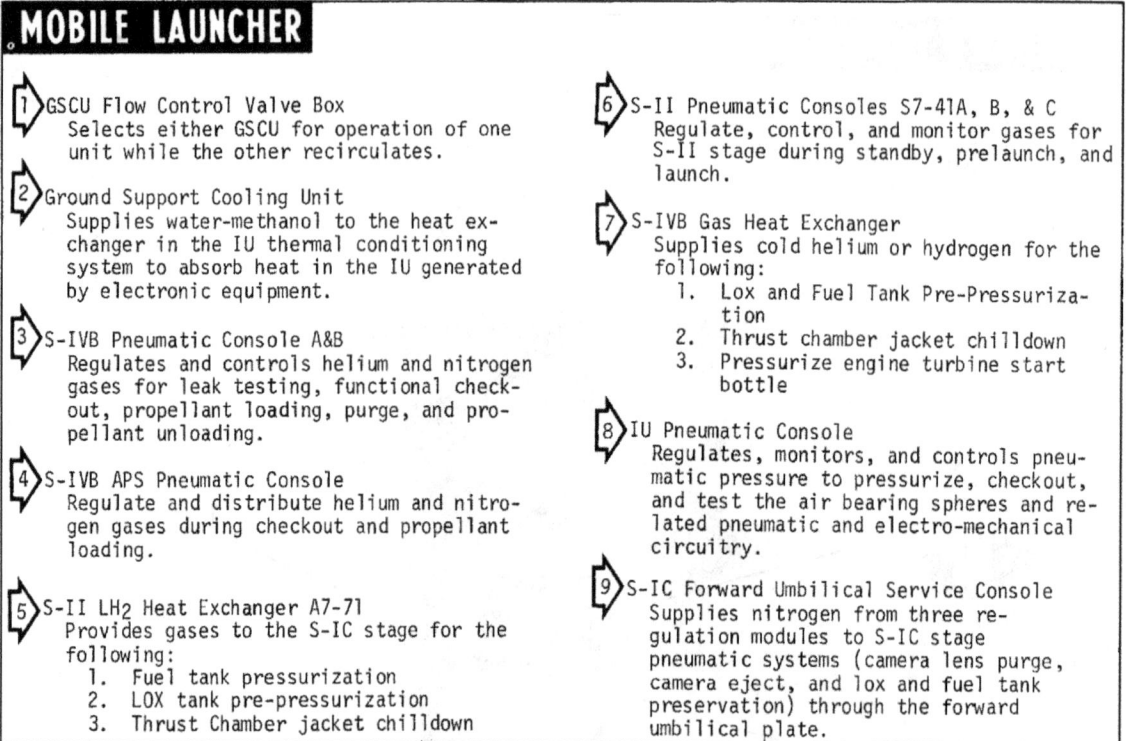

Figure 8-5 (Sheet 2 of 2)

All electrical/mechanical interfaces between vehicle systems and the VAB or the launch site are located through or adjacent to the base structure. A number of permanent pedestals at the launch site provide support for the interface plates and servicing lines.

The base houses such items as the computer systems test sets, digital propellant loading equipment, hydraulic test sets, propellant and pneumatic lines, air conditioning and ventilating systems, electrical power systems, and water systems. Shock-mounted floors and spring supports are provided so that critical equipment receives less than + 0.5 g mechanically-induced vibrations. Electronic compartments within the ML base are provided with acoustical isolation to reduce the overall rocket engine noise level.

The air conditioning and ventilating system for the base provides environmental protection for the equipment during operations and standby. One packaged air-conditioner provides minimal environmental conditioning and humidity control during transit. Fueling operations at the launch area require that the compartments within the structure be pressurized to a pressure of three inches of water above atmospheric pressure and that the air supply originate from a remote area free from contamination.

The primary electrical power supplied to the ML is divided into four separate services: instrumentation, industrial, in-transit and emergency. Instrumentation and industrial power systems are separate and distinct. During transit, power from the crawler-transporter is used for the water/glycol systems, computer air conditioning, threshold lighting, and obstruction lights. Emergency power for the ML is supplied by a diesel-driven generator located in the ground facilities. It is used for obstruction lights, emergency lighting, and for one tower elevator. Water is supplied to the ML at the VAB and at the pad for fire, industrial and domestic purposes and at the refurbishment area for domestic purposes.

Umbilical Tower

The umbilical tower is an open steel structure 380 feet high which provides the support for eight umbilical service arms, one access arm, 18 work and access platforms, distribution equipment for the propellant, pneumatic, electrical and instrumentation subsystems, and other ground support equipment. The distance from the vertical centerline of the tower to the vertical centerline of the vehicle is approximately 80 feet. The distance from the nearest vertical column of the tower to the vertical centerline of the vehicle is approximately 60 feet. Two high speed elevators service 18 landings, from level A of the base to the 340-foot tower level.

The hammerhead crane is located on top of the umbilical tower. The load capacity of the crane is 25 ton when the boom is extended up to 50 feet from the tower centerline. When the boom is extended between 50 to 85 feet the load capacity is 10 ton. The hook can be raised or lowered at 30 feet per minute for a distance of 468 feet. The trolley speed is 110 feet per minute. The crane can rotate 360 degrees at one revolution per minute. Control from each of the 18 levels is available through plug-in controls.

# MOBILE LAUNCHER LEVEL A AND B

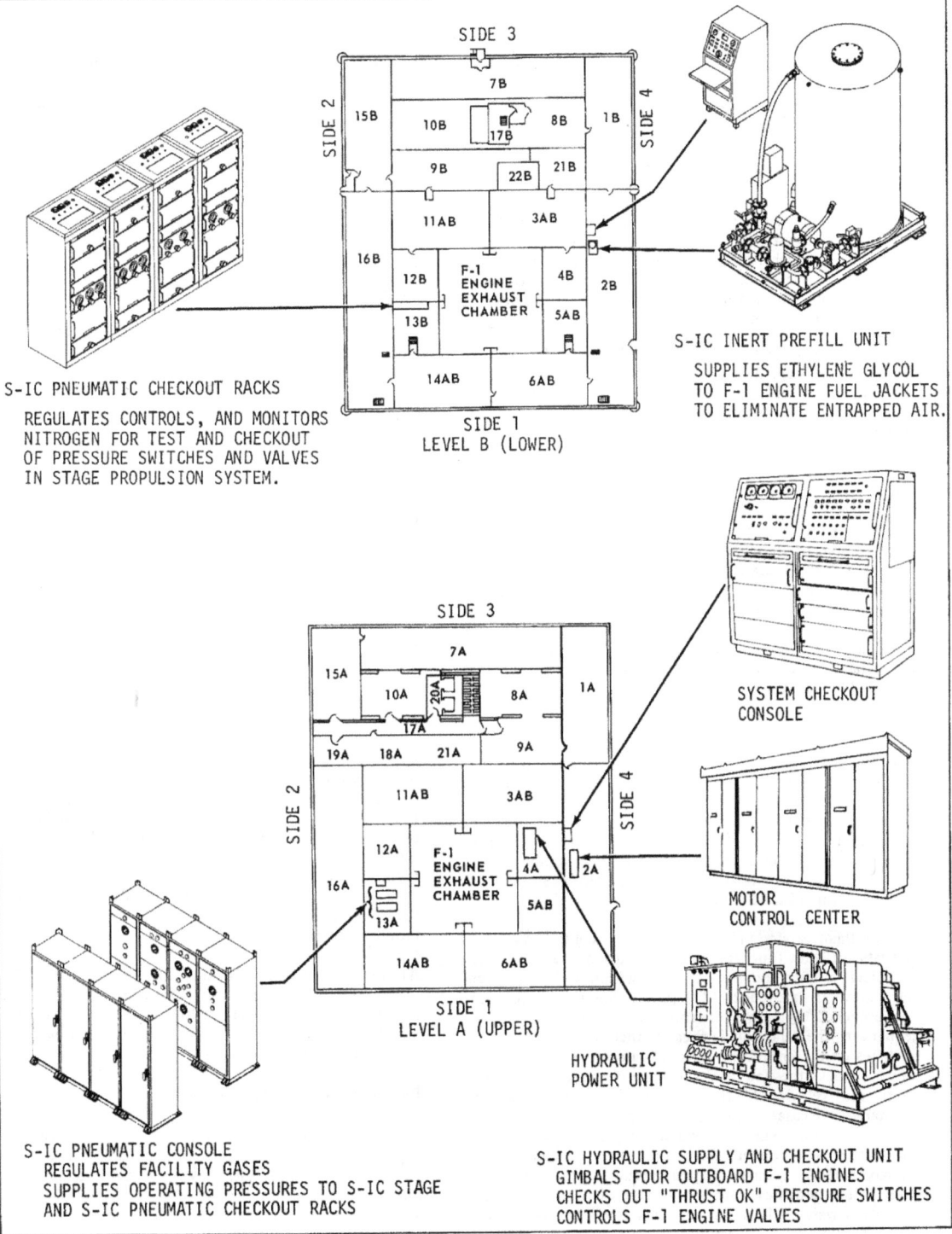

SIDE 3

SIDE 2 | SIDE 4

LEVEL B (LOWER)
SIDE 1

**S-IC PNEUMATIC CHECKOUT RACKS**

REGULATES CONTROLS, AND MONITORS
NITROGEN FOR TEST AND CHECKOUT
OF PRESSURE SWITCHES AND VALVES
IN STAGE PROPULSION SYSTEM.

**S-IC INERT PREFILL UNIT**

SUPPLIES ETHYLENE GLYCOL
TO F-1 ENGINE FUEL JACKETS
TO ELIMINATE ENTRAPPED AIR.

SIDE 3

SIDE 2 | SIDE 4

LEVEL A (UPPER)
SIDE 1

**SYSTEM CHECKOUT CONSOLE**

**MOTOR CONTROL CENTER**

**HYDRAULIC POWER UNIT**

**S-IC PNEUMATIC CONSOLE**
REGULATES FACILITY GASES
SUPPLIES OPERATING PRESSURES TO S-IC STAGE
AND S-IC PNEUMATIC CHECKOUT RACKS

**S-IC HYDRAULIC SUPPLY AND CHECKOUT UNIT**
GIMBALS FOUR OUTBOARD F-1 ENGINES
CHECKS OUT "THRUST OK" PRESSURE SWITCHES
CONTROLS F-1 ENGINE VALVES

Figure 8-6

## Holddown Arms

The four holddown arms (figure 8-7) are mounted on the ML deck 90° apart around the vehicle base. They position and hold the vehicle on the ML during the VAB checkout, movement to the pad, and pad checkout. The arm bases have sufficient strength to support the vehicle before launch and to withstand the dynamic loads caused by engine cutoff in an abort situation. The vehicle base is held with a preloaded force of 700,000 pounds at each arm.

At engine ignition, the vehicle is restrained until proper engine thrust is achieved, at which time a signal from the countdown sequencer causes each of two identical pneumatic systems to release high pressure helium to a separator mechanism in each holddown arm. The unlatching interval for the four arms should not exceed 0.050 seconds. If any of the separators fail to operate in 0.180 seconds, release is effected by detonating an explosive nut link.

Controlled release mechanisms are used to provide a gradual release of the stage at launch, thereby keeping the dynamic loads at launch within the design capability of the vehicle. The controlled release mechanisms (Figure 8-7) restrain the vehicle during the first few inches of travel. Each controlled release mechanism consists of a bracket bolted to the holddown arm base, a tapered pin fastened to the bracket, and a die installed on the vehicle. Upon holddown arm release, each tapered pin is drawn through a die as the vehicle rises through the first six inches. This reduces the diameter of the pin from its maximum to the diameter of the die. The force required to draw the pins through the die decreases linearly from maximum restraint at holddown release to zero restraint when the pins are free of the dies. The vehicle is then free with the pins remaining in the brackets and the dies traveling with the vehicle. There are provisions for as many as 16 mechanisms per vehicle. Twelve mechanisms, using greased pins, will be used for the SA-503 launch.

## Service Arms

The nine service arms provide access to the launch vehicle and support the service lines that are required to sustain the vehicle as described in figure 8-8. The service arms are designated as either preflight or in flight arms. The preflight arms are retracted and locked against the umbilical tower prior to liftoff. The in flight arms retract at vehicle liftoff, after receiving a command signal from the service arm control switches located in the holddown arms.

The in flight service arm launch retract sequence typically consists of the four following operations: arm unlock, umbilical carrier release, carrier withdrawal, and arm retraction and latchback. At T-15 seconds the service arms are unlocked by a signal from the terminal countdown sequencer. When the vehicle rises 3/4-inch, the primary liftoff switches on the holddown arms activate a pneumatic system which unlocks the umbilical carriers and pushes each carrier from the vehicle. If this system fails, the secondary mechanical release mechanism will be actuated when the vehicle rises approximately two inches. If both systems fail the carrier is cammed from the vehicle when it rises approximately 15 inches. Upon carrier ejection, a double pole switch activates both the carrier withdrawal and arm retraction systems. If this switch fails, it will be by-passed by a signal from the secondary liftoff switches when the vehicle rises 18 inches. Line handling devices on the S-IVB forward and aft arms are also activated on carrier ejection. Carrier

withdrawal and arm retraction is accomplished by pneumatic and/or hydraulic systems.

## Tail Service Masts

The three tail service mast (TSM) assemblies, figure 8-7, support service lines to the S-IC stage and provide a means for rapid retraction at vehicle liftoff. The TSM assemblies are located on level 0 of the ML base. Each TSM is a counter-balanced structure which is pneumatically/electrically controlled and hydraulically operated. Retraction of the umbilical carrier and vertical rotation of the mast is accomplished simultaneously to ensure no physical contact between the vehicle and mast. The carrier after retraction rotates into a hood assembly which protects it from the exhaust plume.

## LAUNCH PAD

The launch pad, figure 8-9, provides a stable foundation for the ML during Apollo/Saturn V launch and prelaunch operations and an interface to the ML for ML and vehicle systems. There are presently two pads at LC-39 located approximately three miles from the VAB area. Each launch site is an eight sided polygon measuring approximately 3,000 feet across.

## Launch Pad Structure

The launch pad is a cellular, reinforced concrete structure with a top elevation of 48 feet above sea level (42 feet above grade elevation). The longitudinal axis of the pad is oriented north-south, with the crawlerway and ramp approach from the south.

Located within the fill under the west side of the structure (figure 8-10) is a two-story concrete building to house environmental control and pad terminal connection equipment. On the east side of the structure, within the fill, is a one-story concrete building to house the high-pressure gas storage battery. On the pad surface are elevators, staircases, and interface structures to provide service to the ML and the mobile service structure (MSS). A ramp, with a five percent grade, provides access from the crawlerway. This is used by the C-T to position the ML/Saturn V and the MSS on the support pedestals. The azimuth alignment building is located on the approach ramp in the crawlerway median strip. A flame trench 58 feet wide by 450 feet long, bisects the pad. This trench opens to grade at the north end. The 700,000-pound mobile wedge-type flame deflector is mounted on rails in the trench.

An escape chute is provided to connect the ML to an underground, hardened room. This room is located in the fill area west of the support structure. This is used by astronauts and service crews in the event of a malfunction during the final phase of the countdown.

## Pad Terminal Connection Room

The pad terminal connection room (PTCR) (figure 8-10) provides the terminals for communication and data link transmission connections between the ML or MSS and the launch area facilities and between the ML or MSS and the LCC. This facility also accommodates the electronic equipment that simulates the vehicle and the functions for checkout of the facilities during the absence of the launcher and vehicle.

# HOLDDOWN ARMS/TAIL SERVICE MAST

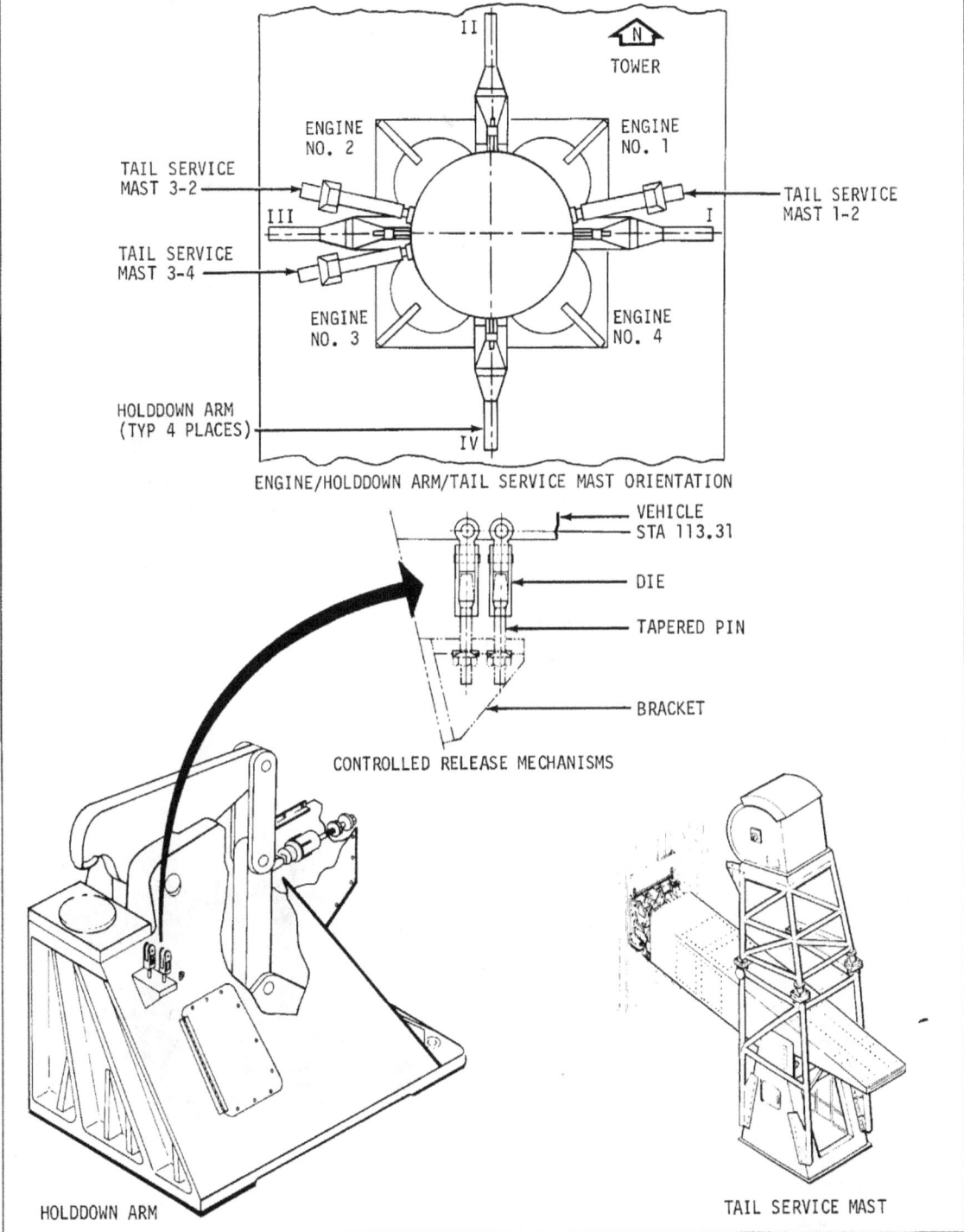

II

TOWER — N

ENGINE NO. 2

ENGINE NO. 1

TAIL SERVICE MAST 3-2

TAIL SERVICE MAST 1-2

III

I

TAIL SERVICE MAST 3-4

ENGINE NO. 3

ENGINE NO. 4

HOLDDOWN ARM (TYP 4 PLACES)

IV

ENGINE/HOLDDOWN ARM/TAIL SERVICE MAST ORIENTATION

VEHICLE STA 113.31

DIE

TAPERED PIN

BRACKET

CONTROLLED RELEASE MECHANISMS

HOLDDOWN ARM

TAIL SERVICE MAST

Figure 8-7

# MOBILE LAUNCHER SERVICE ARMS

1 ⟩ S-IC Intertank (preflight). Provides lox fill and drain interfaces. Umbilical withdrawal by pneumatically driven compound parallel linkage device. Arm may be reconnected to vehicle from LCC. Retract time is 8 seconds. Reconnect time is approximately 5 minutes.

2 ⟩ S-IC Forward (preflight). Provides pneumatic, electrical, and air-conditioning interfaces. Umbilical withdrawal by pneumatic disconnect in conjunction with pneumatically driven block and tackle/lanyard device. Secondary mechanical system. Retracted at T-20 seconds. Retract time is 8 seconds.

3 ⟩ S-II Aft (preflight). Provides access to vehicle. Arm retracted prior to liftoff as required.

4 ⟩ S-II Intermediate (in-flight). Provides $LH_2$ and lox transfer, vent line, pneumatic, instrument cooling, electrical, and air-conditioning interfaces. Umbilical withdrawal systems same as S-IVB Forward with addition of a pneumatic cylinder actuated lanyard system. This system operates if primary withdrawal system fails. Retract time is 6.4 seconds (max).

5 ⟩ S-II Forward (in-flight). Provides $GH_2$ vent, electrical, and pneumatic interfaces. Umbilical withdrawal systems same as S-IVB Forward. Retract time is 7.4 seconds (max).

6 ⟩ S-IVB Aft (in-flight). Provides $LH_2$ and lox transfer, electrical, pneumatic, and air-conditioning interfaces. Umbilical withdrawal systems same as S-IVB Forward. Also equipped with line handling device. Retract time is 7.7 seconds (max).

7 ⟩ S-IVB Forward (in-flight). Provides fuel tank vent, electrical, pneumatic, air-conditioning, and preflight conditioning interfaces. Umbilical withdrawal by pneumatic disconnect in conjunction with pneumatic/hydraulic redundant dual cylinder system. Secondary mechanical system. Arm also equipped with line handling device to protect lines during withdrawal. Retract time is 8.4 seconds (max).

8 ⟩ Service Module (in-flight). Provides air-conditioning, vent line, coolant, electrical, and pneumatic interfaces. Umbilical withdrawal by pneumatic/mechanical lanyard system with secondary mechanical system. Retract time is 9.0 seconds (max).

9 ⟩ Command Module Access Arm (preflight). Provides access to spacecraft through environmental chamber. Arm may be retracted or extended from LCC. Retracted 12° park position until T-4 minutes. Extend time is 12 seconds from this position.

Figure 8-8

# LAUNCH PAD A, LC-39

Figure 8-9

# LAUNCH STRUCTURE EXPLODED VIEW

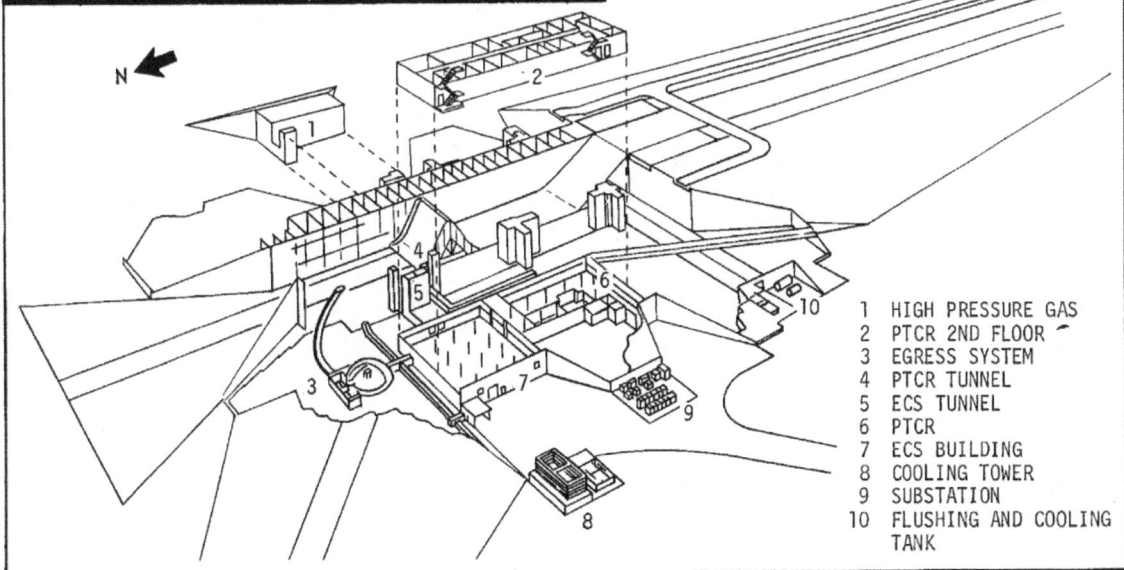

1  HIGH PRESSURE GAS
2  PTCR 2ND FLOOR
3  EGRESS SYSTEM
4  PTCR TUNNEL
5  ECS TUNNEL
6  PTCR
7  ECS BUILDING
8  COOLING TOWER
9  SUBSTATION
10 FLUSHING AND COOLING
   TANK

Figure 8-10

The PTCR is a two-story hardened structure within the fill on the west side of the launch support structure. The launch pedestal and the deflector area are located immediately adjacent to this structure. Each of the floors of this structure measures approximately 136 feet by 56 feet. Entry is made from the west side of the launch support structure at ground level into the first floor area. Instrumentation cabling from the PTCR extends to the ML, MSS, high-pressure gas storage battery area, lox facility, RP-1 facility, LH₂ facility, and azimuth alignment building. The equipment areas of this building have elevated false floors to accommodate the instrumentation and communication cables used for interconnecting instrumentation racks and terminal distributors.

The air conditioning system, located on the PTCR ground floor, provides a controlled environment for personnel and equipment. The air conditioning system is controlled remotely from the LCC when personnel are evacuated for launch. This system provides chilled water for the air handling units located in the equipment compartments of the ML. A hydraulic elevator serves the two floors and the pad level.

Industrial and instrumentation power is supplied from a nearby substation.

### Environmental Control System

The ECS room located in the pad fill west of the pad structure and north of the PTCR (figure 8-10) houses the equipment which furnishes temperature and/or humidity controlled air or nitrogen for space vehicle cooling at the pad.

The ECS room is 96 feet wide by 112 feet long and houses air and nitrogen handling units, liquid chillers, air compressors, a 3000-gallon water/glycol storage tank, and other auxiliary electrical and mechanical equipment.

### High Pressure Gas System

The high-pressure gas storage facility at the pad provides the launch vehicle with high-pressure helium and nitrogen. This facility is an integral part of the east portion of the launch support structure. It is entered from ground elevation on the east side of the pad. The high pressure (6,000 psig) facilities at the pad are provided for high pressure storage of 3,000 cubic feet of gaseous nitrogen and 9,000 cubic feet of gaseous helium.

### Launch Pad Interface Structure

The launch pad interface structure (figure 8-11) provides mounting support pedestals for the ML and MSS, an engine access platform, and support structures for fueling, pneumatic, electric power, and environmental control interfaces.

The ML at the launch pad (as well as the VAB and refurbish area) is supported by six mount mechanisms which are designed to carry vertical and horizontal loading. Four extensible columns, located near each corner of the launcher base exhaust chamber, also support the ML at the launch site. These columns are designed to prevent excessive deflections of the launcher base when the vehicle is fueled and from load reversal in case of an abort between engine ignition and vehicle liftoff.

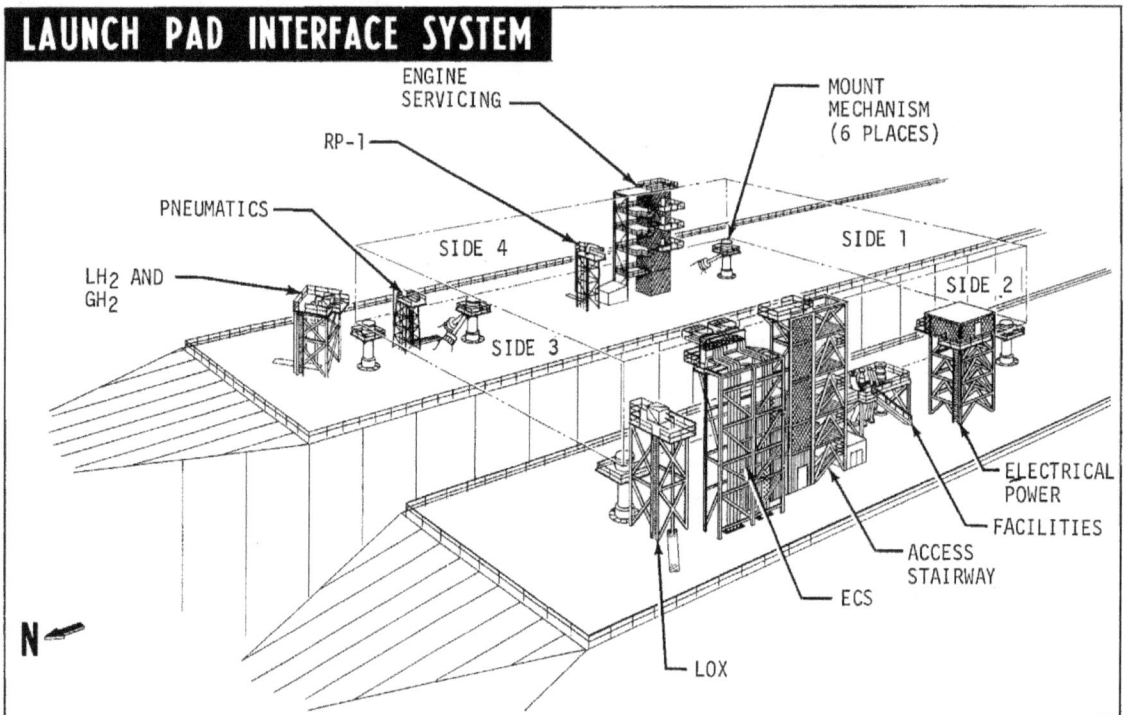

Figure 8-11

The MSS is supported on the launch pad by four mounting mechanisms similar to those used to support the ML.

The engine servicing structure provides access to the ML deck for servicing of the S-IC engines and ML deck equipment.

Interface structures are provided on the east and west portions of the pad structure (figure 8-11) for propellant, pneumatic, power, facilities, environmental control, communications, control, and instrumentation systems.

### Apollo Emergency Ingress/Egress and Escape Systems

The Apollo emergency Ingress/egress and escape systems provide access to and from the Command Module (CM), plus an escape route and safe quarters for the astronauts and service personnel in the event of a serious malfunction prior to launch. Depending upon the time available, the system provides alternate means of escape, i.e., by slide wire or by elevator. Both means utilize the CM access arm as a component.

The slide wire provides the primary means of escape through the use of a cable attached to the LUT, a pulley suspended chair, and a terminal point 2500 feet west of the pad. The cable is attached to the LUT at the 320 foot level (443 feet above ground level), descends to within 20 feet off the ground, and then ascends to the top of the tail tower which is 30 feet high. The chair and slide assembly accelerate to approximately 50 miles per hour. At the low point (1800 feet outward) of descent, a ferrule on the cable activates a braking mechanism, which then brings the slide assembly to a controlled deceleration and stop. The slide wire can accommodate a total of 11 persons from the tower.

The secondary escape and normal egress means are the tower high speed elevators. These move between the 340 foot level of the tower and level A at 600 feet per minute. From level A, egressing personnel move through a vestibule to elevator No. 2, which then takes them down to the bottom of the pad. Armored personnel carriers are available at this point to remove them from the pad area. If the state of emergency does not permit evacuation by vehicle, the personnel may utilize the blast room (figure 8-12).

The blast room is entered from level A by sliding down an escape tube which carries them into the blast room vestibule, commonly called the rubber room. The rubber room consists of a deceleration ramp and is rubber lined to prevent injury to descending personnel. From the rubber room the egress exits into the blast room which can accommodate 20 persons for a period of 24 hours. The blast room is hardened to explosive forces by being mounted on coil springs which reduce the effect of outside acceleration forces to 3 to 5 g's.

Access to the blast room is through blast proof doors or through an emergency hatch in the top. Communication facilities are provided in the room and include an emergency RF link with its antenna built into the ceiling. Final exit from the blast room is through the air intake duct to the air intake facility and them to the pad perimeter. Air velocity in the air duct is decreased to permit personnel to move through the duct.

Emergency ingress to the CM utilizes the tower high speed elevator components and the CM access arm.

### Electrical Power

The electrical power for launch pad A is fed from the 69 kv main substation to Switching Station No. 1, where it is stepped down to 13.8 kv. The 13.8 kv power is then fed to Switching Station No. 2 from where it is distributed to the various substations in the pad area. The output of each of the substations is 480-volts with the exception of the 4160-volt substations supplying power to the fire water booster pump motors and the lox pump motors.

### Fuel System Facilities

The fuel facilities, located in the northeast quadrant of the pad approximately 1,450 feet from pad center, store RP-1 and liquid hydrogen.

The RP-1 facility consists of three 86,000 gallon (577,000 pound) steel storage tanks, a pump house, a circulating pump, a transfer pump, two filter-separators, an 8-inch stainless steel transfer line, RP-1 foam generating building, and necessary valves, piping, and controls. Two RP-1 holding ponds, 150 feet by 250 feet with a water depth of two feet are located north of the launch pad, one on each side of the north-south axis. The ponds retain spilled RP-1 and discharge water to drainage ditches.

The LH$_2$ facility consists of one 850,000 gallon spherical storage tank, a vaporizer/heat exchanger which is used to pressurize the storage tank to 65 psig, a vacuum-jacketed, 10-inch, Invar transfer line, and a burn pond venting system. The internal tank pressure, maintained by circulating LH$_2$ from the tank through the vaporizer and back into the tank, is sufficient to provide the proper flow of LH$_2$ from the storage tank to the vehicle without using a transfer pump. Liquid hydrogen boiloff from the storage and ML areas is directed through vent-piping to bubblecapped headers submerged in the burn pond. The hydrogen is bubbled to the surface of the 100 foot square pond where a hot wire ignition system maintains the burning process.

### LOX System Facility

The lox facility is located in the northwest quadrant of the pad area, approximately 1,450 feet from the center of the pad. The facility consists of one 900,000 gallon spherical storage tank, a lox vaporizer to pressurize the storage tank, main fill and replenish pumps, a drain basin for venting and dumping of lox, and two transfer lines.

### Gaseous Hydrogen Facility

This facility is located on the pad perimeter road northwest of the liquid hydrogen facility. The facility provides GH$_2$ at 6,000 psig to the launch vehicle. The facility consists of four storage tanks having a total capacity of 800 cubic feet, a flatbed trailer on which are mounted liquid hydrogen tanks and a liquid-to-gas converter, a transfer line, and necessary valves and piping.

### Azimuth Alignment Building

The azimuth alignment building is located in the approach ramp to the launch structure in the median of the crawlerway about 700 feet from the ML positioning pedestals. The building houses the auto-collimator theodolite which senses, by a light source, the rotational output of the stable

# EGRESS SYSTEM

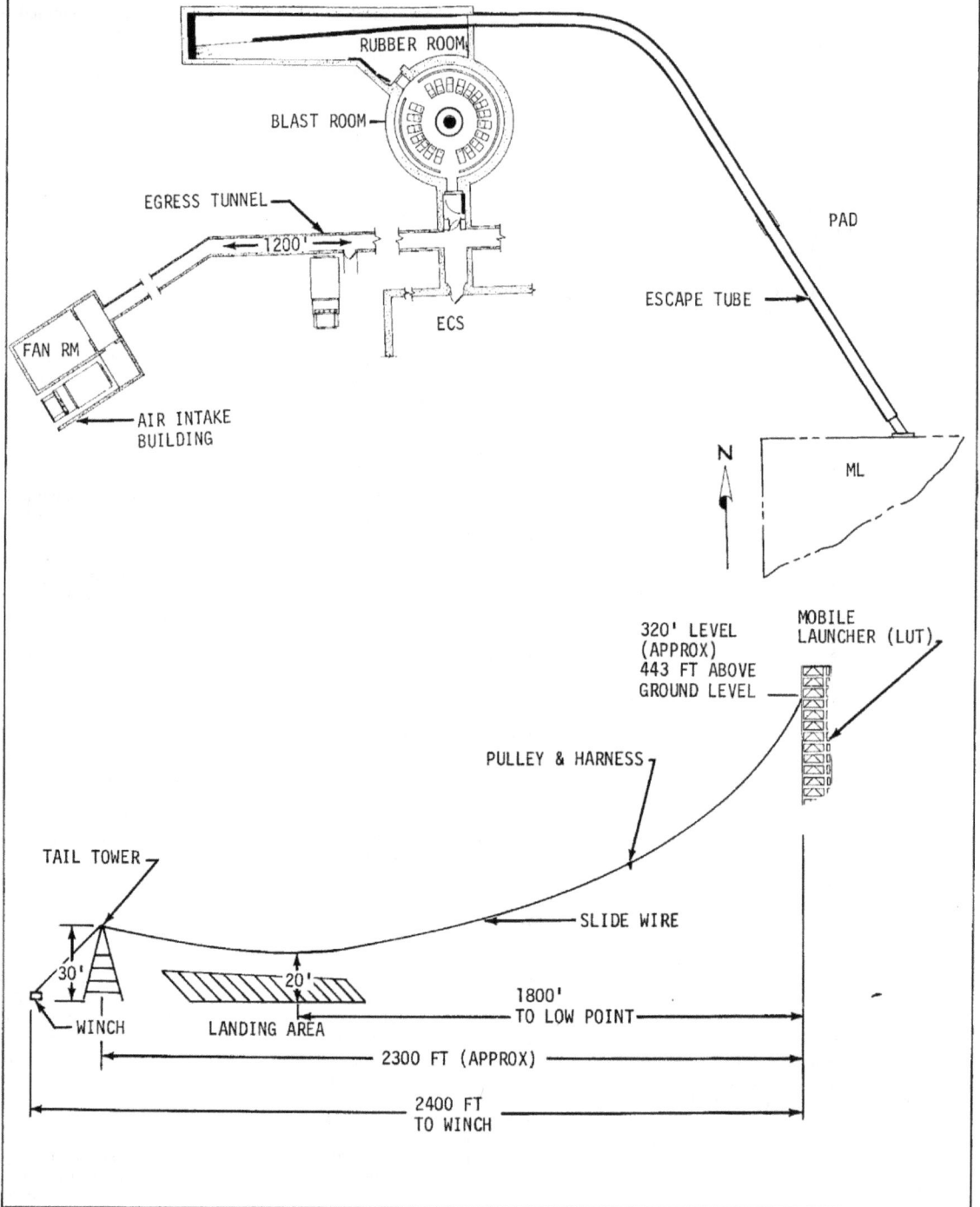

RUBBER ROOM

BLAST ROOM

EGRESS TUNNEL

←1200'→

FAN RM

AIR INTAKE
BUILDING

ECS

PAD

ESCAPE TUBE →

N

ML

320' LEVEL
(APPROX)
443 FT ABOVE
GROUND LEVEL

MOBILE
LAUNCHER (LUT)

PULLEY & HARNESS

TAIL TOWER

SLIDE WIRE

30'

20'

WINCH

LANDING AREA

1800'
TO LOW POINT

2300 FT (APPROX)

2400 FT
TO WINCH

Figure 8-12

platform. A short pedestal, with a spread footing isolated from the building, provides the mounting surface for the theodolite.

### Photographic Facilities

These facilities support photographic camera and closed circuit television equipment to provide real-time viewing and photographic documentation coverage. There are six camera sites in the launch pad area, each site containing an access road, five concrete camera pads, a target pole, communication boxes, and a power transformer with a distribution panel and power boxes. These sites cover prelaunch activities and launch operations from six different angles at a radial distance of approximately 1,300 feet from the launch vehicle. Each site has four engineering sequential cameras and one fixed, high speed, metric camera (CZR). A target pole for optical alignment of the CZR camera is located approximately 225 feet from the CZR pad and is approximately 86 feet high.

### Pad Water System Facilities

The pad water system facilities furnish water to the launch pad area for fire protection, cooling, and quenching. Specifically, the system furnishes water for the industrial water system, flame deflector cooling and quench, ML deck cooling and quench, ML tower fogging and service arm quench, sewage treatment plant, Firex water system, lox and fuel facilities, ML and MSS fire protection, and all fire hydrants in the pad area. The water is supplied from three 6-inch wells, each 275 feet deep. The water is pumped from the wells through a desanding filter and into a 1,000,000 gallon reservoir.

### Air Intake Building

This building houses fans and filters for the air supply to the PTCR, pad cellular structure, and the ML base. The building is located west of the pad, adjacent to the perimeter road.

### Flame Deflector

There are two flame deflectors provided at each pad; one for use and the other held in reserve. Their normal parking position is north of the launch support structure within the launch pad area. The flame deflector protects the boattail section of the Saturn V launch vehicle and the launch stand from hot gases, high pressures, and flame generated by the launch vehicle during the period of engine ignition and liftoff.

## MOBILE SERVICE STRUCTURE

The mobile service structure (figure 8-13) provides access to those portions of the space vehicle which cannot be serviced from the ML while at the launch pad. During nonlaunch periods, the MSS is located in a parked position along side of the crawlerway, 7,000 feet from the nearest launch pad. The MSS is transported to the launch site by the C-T. It is removed from the pad a few hours prior to launch and returned to its parking area.

The MSS is approximately 402 feet high, measured from ground level, and weighs 12 million pounds. The tower

**MOBILE SERVICE STRUCTURE**

Figure 8-13

structure rests on a base 135 feet by 135 feet. The top of the MSS base is 47 feet above grade. At the top, the tower is 87 feet by 113 feet.

The MSS is equipped with systems for air conditioning, electrical power, communications networks, fire protection, nitrogen pressurization, hydraulic pressure, potable water, and spacecraft fueling.

The structure contains five work platforms which provide access to the space vehicle. The outboard sections of the platforms are actuated by hydraulic cylinders to open and accept the vehicle and to close around it to provide access to the launch vehicle and spacecraft. The three upper platforms are fixed but can be relocated as a unit to meet changing vehicle configurations. The uppermost platform is open, with a chain-link fence for safety. The two platforms immediately below are enclosed to provide environmental control to the spacecraft. The two lowest platforms can be adjusted vertically to serve different parts of the vehicle. Like the uppermost platform, they are open with a chain-link fence for safety.

## CRAWLER-TRANSPORTER

The crawler-transporter (figure 8-14) is used to transport the mobile launcher and the mobile service structure. The ML,

# CRAWLER TRANSPORTER

Figure 8-14

with the space vehicle, is transported from the vehicle assembly building to the launch pad. The MSS is transported from its parking area to and from the launch pad. After launch, the ML is transported to the refurbishment area and subsequently back to the VAB. The C-T is capable of lifting, transporting, and lowering the ML or the MSS, as required, without the aid of auxiliary equipment. The C-T supplies limited electric power to the ML and the MSS during transit.

The C-T consists of a rectangular chassis which is supported through, a suspension system by four dual-tread crawler-trucks. The overall length is 131 feet and the overall width is 114 feet. The unit weighs approximately 6 million pounds. The C-T is powered by self-contained, diesel-electric generator units. Electric motors in the crawler-trucks propel the vehicle. Electric motor-driven pumps provide hydraulic power for steering and suspension control. Air conditioning and ventilation are provided where required.

The C-T can be operated with equal facility in either direction. Control cabs are located at each end and their control function depends on the direction of travel. The leading cab, in the direction of travel, will have complete control of the vehicle. The rear cab will, however, have override controls for the rear trucks only.

Maximum C-T unloaded speed is 2 mph, 1 mph with full load on level grade, and 0.5 mph with full load on a five percent grade. It has a 500-foot minimum turning radius and can position the ML or the MSS on the facility support pedestals within ± two inches.

## CONVERTER/COMPRESSOR FACILITY

The converter/compressor facility (CCF) converts liquid nitrogen to low pressure and high pressure gaseous nitrogen and compresses gaseous helium to 6,000 psig. The gaseous nitrogen and helium are then supplied to the storage facilities at the launch pad and at the VAB. The CCF is located on the north side of the crawlerway, approximately at the mid-point between the VAB and the main crawlerway junction to launch pads A and B.

The facility includes a 500,000 gallon, liquid nitrogen, Dewar storage tank, tank vaporizers, high pressure liquid nitrogen pump and vaporizer units, high pressure helium compressor units, helium and nitrogen gas driver/purifiers, rail and truck

transfer facilities, and a data link transmission cable tunnel.

The 500,000-gallon storage tank for the liquid nitrogen is located adjacent to the equipment building that houses the evaporators for conversion of the liquid nitrogen to high-pressure gas. The liquid nitrogen is transferred to the vaporizing compressors by pressurizing the storage tank. After vaporizing and compressing to 150 psig or 6,000 psig, the gaseous nitrogen is piped to the distribution lines supplying the VAB area (6,000 psig) and the pad (150 psig and 6,000 psig).

The gaseous helium is stored in tube-bank rail cars. These are then connected to the facility via a common manifold and a flexible one-inch inside diameter high-pressure line. The helium passes through the CCF helium compressors which boost its pressure from the tube-bank storage pressure to 6,000 psig after which it is piped to the VAB and pad high-pressure storage batteries.

Controls and displays are located in the CCF. Mass flow rates of high-pressure helium, high-pressure nitrogen, and low-pressure nitrogen gases leaving the CCF are monitored on panels located in the CCF via cableway ducts running between the CCF and the VAB, LCC, and launch pad.

## ORDNANCE STORAGE AREA

The ordnance storage area serves LC-39 in the capacity of laboratory test area and storage area for ordnance items. This facility is located on the north side of the crawlerway and approximately 2,500 feet north-east of the VAB. This remote site was selected for maximum safety.

The ordnance storage installation, enclosed by a perimeter fence, is comprised of three archtype magazines, two storage buildings, one ready-storage building, an ordnance test building and a guard service building. These structures are constructed of reinforced concrete, concrete blocks, and over-burdened where required. This facility contains approximately 10,000 square feet of environmentally controlled space. It provides for storage and maintenance of retrorockets, ullage rockets, explosive separation devices, escape rockets, and destruct packages. It also includes an area to test the electro-explosive devices that are used to initiate or detonate ordnance items. A service road from this facility connects to Saturn Causeway.

## VEHICLE ASSEMBLY AND CHECKOUT

The vehicle stages and the instrument unit (IU) are, upon arrival at KSC, transported to the VAB by special carriers. The S-IC stage is erected on a previously positioned mobile launcher (ML) in one of the checkout bays in the high bay area. Concurrently, the S-II and S-IVB stages and the IU are delivered to preparation and checkout cells in the low bay area for inspection, checkout, and pre-erection preparations. All components of the launch vehicle, including the Apollo spacecraft and launch escape system, are then assembled vertically on the ML in the high bay area.

Following assembly, the space vehicle is connected to the LCC via a high speed data link for integrated checkout and a simulated flight test. When checkout is completed, the crawler-transporter (C-T) picks up the ML, with the assembled space vehicle, and moves it to the launch site over the crawlerway.

At the launch site, the ML is emplaced and connected to system interfaces for final vehicle checkout and launch monitoring. The mobile service structure (MSS) is transported from its parking area by the C-T and positioned on the side of the vehicle opposite the ML. A flame deflector is moved on its track to its position beneath the blast opening of the ML to deflect the blast from the S-IC stage engines. During the prelaunch checkout, the final system checks are completed, the MSS is removed to the parking area, propellants are loaded and various items of support equipment are removed from the ML and the vehicle is readied for launch. After vehicle launch, the C-T transports the ML to the parking area near the VAB for refurbishment.

## TEST SYSTEM

A computer controlled automatic checkout system is used to accomplish the VAB (high bay) and pad testing. An RCA-110A computer and the equipment necessary to service and check out the launch vehicle are installed on the ML. Also an RCA-110A computer and the display and control equipment necessary to monitor and control the service and checkout operations are installed in the LCC. The computers operate in tandem through a data link with the computer in the ML receiving commands from and transmitting data to the computer in the LCC. The physical arrangement of the LCC and the ML are illustrated in figures 8-15 and 8-6 respectively.

### Test System Operation

Test system operation for Saturn V launch vehicle checkout is conducted from the firing room (see figure 8-16). During prelaunch operations, each stage is checked out utilizing the stage control and display console. Each test signal is processed through the computer complex, and is sent to the vehicle. The response signal is sent from the vehicle, through the computer complex, and the result is monitored on the display console. The basic elements of the test system and their functional relationship are shown in figure 8-17.

A switch on the control console can initiate individual operation of a system component or call up a complete test routine from the computer. A CRT is also provided for test conduction and evaluation.

The insertion of a plastic coded card key, prior to console operation, is a required precaution against improper program callup. Instructions, interruptions, and requests for displays are entered into the system by keying in proper commands at the console keyboards.

A complete test routine is called up by initiating a signal at the control panel. The signal is sent to the patch distributor located in the LCC and is routed to the appropriate signal conditioning equipment where the signal is prepared for acceptance by the LCC computer complex. The LCC computer communicates with the ML computer to call up the test routine. The ML computer complex sends the signal to the ML signal conditioning equipment and then to the stage relay rack equipment. The signal is then routed to the terminal distribution equipment and through the crossover distributor to interrogate the vehicle sensors. The sensor outputs are sent back to the ML computer complex for evaluation. The result is then sent to the LCC computer complex which routes the result to the stage console for display. Manual control of vehicle functions is provided at the control consoles. This control bypasses the computers

and is sent to the vehicle by means of hardwire. The result is also sent back to the display console by hardwire.

The digital data acquisition system (DDAS) collects the vehicle and support equipment responses to test commands, formats the test data for transmission to the ML and LCC, and decommutates the data for display in the ML and LCC. Decommutated test data is also fed to ML and LCC computer for processing and display, and for computer control of vehicle checkout. The DDAS consists of telemetry equipment, data transmission equipment, and ground receiving stations to perform data commutation, data transmission, and data decommutation.

The digital event evaluators are used to monitor the status of input lines and generate a time tagged printout for each detected change in input status. High speed printers in the LCC are connected to each DEE to provide a means for real time or post-test evaluation of discrete data. Two systems (DEE-3 and DEE-6) are used to monitor discrete events.

The DEE-3 is located in the PTCR with a printer located in the LCC. It monitors 768 inputs associated with propellant loading, environmental control, water control, and DDAS.

The DEE-6 is located in the ML base with a printer and remote control panel in the LCC. It monitors up to 4320 discrete signals from the vehicle stage umbilicals, pad and tower ground support equipment, and the DDAS.

The computer complex consists of two RCA-110A general purpose computers and peripheral equipment. This equipment includes a line printer, card reader, card punch, paper tape reader, and magnetic tape transports. The peripheral equipment provides additional bulk storage for the computer, acts as an input device for loading test routines into the computer memory, and as an output device to record processed data. One computer is located in the ML base and the other in the LCC behind the firing room. The computers are connected by underground hardwire. The computer system uses a tandem philosophy of checkout and control. The LCC computer is the main control for the system. It accepts control inputs from test personnel at the consoles in the firing room as well as inputs from tape storage and transmits them as test commands to the ML computer. The ML computer has the test routines stored in its memory banks. These routines are called into working memory and sent as discrete signals to the launch vehicle in response to the commands received from the LCC computer. The ML computer reports test routine status, data responses and results of test to the LCC computer. It is through this link that the control equipment and personnel in the firing room are informed of the test progress.

The propellant tanking computer system (PTCS) determines and controls the quantities of fuel and oxidizer on board each stage. Optimum propellant levels are maintained and lox and $LH_2$ are replenished as boiloff occurs during the countdown. The propellant tanking operation is monitored on the PTCS control panel.

Visual surveillance of launch vehicle checkout is provided to the launch management team and for distribution to MSC and MSFC through the operational television system (OTV). Sixty cameras provide this capability, 27 of which are located on the ML, 15 in the pad area, 12 on the MSS and 6 in the LCC. Any camera may be requested for viewing on the 10 x 10 foot screens in the firing room.

# LCC FACILITY LAYOUT

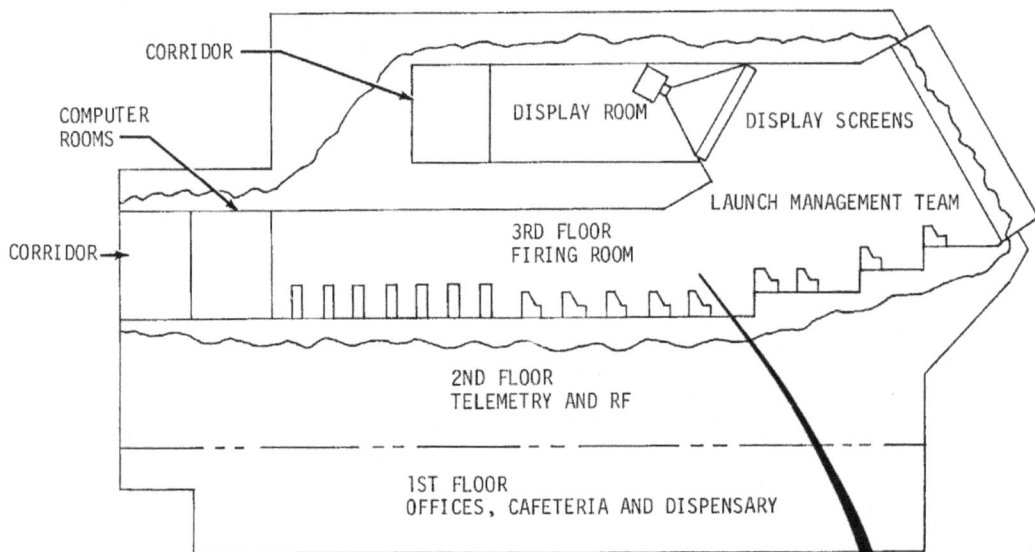

CORRIDOR

COMPUTER
ROOMS

CORRIDOR

DISPLAY ROOM

DISPLAY SCREENS

LAUNCH MANAGEMENT TEAM

3RD FLOOR
FIRING ROOM

2ND FLOOR
TELEMETRY AND RF

1ST FLOOR
OFFICES, CAFETERIA AND DISPENSARY

Figure 8-15

# FIRING ROOM (TYPICAL)

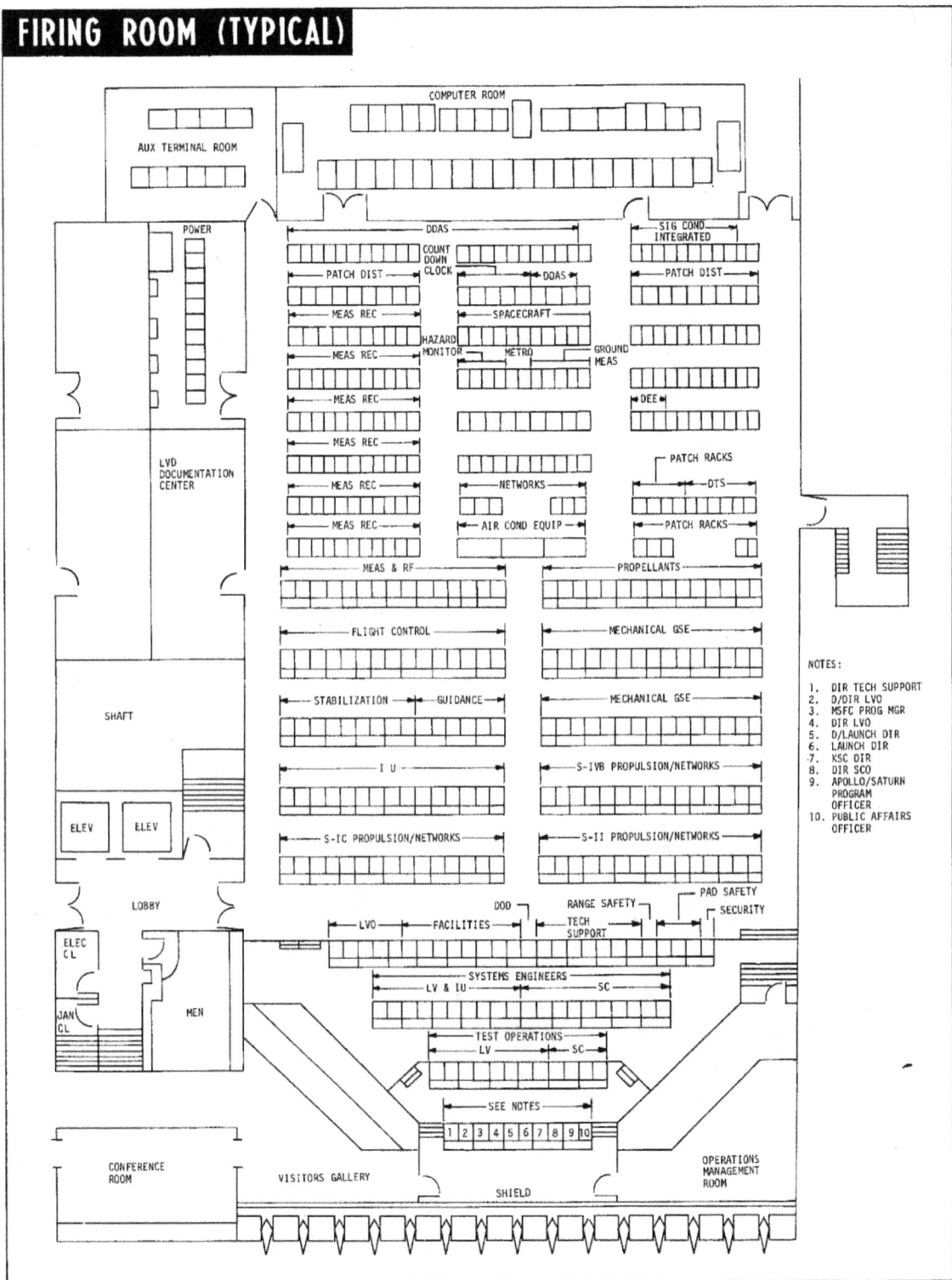

COMPUTER ROOM

AUX TERMINAL ROOM

POWER

SIG COND INTEGRATED

DDAS

COUNT DOWN CLOCK

PATCH DIST          DDAS          PATCH DIST

MEAS REC          SPACECRAFT

HAZARD MONITOR    METRO    GROUND MEAS

MEAS REC

LVD DOCUMENTATION CENTER

MEAS REC          DEE

MEAS REC          PATCH RACKS

MEAS REC          NETWORKS          DTS

MEAS REC          AIR COND EQUIP          PATCH RACKS

MEAS & RF          PROPELLANTS

FLIGHT CONTROL          MECHANICAL GSE

SHAFT

STABILIZATION    GUIDANCE          MECHANICAL GSE

I U          S-IVB PROPULSION/NETWORKS

ELEV        ELEV

S-IC PROPULSION/NETWORKS          S-II PROPULSION/NETWORKS

LOBBY

PAD SAFETY

DOD    RANGE SAFETY

ELEC CL          LVO    FACILITIES    TECH SUPPORT    SECURITY

JAN CL          SYSTEMS ENGINEERS

MEN          LV & IU          SC

TEST OPERATIONS
LV          SC

SEE NOTES

1 2 3 4 5 6 7 8 9 10

CONFERENCE ROOM          OPERATIONS MANAGEMENT ROOM

VISITORS GALLERY          SHIELD

NOTES:
1. DIR TECH SUPPORT
2. D/DIR LVO
3. MSFC PROG MGR
4. DIR LVO
5. D/LAUNCH DIR
6. LAUNCH DIR
7. KSC DIR
8. DIR SCO
9. APOLLO/SATURN PROGRAM OFFICER
10. PUBLIC AFFAIRS OFFICER

Figure 8-16

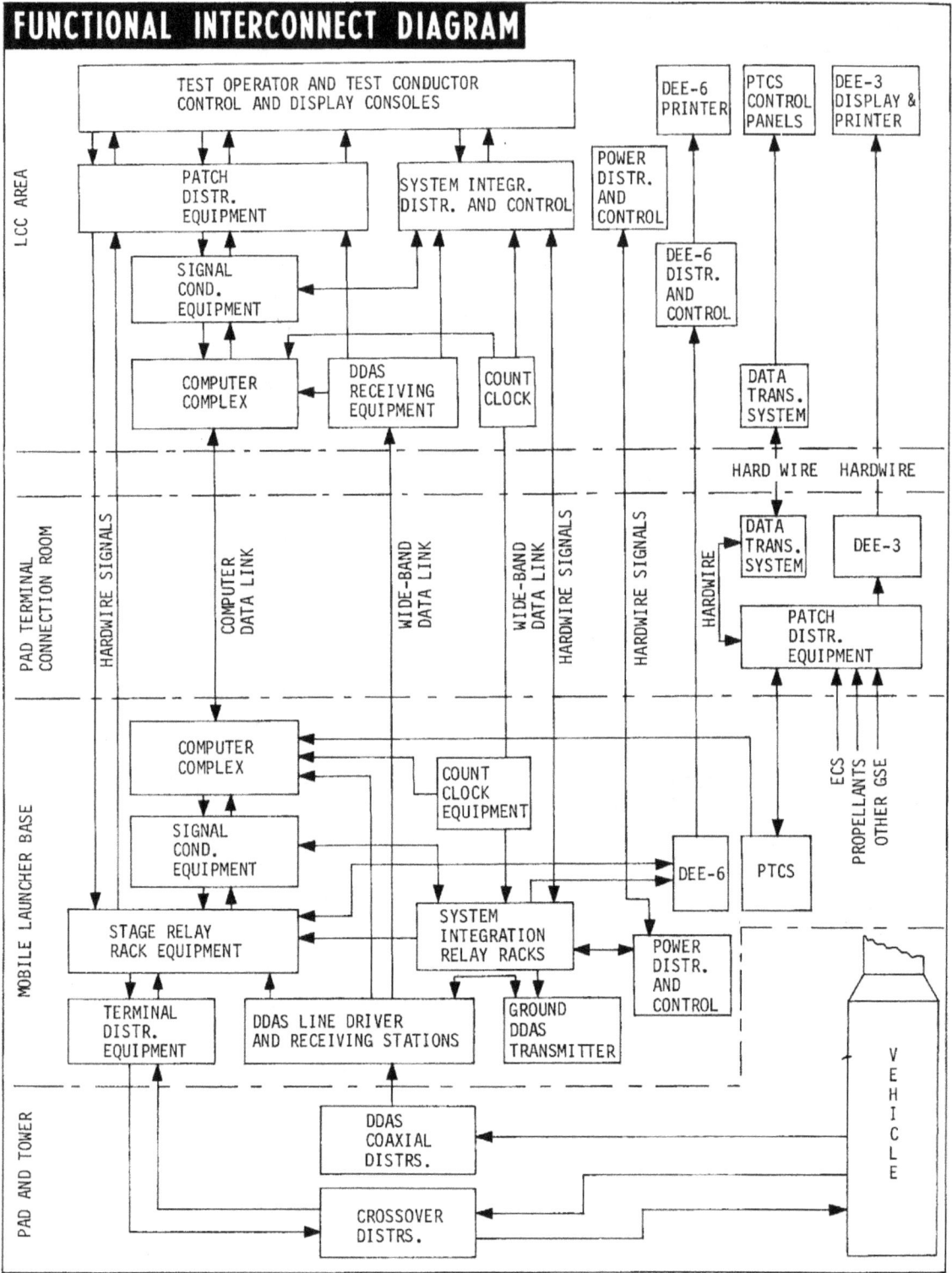

# FUNCTIONAL INTERCONNECT DIAGRAM

**LCC AREA**

TEST OPERATOR AND TEST CONDUCTOR
CONTROL AND DISPLAY CONSOLES

DEE-6
PRINTER

PTCS
CONTROL
PANELS

DEE-3
DISPLAY &
PRINTER

PATCH
DISTR.
EQUIPMENT

SYSTEM INTEGR.
DISTR. AND CONTROL

POWER
DISTR.
AND
CONTROL

SIGNAL
COND.
EQUIPMENT

DEE-6
DISTR.
AND
CONTROL

COMPUTER
COMPLEX

DDAS
RECEIVING
EQUIPMENT

COUNT
CLOCK

DATA
TRANS.
SYSTEM

**PAD TERMINAL CONNECTION ROOM**

HARD WIRE    HARDWIRE

HARDWIRE SIGNALS

COMPUTER DATA LINK

WIDE-BAND DATA LINK

WIDE-BAND DATA LINK

HARDWIRE SIGNALS

HARDWIRE SIGNALS

HARDWIRE

DATA
TRANS.
SYSTEM

DEE-3

PATCH
DISTR.
EQUIPMENT

**MOBILE LAUNCHER BASE**

COMPUTER
COMPLEX

COUNT
CLOCK
EQUIPMENT

ECS

PROPELLANTS

OTHER GSE

SIGNAL
COND.
EQUIPMENT

DEE-6

PTCS

STAGE RELAY
RACK EQUIPMENT

SYSTEM
INTEGRATION
RELAY RACKS

POWER
DISTR.
AND
CONTROL

TERMINAL
DISTR.
EQUIPMENT

DDAS LINE DRIVER
AND RECEIVING STATIONS

GROUND
DDAS
TRANSMITTER

**PAD AND TOWER**

DDAS
COAXIAL
DISTRS.

CROSSOVER
DISTRS.

VEHICLE

Figure 8-17

Certain major events may be observed by members of the launch management team who occupy the first four rows in the firing room. The significant launch vehicle events which are displayed on the 10 x 10 foot screen are shown in figure 8-18.

## PRELAUNCH SEQUENCE

The prelaunch sequence of events (see figure 8-19) take place first in the VAB and then at the pad. The VAB events are V-times and are referenced to completion of VAB activities. The pad events are T-times and are referenced to liftoff.

Prior to VAB activities, the stages and components are received at KSC. The stages and components are unloaded, transported to the VAB, inspected, and erected in the applicable checkout bay.

### VAB Activities

The VAB activities are the assembly and checkout activities which are completed in two major areas of the VAB - the high bay and the low bay. These activities require approximately 250 clock hours and include phases A and B as illustrated in figure 8-19.

Phase A .includes the time period V-253 to V-115 and encompasses the vehicle assembly and checkout activities accomplished prior to spacecraft assembly and installation. Phase B includes the remaining activities which are completed in the VAB.

Low Bay Activities. The low bay activities include receipt and inspection of the S-II stage, S-IVB stage and IU, and the assembly and checkout of the S-II and S-IVB stages.

## MAJOR EVENTS

| | | | | |
|---|---|---|---|---|
| R F SILENCE | S-IVB LOX TANK PRESSURIZED | S-IC LOX TANK PRESSURIZED | COMMIT | S-IVB ENGINE START |
| S-II PREP COMPLETE | S-IC FUEL TANK PRESSURIZED | S-IC PROPELLANTS PRESSURIZED | LIFTOFF | S-IVB CUTOFF |
| S-IVB PREP COMPLETE | S-IVB LH2 TANK PRESSURIZED | S-IC INTERTANK UMB DISCONNECTED | AUTOMATIC ABORT ENABLED | S/C SEPARATION |
| IU READY | S-IVB PROPELLANTS PRESSURIZED | S-IC FORWARD UMB DISCONNECTED | S-IC CUTOFF | PAD ABORT REQUEST |
| S/C READY | S-II LH2 TANK PRESSURIZED | S-II AFT UMB DISCONNECTED | S-IC/S-II SEPARATION | R F ABORT REQUEST |
| E D S READY | S-II PROPELLANTS PRESSURIZED | S-IC INTERTANK UMB RETRACTED | S-II ENGINE START | |
| RANGE SAFE | S-IC ON INTERNAL POWER | S-IC FORWARD UMB RETRACTED | S-II SECOND PLANE SEPARATION | |
| S-IC PREP COMPLETE | S-II ON INTERNAL POWER | S-II AFT UMB RETRACTED | LET JETTISON | |
| LAUNCH SEQUENCE START | S-IVB ON INTERNAL POWER | READY FOR S-IC IGNITION | S-II CUTOFF | |
| S-II LOX TANK PRESSURIZED | IU ON INTERNAL POWER | S-IC IGNITION | S-II/S-IVB SEPARATION | EVENT SYSTEM CALIBRATING |

Figure 8-18

The S-II stage is brought into the low bay area and positioned on the checkout dolly and access platforms are installed. An insulation leak check, J-2 engine leak check, and propellant level probes electrical checks are made.

The S-IVB stage is brought into the low bay area and positioned on the checkout dolly and access platforms are installed. A fuel tank inspection, J-2 engine leak test, hydraulic system leak check, and propellant level sensor electrical checks are made.

# PHASE BREAKDOWN FOR SEQUENCE OF EVENTS

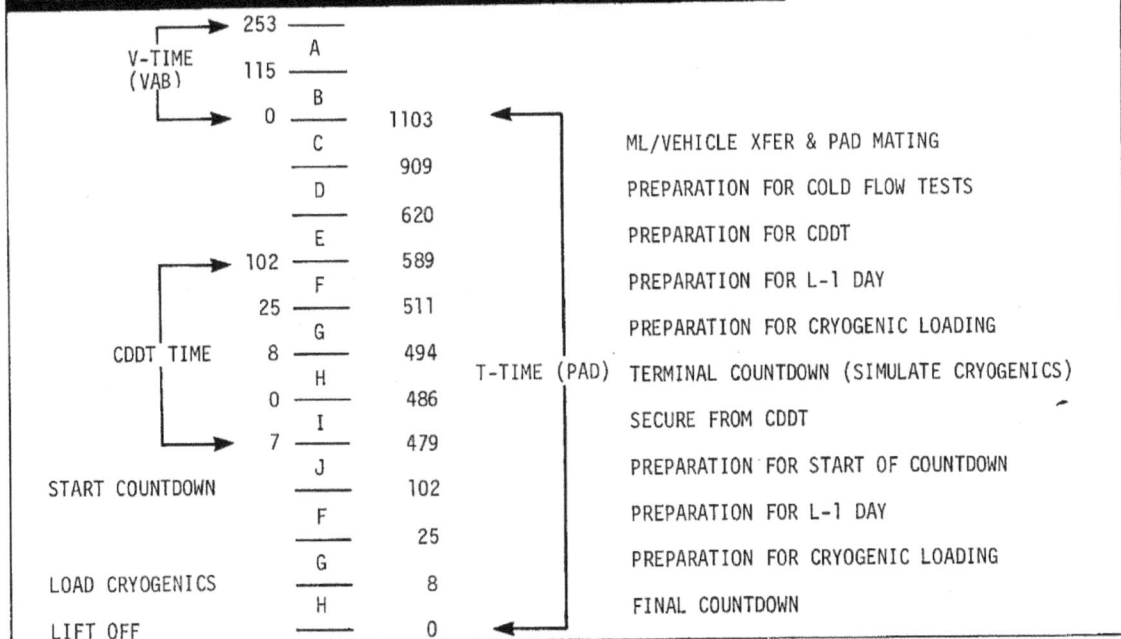

| | | |
|---|---|---|
| V-TIME (VAB) | 253 — A | |
| | 115 — B | |
| | 0 — C | 1103 — ML/VEHICLE XFER & PAD MATING |
| | — D | 909 — PREPARATION FOR COLD FLOW TESTS |
| | — E | 620 — PREPARATION FOR CDDT |
| CDDT TIME | 102 — F | 589 — PREPARATION FOR L-1 DAY |
| | 25 — G | 511 — PREPARATION FOR CRYOGENIC LOADING |
| | 8 — H | 494 — TERMINAL COUNTDOWN (SIMULATE CRYOGENICS) |
| | 0 — I | 486 — SECURE FROM CDDT |
| | 7 — J | 479 — PREPARATION FOR START OF COUNTDOWN |
| START COUNTDOWN | — F | 102 — PREPARATION FOR L-1 DAY |
| | — G | 25 — PREPARATION FOR CRYOGENIC LOADING |
| LOAD CRYOGENICS | — H | 8 — FINAL COUNTDOWN |
| LIFT OFF | — | 0 |

T-TIME (PAD)

Figure 8-19

## Instant Cutoff Interlocks

The TCS may be terminated automatically by any one of the following six cutoff interlocks:

1. S-IC stage logic cutoff. The cutoff signal is caused by (a) the presence of an engine shutdown signal on the stop solenoids from internal sources; or (b) a malfunction in either of the redundant range safety command destruct systems. The range safety command receiver system check is enabled from T-17 seconds to T-50 milliseconds (ms).

2. S-IC main fuel valve failure. The cutoff signal is caused by (a) both main fuel valves on any engine open prior to hypergol rupture or (b) on any engine, one main fuel valve open while the other valve remains closed (to prevent lox-rich condition in the engines). For main fuel valve failure, engines will shut down in a 3-2 sequence with the first three engines shutting down at cutoff and the other two 100 ms later. If the failure occurs in engine 2 or 4, a preferred engine shutdown sequence will occur, i.e., 2, 4, and 5 followed by 1 and 3. If the failure occurs in engine 1, 3, or 5, the normal engine shutdown sequence will occur, i.e., 1, 3, and 5 followed by 2 and 4. If cutoff is caused by other than main fuel valve failure, the normal shutdown sequence will occur.

3. Sequencer power supply failure. The cutoff signal is caused by an out-of-tolerance value of the voltage supply. A new TCS under development having a battery backup will eventually eliminate this interlock.

4. S-IC voltage failure. The cutoff signal is caused by improper voltage output from either the stage main bus (+ 1D11) or the stage instrumentation bus (+1D21).

5. Emergency detection system (EDS) failure. The EDS failure interlock is enabled from T-8.9 seconds to T-50 ms. A cutoff signal is caused by one of the three manual cutoff commands from the spacecraft or loss of one of the three EDS voting logic buses.

6. IU failure cutoff. This interlock is enabled from T-8.9 seconds to T-50 ms. During this period, a loss of IU ready to launch will initiate cutoff. The IU ready to launch interlock monitors the IU power systems, the flight computer, and the presence of the S-IC ignition command.

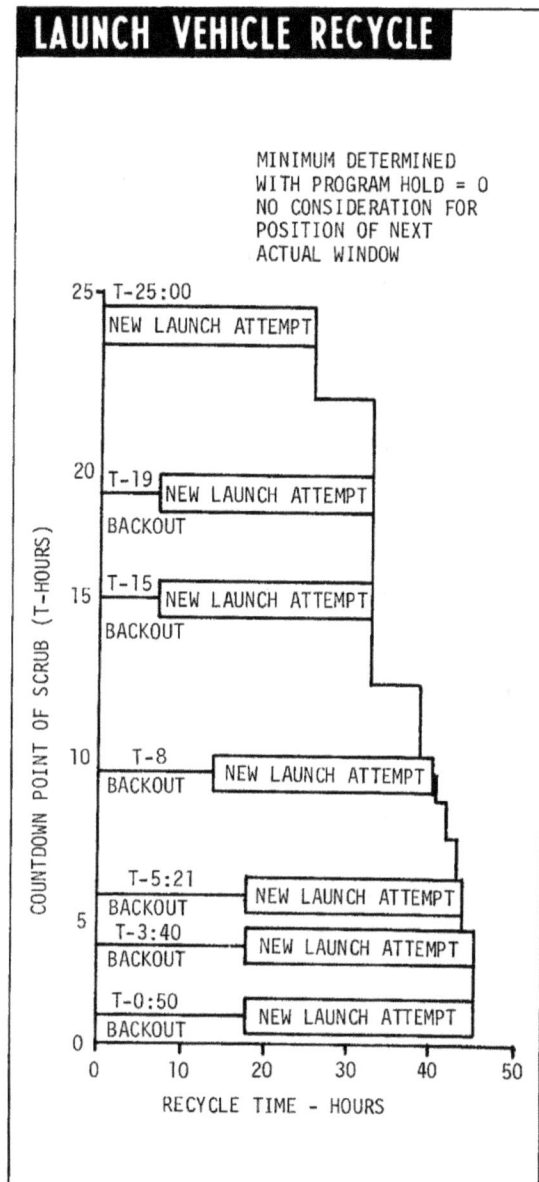

Figure 8-21

# SECTION IX

# MISSION CONTROL MONITORING

## TABLE OF CONTENTS

## INTRODUCTION

Mission control monitoring provides, at the various operational levels, the information required to control, direct and evaluate the mission from prelaunch checkout through recovery. The monitoring function during vehicle flight includes space vehicle tracking, receipt and evaluation of flight crew and launch vehicle status, transmittal of up-data commands to the onboard computer and voice communications with the flight crew. The facilities used in the accomplishment of the monitoring function include an assembly, checkout and launch facility, a central flight control facility, a worldwide network of monitoring stations and a real-time display system.

Associated with the flight crew in mission control operations are the following organizations and facilities:

1.  Mission Control Center (MCC), Manned Spacecraft Center, Houston, Texas. The MCC contains the communication, computer, display, and command systems to enable the flight controllers to effectively monitor and control the space vehicle.

2.  Kennedy Space Center, Cape Kennedy, Florida. The space vehicle is assembled and launched from this facility.
    Prelaunch, launch, and powered flight data are collected by the Central Instrumentation Facility (CIF) at KSC from the launch pads, CIF receivers, Merritt Island Launch Area (MILA), and the downrange Air Force Eastern Test Range (AFETR) stations. This data is transmitted to MCC via the Apollo Launch Data System (ALDS). Also located at KSC is the Impact Predictor (IP).

3.  Goddard Space Flight Center (GSFC), Greenbelt, Maryland. GSFC manages and operates the Manned Space Flight Network (MSFN) and the NASA Communications (NASCOM) networks. During flight, the MSFN is under operational control of the MCC.

4.  George C. Marshall Space Flight Center (MSFC), Huntsville, Alabama. MSFC, by means of the Launch Information Exchange Facility (LIEF) and the Huntsville Operations Support Center (HOSC), provides launch vehicle systems real-time support to KSC and MCC for preflight, launch, and flight operations.

A block diagram of the basic flight control interfaces is shown in figure 9-1.

## VEHICLE FLIGHT CONTROL CAPABILITY

Flight operations are controlled from the MCC. The MCC is staffed by flight control personnel who are trained and oriented on one program and mission at a time. The flight control team members perform mission planning functions and monitor flight preparations during preflight periods. Each member becomes and operates as a specialist on some aspect of the mission.

### MCC ORGANIZATION

The MCC has two control rooms for flight control of manned space flight missions. Each control room, called a Mission Operations Control Room (MOCR), is used independently of the other and is capable of controlling individual missions. The control of one mission involves one MOCR and a designated team of flight controllers. Staff Support Rooms (SSR's), located adjacent to the MOCR are manned by flight control specialists who provide detailed support to the MOCR. Figure 9-2 outlines the organization of the MCC for flight control and briefly describes key responsibilities. Information flow within the MOCR is shown in figure 9-3.

The consoles within the MOCR and SSR's permit the necessary interface between the flight controllers and the spacecraft. The displays and controls on these consoles and other group displays provide the flight controllers with the capability to monitor and evaluate data concerning the mission.

Problems concerning crew safety and mission success are identified to flight control personnel in the following ways: **IX**

1.  Flight crew observations,

2.  Flight controller real-time observations,

3.  Review of telemetry data received from tape recorder playback,

4.  Trend analysis of actual and predicted values,

5.  Review of collected data by systems specialists,

6.  Correlation and comparison with previous mission data,

7.  Analysis of recorded data from launch complex testing.

The facilities at the MCC include an input/output processor designated as the Command, Communications and Telemetry System (CCATS) and a computational facility, the Real-Time Computer Complex (RTCC). Figure 9-4 shows the MCC functional configuration.

The CCATS consists of three Univac 494 general purpose computers. Two of the computers are configured so that either may handle all of the input/output communications for two complete missions. One of the computers acts as a dynamic standby. The third computer is used for nonmission activities.

The RTCC is a group of five IBM 360 large scale, general purpose computers. Any of the five computers may be designated as the mission operations computer (MOC). The MOC performs all the required computations and display formatting for a mission. One of the remaining computers will be a dynamic standby. Another pair of computers may be used for a second mission or simulation.

SPACE VEHICLE TRACKING

From liftoff of the launch vehicle to insertion into orbit, accurate position data are required to allow the Impact Predictor (IP) and the RTCC to compute a trajectory and an orbit. These computations are required by the flight controllers to evaluate the trajectory, the orbit, and/or any abnormal situations to ensure safe recovery of the astronauts. The launch tracking data are transmitted from the AFETR sites to the IP and then to the RTCC via high-speed data communications circuits at the rate of ten samples per second (s/s). The IP also generates a state vector smooth sample which is transmitted to the RTCC at a rate of two s/s. (A state vector is defined as spacecraft inertial position and inertial rate of motion at an instant of time.) The message from the IP to the RTCC alternately contains one smoothed vector, then five samples of best radar data. Low speed tracking data are also transmitted via teletype (TTY) to MCC,

at a rate of one sample per six seconds, from all stations actively tracking the spacecraft. Figure 9-5 shows data flow from liftoff to orbital insertion.

As the launch vehicle is boosting the spacecraft to an altitude and velocity that will allow the spacecraft to attain earth orbit, the trajectory is calculated and displayed on consoles and plotboards in the MOCR and SSR's. Also displayed are telemetry data concerning status of launch vehicle and spacecraft systems. If the space vehicle deviates excessively from the nominal flight path, or if any critical vehicle condition exceeds tolerance limits, or if the safety of the astronauts or range personnel is endangered, a decision is made to abort the mission.

During the orbit phase of a mission, all stations that are actively tracking the spacecraft will transmit the tracking data through GSFC to the RTCC by teletype, at a frequency of one sample every six seconds. If a thrusting maneuver is performed by the spacecraft, high-speed tracking data at the rate of five s/s is transmitted in addition to the teletype data.

Any major maneuver during a mission is planned to occur during or just prior to acquisition by a tracking station that can relay high-speed tracking data to the MCC. This is to ensure that data is available for the calculation of the new spacecraft orbit and ephemeris.

Approximately 25 minutes prior to anticipated spacecraft acquisition by a tracking station, a message giving time, antenna position coordinates, and range is dispatched to that station. This information is computed from the ephemeris and is used by station personnel to pre-position the antenna and enable spacecraft acquisition with minimum delay.

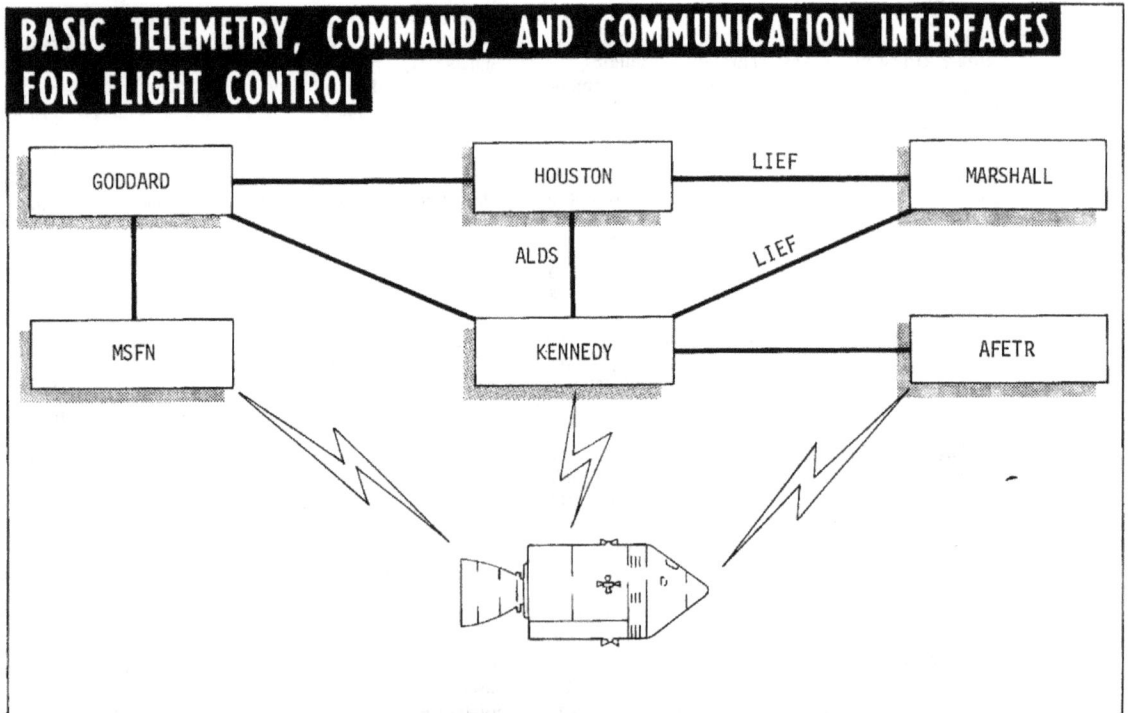

BASIC TELEMETRY, COMMAND, AND COMMUNICATION INTERFACES FOR FLIGHT CONTROL

Figure 9-1

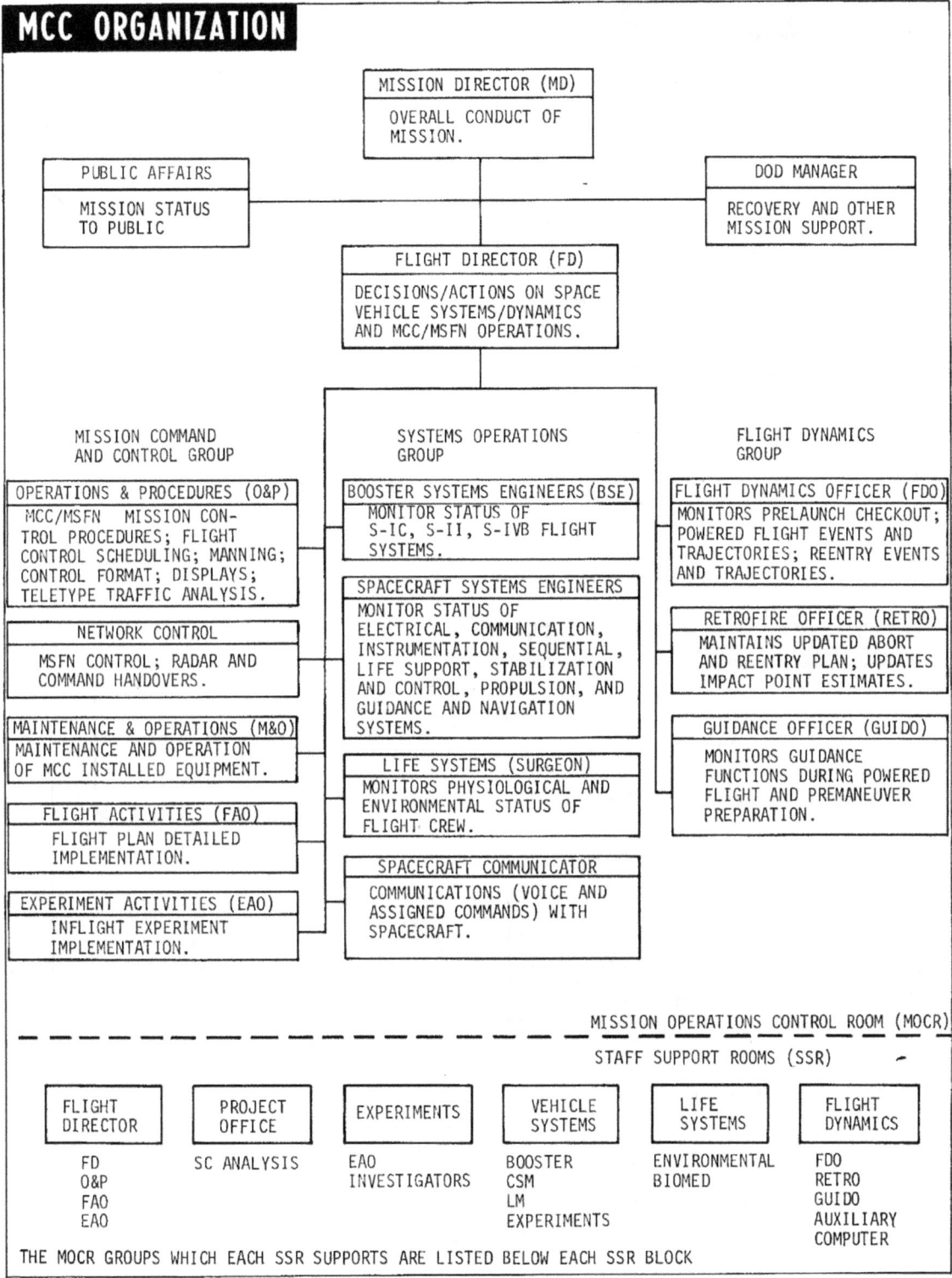

# MCC ORGANIZATION

**MISSION DIRECTOR (MD)**
OVERALL CONDUCT OF MISSION.

**PUBLIC AFFAIRS**
MISSION STATUS TO PUBLIC

**DOD MANAGER**
RECOVERY AND OTHER MISSION SUPPORT.

**FLIGHT DIRECTOR (FD)**
DECISIONS/ACTIONS ON SPACE VEHICLE SYSTEMS/DYNAMICS AND MCC/MSFN OPERATIONS.

### MISSION COMMAND AND CONTROL GROUP

**OPERATIONS & PROCEDURES (O&P)**
MCC/MSFN MISSION CONTROL PROCEDURES; FLIGHT CONTROL SCHEDULING; MANNING; CONTROL FORMAT; DISPLAYS; TELETYPE TRAFFIC ANALYSIS.

**NETWORK CONTROL**
MSFN CONTROL; RADAR AND COMMAND HANDOVERS.

**MAINTENANCE & OPERATIONS (M&O)**
MAINTENANCE AND OPERATION OF MCC INSTALLED EQUIPMENT.

**FLIGHT ACTIVITIES (FAO)**
FLIGHT PLAN DETAILED IMPLEMENTATION.

**EXPERIMENT ACTIVITIES (EAO)**
INFLIGHT EXPERIMENT IMPLEMENTATION.

### SYSTEMS OPERATIONS GROUP

**BOOSTER SYSTEMS ENGINEERS (BSE)**
MONITOR STATUS OF S-IC, S-II, S-IVB FLIGHT SYSTEMS.

**SPACECRAFT SYSTEMS ENGINEERS**
MONITOR STATUS OF ELECTRICAL, COMMUNICATION, INSTRUMENTATION, SEQUENTIAL, LIFE SUPPORT, STABILIZATION AND CONTROL, PROPULSION, AND GUIDANCE AND NAVIGATION SYSTEMS.

**LIFE SYSTEMS (SURGEON)**
MONITORS PHYSIOLOGICAL AND ENVIRONMENTAL STATUS OF FLIGHT CREW.

**SPACECRAFT COMMUNICATOR**
COMMUNICATIONS (VOICE AND ASSIGNED COMMANDS) WITH SPACECRAFT.

### FLIGHT DYNAMICS GROUP

**FLIGHT DYNAMICS OFFICER (FDO)**
MONITORS PRELAUNCH CHECKOUT; POWERED FLIGHT EVENTS AND TRAJECTORIES; REENTRY EVENTS AND TRAJECTORIES.

**RETROFIRE OFFICER (RETRO)**
MAINTAINS UPDATED ABORT AND REENTRY PLAN; UPDATES IMPACT POINT ESTIMATES.

**GUIDANCE OFFICER (GUIDO)**
MONITORS GUIDANCE FUNCTIONS DURING POWERED FLIGHT AND PREMANEUVER PREPARATION.

MISSION OPERATIONS CONTROL ROOM (MOCR)

- - - - - - - - - - - - - - - - - - - - - - - - - - - - - - - - - - - - - - - - -

STAFF SUPPORT ROOMS (SSR)

| FLIGHT DIRECTOR | PROJECT OFFICE | EXPERIMENTS | VEHICLE SYSTEMS | LIFE SYSTEMS | FLIGHT DYNAMICS |
|---|---|---|---|---|---|
| FD O&P FAO EAO | SC ANALYSIS | EAO INVESTIGATORS | BOOSTER CSM LM EXPERIMENTS | ENVIRONMENTAL BIOMED | FDO RETRO GUIDO AUXILIARY COMPUTER |

THE MOCR GROUPS WHICH EACH SSR SUPPORTS ARE LISTED BELOW EACH SSR BLOCK

Figure 9-2

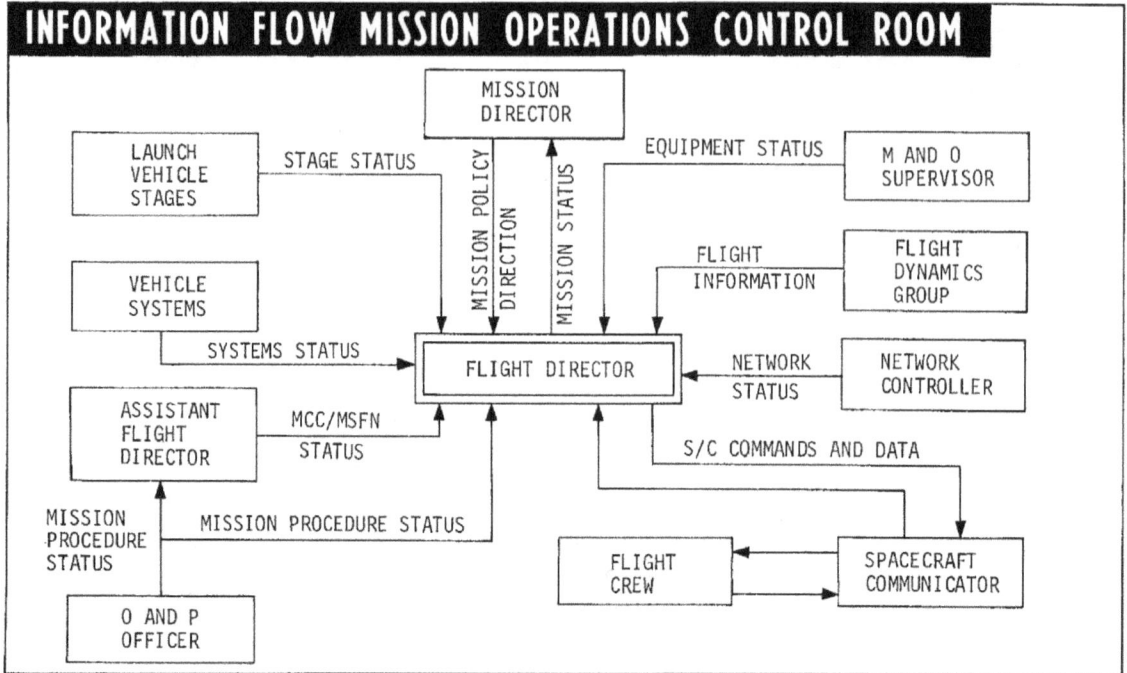

Figure 9-3

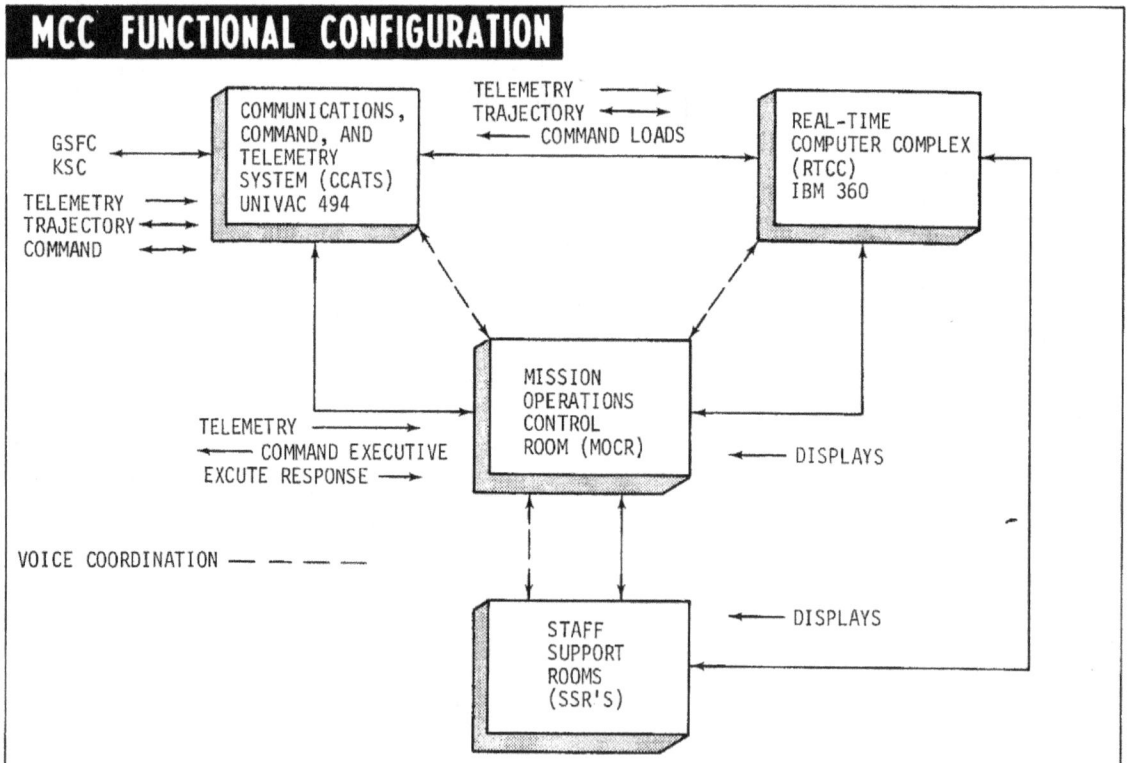

Figure 9-4

## COMMAND SYSTEM

The Apollo ground command systems have been designed to work closely with the telemetry and trajectory systems to provide a method of "closed-loop" command which includes the astronauts and flight controllers as vital links in the commanding operation. For example, analysis of spacecraft data by flight controllers results in a command to alter an observed condition. The effects of the command will be observed in subsequent data presented to the flight control team. This verifies the correct execution of the command and closes the loop. In some cases, such as maneuvering the spacecraft, the command may not be immediately executed, but instead, the astronaut will maneuver the spacecraft at the optimum time specified by the command. The resulting telemetry and trajectory will reflect the maneuver and close the commanding loop.

To prevent spurious commands from reaching the space vehicle, switches on the Command Module console block uplink data from the onboard computers. At the appropriate times, the flight crew will move the switches from the BLOCK to ACCEPT position and thus permit the flow of uplink data.

With a few exceptions, commands to the space vehicle fall into two categories, real-time commands and command loads (also called computer loads, computer update, loads, or update). Among the exceptions is the "clock word" command. This command is addressed to the onboard timing system and is used when the downlink (telemetry) time word and the ground timing system are out of tolerance.

### Real-Time Commands

Real-time commands are used to control space vehicle systems or subsystems from the ground. The execution of a real-time command results in immediate reaction by the affected system. Real-time commands correspond to unique space vehicle hardware and therefore require careful pre-mission planning to yield commands which provide alternate systems operation in the event of an anticipated failure. Pre-mission planning also includes commands necessary to initiate mission contingency plans. Descriptions of several real-time commands used by the Booster System Engineer follow.

The ALTERNATE SEQUENCE real-time commands permit the onboard Launch Vehicle Digital Computer (LVDC) to deviate from its normal program and enter a predefined alternate sequence of program steps. For to deviate from its normal program and enter a predefined, alternate sequence of program steps. For switch selector sequence would jettison the S-II stage and the S-IVB stage would be used to attain switch selector sequence would jettison the S-II stage and the S-IVB stage would be used to attain a parking orbit.

The SEQUENCE INHIBIT real-time command provides the capability to inhibit a programed sequence, usually a maneuver. Each sequence must be separately inhibited with the command being processed immediately after LVDC acceptance. The maneuvers may be inhibited in any random order required during the mission. If an update for a particular command is received after the inhibit for that command, the inhibit is removed and the maneuver will occur at the update time specified.

Other examples of real-time commands are: $LH_2$ VENT CLOSED, $LH_2$ VENT OPEN, LOX VENT OPEN, TERMINATE, SET ANTENNA OMNI, SET ANTENNA LO-GAIN, and SET ANTENNA HI-GAIN.

Real-time commands are stored prior to the mission in the Command Data Processor (CDP) at the applicable command site. The CDP, a Univac 642B general purpose digital computer, is programed to format, encode, and output commands when a request for uplink is generated.

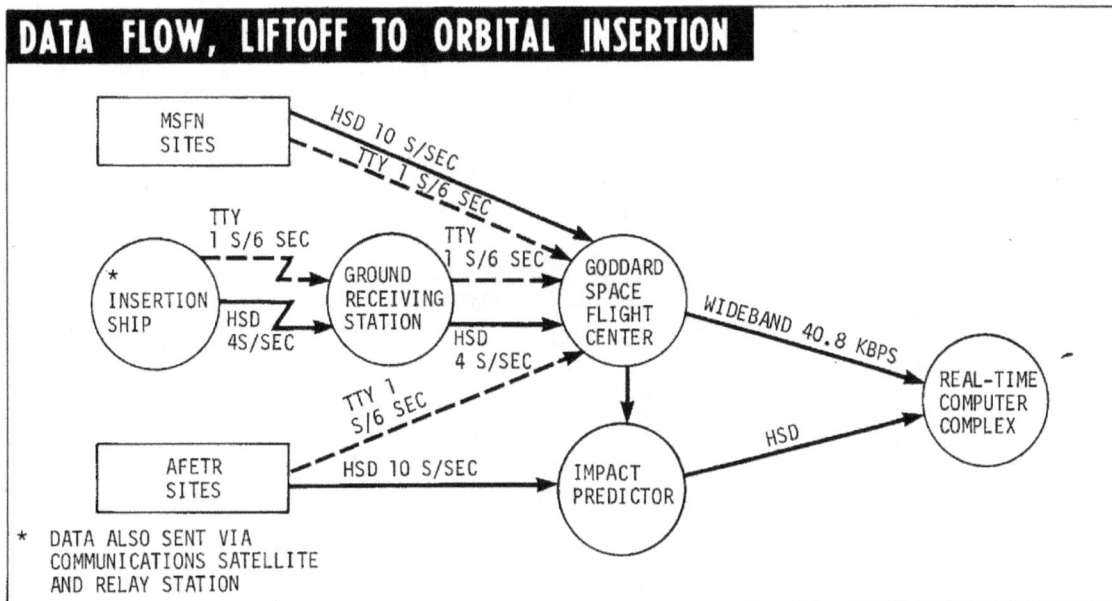

Figure 9-5

## Command Loads

Command loads are generated by the real-time computer complex on request of flight controllers. Command loads are based on the latest available telemetry and/or trajectory data. Due to the nature of these commands, the data structure cannot be determined prior to the mission but must be calculated as a result of real-time data. A command load, for example, may define the exact conditions under which a thrust may be applied that will change a faulty orbit to the desired orbit.

The RTCC operating personnel take data supplied by the flight controllers requesting the command load, and by selecting the appropriate computer program, cause the computer to "make up" a command load. When the load is "ready", it is reviewed by the responsible flight controller via the display system. When the load is approved it is transferred via NASCOM, in the form of high-speed data and/or teletype messages, to the appropriate site and stored in the command data processor. The CCATS will retain the load in memory where it is available for re-transmission should difficulties be encountered in the transfer procedure. When the command load is properly stored in the site's command data processor, a load validation message is sent to the CCATS and to the flight controller.

Flight controllers typically required to generate a command load include the Booster Systems Engineer (BSE), the Flight Dynamics Officer (FDO), the Guidance Officer (GUIDO), and the Retrofire Officer (RETRO).

Prior to the acquisition of the space vehicle by a site, the flight controllers requiring command capability during the pass indicate their requirements to the Real-Time Command Controller (R/T CMD) in the CCATS area. The R/T CMD enables the circuitry to permit the command function for the console/site combinations requested.

When the space vehicle has been acquired by the site, it is announced over one of the voice coordination loops and each flight controller executes his commands according to the priority assigned by real-time decision. Disposition of each command is indicated by the indicator lights on the command panel. These indicator lights are operated by CCATS in response to the verification and/or reject messages received from the site. Typical Command loads (BSE) are described below.

The SECTOR DUMP command causes the LVDC to telemeter the entire contents of one memory sector, or a series of memory sectors within the same memory module. For example: this command is used to telemeter the memory sector in which the navigation update parameters are stored. The real-time TERMINATE command may by used to halt a sector dump before the last block of data is telemetered. The SECTOR DUMP command applies to the orbital phase.

The NAVIGATION UPDATE command permits loading of six navigation parameters and an execution time into the LVDC.

The SEQUENCE INITIATE UPDATE command permits update of stored values for the time of initiation of each of the pre-programmed maneuvers specified for the mission. The pre-stored values will be adjusted immediately upon the receipt of the update command.

Other examples of command loads are: TIME BASE UPDATE, SLV PRELAUNCH TARGET UPDATE, and SLV ORBIT TARGET UPDATE.

## DISPLAY AND CONTROL SYSTEM

MCC is equipped with facilities which provide for the input of data from the MSFN and KSC over a combination of high-speed data, low-speed data, wide-band data, teletype, and television channels. This data is computer processed for display to the flight controllers. With this displayed data, detailed mission control by the MOCR and detailed support in the various specialty areas by the SSR's are made possible.

### Display System

Several methods of displaying data are used including television (projection TV, group displays, closed circuit TV, and TV monitors), console digital readouts, and event lights. The display and control system interfaces with the RTCC and includes computer request, encoder multiplexer, plotting display, slide file, digital-to-TV converter, and telemetry event driver equipments (see figure 9-6).

The encoder multiplexer receives the display request from the console keyboard and encodes it into digital format for transmission to the RTCC.

The converter slide file data distributor routes slide selection data from the RTCC to reference slide files and converter slide files, receives RTCC control data signals required to generate individual console television displays and large scale projection displays; and distributes control signals to a video switching matrix to connect an input video channel with an output television viewer or projector channel.

The digital-to-television conversion is accomplished by processing the digital display data into alphanumeric, special symbol, and vector displays for conversion into video signals. This process produces analog voltages which are applied to the appropriate element of a character-shaped beam cathode ray tube. The resultant display image on the face of the cathode ray tube is optically mixed with the slide file image and viewed by a television camera, which transmits the mixed images to the TV monitors and projectors. The digital-to-TV data consists of preprogramed computer-generated dynamic data formats which are processed and combined with background data on film slides.

### Control System

A control system is provided for flight controllers to exercise their respective functions for mission control and technical management. This system is comprised of different groups of consoles with television monitors, request keyboards, communications equipment, and assorted modules. These units are assembled as required, to provide each operational position in the MOCR with the control and display capabilities required for the particular mission. The console components are arranged to provide efficient operation and convenience for the flight controller. The console configuration for the Booster System Engineer is shown in figure 9-7. Brief descriptions of the console modules are contained in the following paragraphs.

The event indicator modules display discrete mission events, system modes, equipment modes, and vehicle system status. Each BSE event module consists of 18 bilevel indicators which permit a total of 36 event positions per module.

Another type of event module has the capability to display up to 72 events. The signals affecting the lights are telemetry inputs from the space vehicle.

The ground elapsed time module shows the elapsed time from liftoff.

The communications module provides rapid access to internal, external, and commercial voice communications circuits. Flight controllers may monitor as well as talk over these circuits.

The command module provides the flight controller with the means to select and initiate real-time commands and command loads for transmission to the space vehicle. The module also indicates receipt/rejection of commands to the space vehicle and verification of proper storage of command loads in a site command data processor. The modules are made up of pushbutton indicators (PBI's) which are labeled according to their function.

The status report module (SRM) provides flight controllers in the MOCR with the means to report mission status to the flight director and assistant flight director and to review systems status in the SSR's.

The abort request indicator provides the capability, by toggle switch action, to indicate an abort condition. This produces a priority command to the spacecraft.

The manual select keyboard (MSK) permits the flight controller to select a TV channel, a computer-generated display format, or a reference file for viewing on the TV monitor. The desired item is selected by use of a PBI three-mode switch (TV channel, reference file, display request) and a select-number thumbwheel encoder. BSE consoles may obtain a hardcopy of a display by use of a PBI. The hardcopy is delivered to the console via a pneumatic tube system.

The summary message enable keyboard (SMEK) permits the flight controller to instruct the RTCC to strip out selected data from telemetry inputs and to format this data into digital-to-TV summary displays. The SMEK is also used to instruct the RTCC to convert data into specific teletype messages to designated MSFN sites. The module contains appropriately labeled PBI's.

The TV monitor module provides viewing of computer-generated displays, reference file data, closed circuit TV within the MCC and KSC, and commercial TV.

Console modules not illustrated in figure 9-7 but used in MOCR and SSR consoles are described below.

The analog meter module displays parameters in engineering units. Up to 15 measurements, determined prior to the mission, can be displayed. There are movable markers on each meter which are manually set to show the nominal value

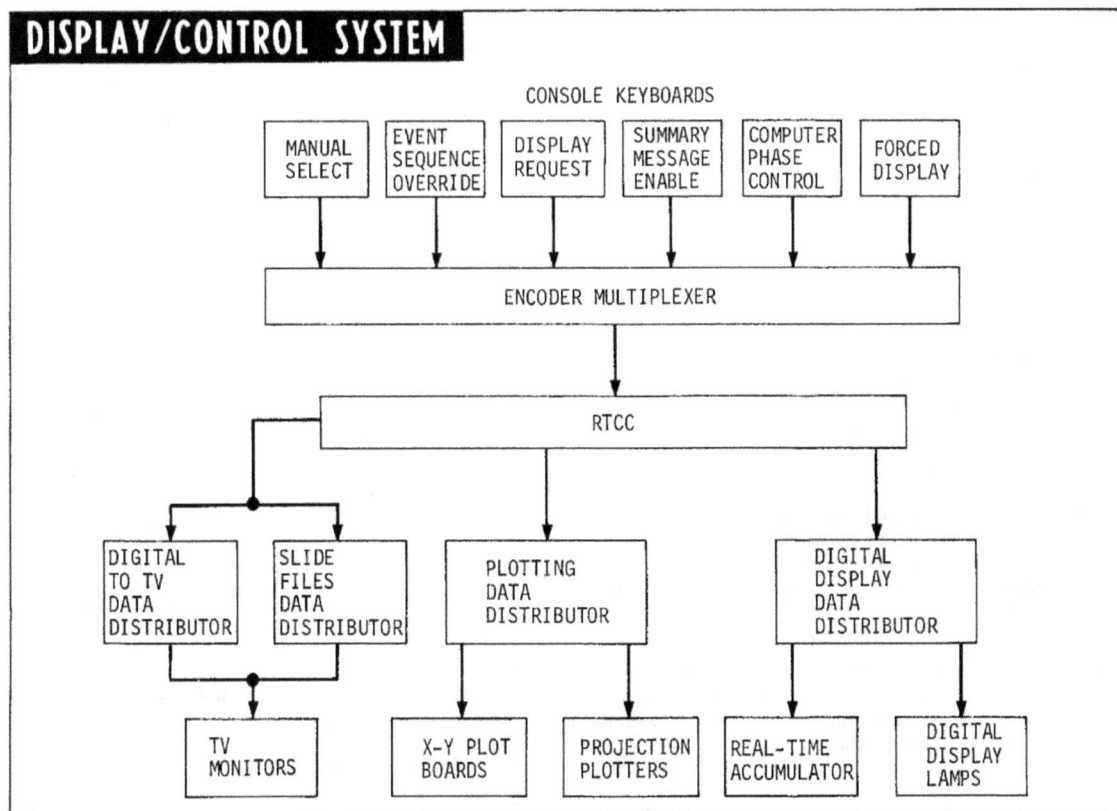

Figure 9-6

of the parameter as well as the upper and lower limits. When a parameter exceeds the established tolerance, a red warning light at the bottom of the meter is lit. The light extinghishes when the parameter returns to tolerance or the exceeded limit is manually extended.

The forced display module (FDK) indicates to the flight controller a violation of preprogrammed limits of specific analog parameters as determined from the incoming data. When an out-of-tolerance condition occurs, the appropriate PBI in the module illuminates as a warning. When the flight controller acknowledges the lit PBI by depressing it, a readout indicator on the FDK will display a four digit code which identifies the display format on which the out of tolerance parameter appears.

The display request keyboard (DRK) provides a fast means of requesting the RTCC for a specific display format. The display is called up by depressing the appropriately labeled PBI. This keyboard provides the same capability as MSK in the "display request mode" except that the callup is faster in that thumbwheel selection is not required.

## CONTINGENCY PLANNING AND EXECUTION

Planning for a mission begins with the receipt of mission requirements and objectives. The planning activity results in specific plans for prelaunch and launch operations, preflight training and simulation, flight control procedures, flight crew activities, MSFN and MCC support, recovery operations, data acquisition and flow, and other mission related operations. Simulations are planned and performed to test procedures and train flight control and flight crew teams in normal and contingency operations. The simulation and training exercises result in a state of readiness for the mission. Mission documentation covering all aspects of the mission is developed and tested during the planning and training period. Included in this documentation are the mission rules.

## MISSION RULES

Mission rules are a compilation of rules governing the treatment of contingency situations. The purpose of the mission rules is to outline preplanned actions to assist in making rapid real-time decisions during prelaunch, flight, and recovery operations. The mission rules are based upon the mission objectives and on the objective of maintaining a high degree of confidence in crew safety during mission implementation. The mission rules categorize the degree of importance assigned to space vehicle/operational support elements as follows:

A mandatory item (M) is a space vehicle element or operational support element that is essential for accomplishment of the primary mission, which includes prelaunch, flight, and recovery operations that ensure crew safety and effective operational control as well as the attainment of the primary mission objectives.

A highly desirable item (HD) is a space vehicle or operational support element that supports and enhances the accomplishment of the primary mission or is essential for accomplishment of the secondary mission objectives.

Redline values are the maximum and/or minimum limits of a critical parameter (redline function) necessary to describe vehicle, system, and component performance and operation. Redline functions are mandatory items.

## Launch Mission Rules

Launch mission rules cover the following listed items and other information as appropriate:

1. Mandatory and highly desirable onboard instrumentation required to collect data for flight control purposes or postflight evaluation.

2. Mandatory and highly desirable onboard instrumentation required to verify that the space vehicle is ready for launch.

3. Redline values defining upper and lower limits of parameters such as pressure, temperature, voltage, current, and operating time, for any system/subsystem essential to mission success.

4. Mandatory and highly desirable range and instrumentation support required to prepare and launch the space vehicle and accomplish postflight analysis.

5. Range safety requirements and instrumentation as established by the Air Force Eastern Test Range.

6. Wind and weather restrictions on the launch.

7. Long-range camera coverage required for the launch.

8. Launch window definition and launch window rules pertaining to launch operations.

9. The space vehicle functional sequence.

10. The time span before launch during which manual cutoff will not be attempted.

## Flight Mission Rules

Flight mission rules cover the following listed items and also medical decision rules for appropriate mission go/no-go points:

1. Mandatory and highly desirable instrumentation for control of the space vehicle after liftoff.

2. Space vehicle nominal and non-nominal subsystem performance in accordance with alternate mission capability.

3. Trajectory and guidance.

4. Mandatory and highly desirable items of mission support in the MCC.

5. Mandatory and highly desirable range and MSFN support required to support the mission after liftoff, and for subsequent analysis and evaluation.

6. Rules relating to the human or medical aspects of manned flight.

7. Recovery restrictions.

8. Launch window rules pertaining to such items as time of liftoff, launch azimuth, recovery, and spacecraft performance limitations.

## VEHICLE MISSION ABORT CAPABILITY

Section III of this manual, dealing with the emergency detection system, describes the manual and automatic capabilities for mission abort designed into the Saturn-Apollo system. Also described in Section III are the abort modes and limits, and the emergency procedures related to mission abort.

Time critical aborts must be initiated onboard the spacecraft because sufficient time is not available for response by the MCC ground based flight controllers.

The detection of slowly diverging conditions which may result in an abort is the prime responsibility of MCC. In the event such conditions are discovered, MCC requests abort of the mission or, circumstances permitting, sends corrective commands to the vehicle or requests corrective flight crew action.

In the event of a non-catastrophic contingency, MCC recommends alternate flight procedures, and mission events are rescheduled to derive maximum benefit from the modified mission.

## ABORT GROUND RULES

Flight crew safety shall take precedence over the accomplishment of mission objectives.

The Command Pilot of a manned mission may initiate such inflight action as he deems necessary for crew safety.

The Launch Operations Manager may send an abort request signal from the time the launch escape system (LES) is armed until the space vehicle reaches sufficient altitude to clear the top of the umbilical tower.

From liftoff to tower clear, the Launch Director and Flight Director have concurrent responsibility for sending an abort request.

Control of the space vehicle passes from the Launch Director to the Flight Director when the space vehicle clears the top of the tower.

In the Mission Control Center, the Flight Director, Flight Dynamics Officer, and Booster Systems Engineer have the capability to send an abort request signal.

Where possible, all manual abort requests from the ground during flight will be based on two independent indications of the failure.

## LAUNCH VEHICLE CONTINGENCIES/REACTIONS

Malfunctions which could result in loss of the space vehicle are analyzed and a mission rule is developed to respond to the malfunction. The contingencies and corresponding reactions are incorporated into the premission simulations and the mission rules are refined as required. Typical contingencies/reactions on which mission rules are based are listed in figure 9-8.

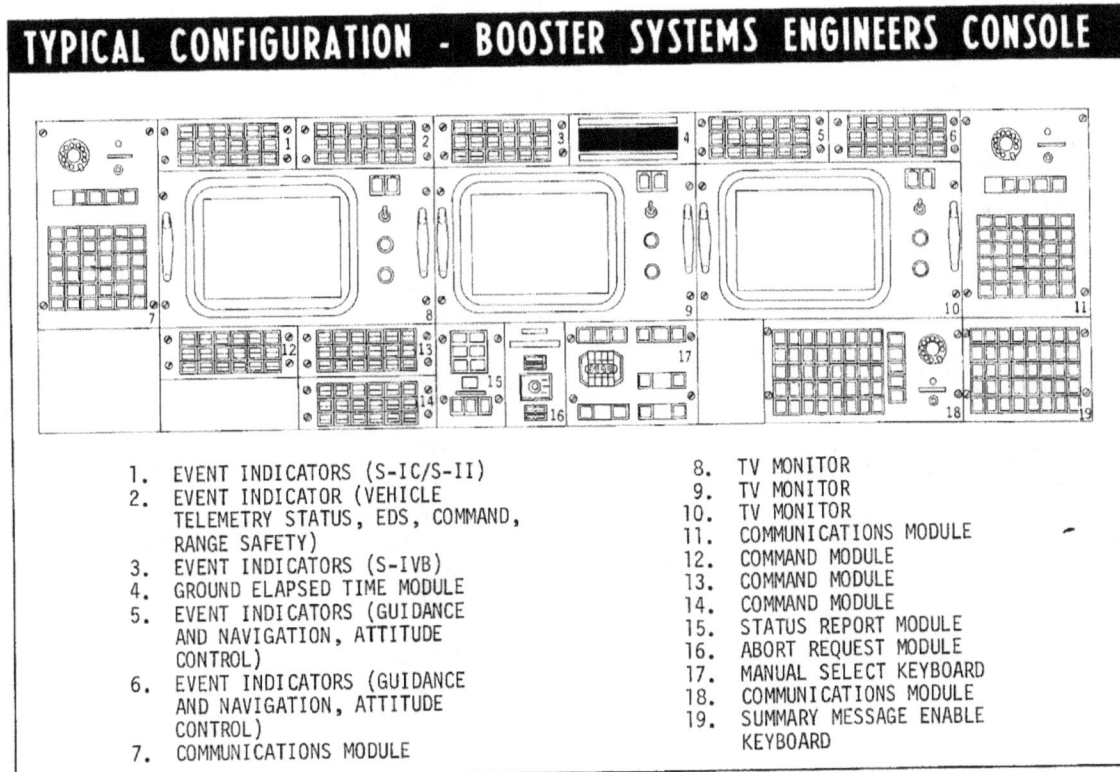

## TYPICAL CONFIGURATION - BOOSTER SYSTEMS ENGINEERS CONSOLE

1. EVENT INDICATORS (S-IC/S-II)
2. EVENT INDICATOR (VEHICLE TELEMETRY STATUS, EDS, COMMAND, RANGE SAFETY)
3. EVENT INDICATORS (S-IVB)
4. GROUND ELAPSED TIME MODULE
5. EVENT INDICATORS (GUIDANCE AND NAVIGATION, ATTITUDE CONTROL)
6. EVENT INDICATORS (GUIDANCE AND NAVIGATION, ATTITUDE CONTROL)
7. COMMUNICATIONS MODULE
8. TV MONITOR
9. TV MONITOR
10. TV MONITOR
11. COMMUNICATIONS MODULE
12. COMMAND MODULE
13. COMMAND MODULE
14. COMMAND MODULE
15. STATUS REPORT MODULE
16. ABORT REQUEST MODULE
17. MANUAL SELECT KEYBOARD
18. COMMUNICATIONS MODULE
19. SUMMARY MESSAGE ENABLE KEYBOARD

Figure 9-7

| TYPICAL INFLIGHT CONTINGENCIES/REACTIONS | | | | |
|---|---|---|---|---|
| CONTINGENCY | FLIGHT TIME MIN:SEC | EFFECT | SENSORS OR DISPLAYS | MISSION RULE DATA |
| Loss of thrust for single control engine | 0:00- 0:00.75 | Collision with hold-down post | Crew<br>1. Thrust OK light<br>2. Physiological<br>Ground<br>1. Thrust OK light<br>2. Thrust chamber pressure<br>3. Longitudinal acceleration<br>4. Visual | No automatic abort<br>Manual abort by LES with two cues<br>1. Thrust OK light<br>2. Physiological<br>3. Abort request light |
| Loss of thrust for any single engine | 0-0.03 0:00- 00:03 | Pad fallback | Crew<br>1. Thrust OK light<br>2. Physiological<br>Ground<br>1. Thrust OK light<br>2. Thrust chamber<br>3. Longitudinal acceleration<br>4. Visual | No automatic abort<br>Manual abort by LES with two cues<br>1. Thrust OK light<br>2. Physiological<br>3. Abort request light |
| Loss of thrust for any single tower side engine (Engine No. 1 or No. 2) | 0:00.75- 0:04.5 | Tower collision | Crew<br>1. Thrust OK light<br>2. Physiological<br>Ground<br>1. Thrust OK light<br>2. Thrust chamber pressure<br>3. Longitudinal acceleration<br>4. Visual | No automatic abort available.<br>Manual abort by LES with two cues<br>1. Thrust OK light<br>2. Physiological<br>3. Abort request light |
| Loss of thrust for single control engine | 0:00- 0:10 | Possible false abort prior to 50 seconds if not covered by above contingencies | Crew<br>1. Thrust OK light<br>2. Q-Ball may exceed 3.2 psi (prior to 50 seconds)<br>3. Physiological<br>Ground<br>1. Thrust OK light<br>2. Thrust chamber pressure<br>3. Longitudinal acceleration<br>4. Visual | Recommendation required to assure no false abort |
| Loss of thrust any single engine | 0:10- 1:13 | No effects | Crew<br>1. Thrust OK light<br>2. Physiological<br>Ground<br>1. Thrust OK light<br>2. Thrust chamber pressure<br>3. Longitudinal acceleration<br>4. Visual | No abort required |

Figure 9-8 (Sheet 1 of 9)

| TYPICAL INFLIGHT CONTINGENCIES/REACTIONS (CONTINUED) | | | | |
|---|---|---|---|---|
| CONTINGENCY | FLIGHT TIME MIN:SEC | EFFECT | SENSORS OR DISPLAYS | MISSION RULE DATA |
| Loss of thrust any single control engine | 1:08-1:18 | Possible loss of control of vehicle resulting in structural breakup within 0.6 seconds after abort | Crew<br>1. Attitude rate exceeds 4°/second<br>2. Thrust OK light<br>3. Q-Ball exceeds 3.2 psi<br>Ground<br>1. Thrust OK light<br>2. Thrust chamber pressure<br>3. Longitudinal acceleration | Automatic abort when attitude rate is exceeded.<br>Manual abort by LES with two cues<br>1. Thrust OK light<br>2. Q-Ball limits exceeded |
| Loss of thrust any single engine | 1:18-2:30 | No effect | Crew<br>1. Thrust OK light<br>2. Physiological<br>Ground<br>1. Thrust OK light<br>2. Thrust chamber pressure<br>3. Longitudinal acceleration | No abort required |
| | 2:30-8:54 | No effect | Crew<br>1. Thrust OK light<br>2. Physiological<br>Ground<br>1. Thrust OK light<br>2. Thrust chamber pressure<br>3. Longitudinal acceleration | No abort required |
| | S-IVB Burn | Complete loss of thrust | Crew<br>1. Thrust OK light<br>2. Physiological<br>Ground<br>1. Thrust OK light<br>2. Thrust chamber pressure<br>3. Longitudinal acceleration | Abort by SPS |
| Loss of thrust any two adjacent engines | 0:00-2:30 | Loss of control resulting in structural breakup | Crew<br>1. Thrust OK lights<br>2. Physiological (if shutdown occurs for 2 or more engines at the same time the effects may be the same as a single engine shutdown)<br>Ground<br>1. Thrust OK light<br>2. Thrust chamber pressure<br>3. Longitudinal acceleration | Automatic abort by LES with loss of thrust for any two engines prior to 2 min 5 sec.<br>Manual abort after 2 min 5 sec. with two engine out lights. |
| Loss of thrust any control engine and center engine | 0:00-1:20 | | | |
| Loss of thrust any opposite control engine | 0:00-1:00 | | | |

Figure 9-8 (Sheet 2 of 9)

| | | TYPICAL INFLIGHT CONTINGENCIES/REACTIONS (CONTINUED) | | |
|---|---|---|---|---|
| CONTINGENCY | FLIGHT TIME MIN:SEC | EFFECT | SENSORS OR DISPLAYS | MISSION RULE DATA |
| One actuator inoperative (null position) | S-IVB boost | Loss of control | Crew<br>1.  No defined cues<br>Ground<br>1.  FDO limits exceeded | No automatic abort. Manual abort by SPS when FDO limits exceeded |
| Any single actuator hardover (pitch or yaw) | 0:00-0:00.32 | Collision with holddown post. | Crew<br>1.  Physiological<br>Ground<br>1.  Visual<br>2.  Loss of hydraulic pressure<br>3.  Engine position | No automatic abort. Manual abort upon contact with holddown posts. |
| Any single actuator hardover (yaw positive only) | 0:00.32-0:02.1 | Tower collision | Crew<br>1.  Roll rate will exceed 4°/second<br>2.  Physiological<br>Ground<br>1.  Actuator position<br>2.  Loss of hydraulic pressure<br>3.  Visual | No automatic abort. Manual abort by LES with two cues<br>1.  Roll attitude error limits exceeded<br>2.  Abort request light |
| Any single actuator hardover (pitch or yaw) | 0:00-0:58 | No effect unless covered by above | Crew<br>1.  Roll attitude error exceeds 5°<br>2.  Q-Ball pressure exceeded<br>Ground<br>1.  Actuator position<br>2.  Loss of hydraulic pressure<br>3.  Visual | No abort required. |
| Any single actuator hardover (pitch or yaw) | 0:58-1:23 | Possible loss of control resulting in structural breakup | Crew<br>1.  Attitude rate exceeds 4°/second<br>2.  Roll attitude error exceeds 5°<br>3.  Q-Ball exceeds 3.2 psi<br>Ground<br>1.  Actuator position<br>2.  Loss of hydraulic pressure | Automatic abort by LES when attitude rate limit is exceeded. Manual abort by LES with two cues<br>1.  Roll attitude limit exceeded<br>2.  Q-Ball limits exceeded<br>3.  Engine position by ground |
| Any single actuator hardover | 1:13-2:30 | No effect | Crew<br>1.  Roll attitude error<br>2.  Q-Ball pressure exceeded<br>Ground<br>1.  Actuator position<br>2.  Loss of hydraulic pressure | No abort required |

Figure 9-8 (Sheet 3 of 9)

| | | TYPICAL INFLIGHT CONTINGENCIES/REACTIONS (CONTINUED) | | |
|---|---|---|---|---|
| CONTINGENCY | FLIGHT TIME MIN:SEC | EFFECT | SENSORS OR DISPLAYS | MISSION RULE DATA |
| | 2:30-8:54 | No effect | Crew<br>1. No defined cues<br>Ground<br>1. Actuator position<br>2. Loss of hydraulic pressure | No abort required |
| | S-IVB Burn Phase | Loss of control | Crew<br>1. Attitude rate exceeds 10°/second<br>2. L/V overrate light<br>Ground<br>1. Actuator position<br>2. Loss of hydraulic pressure | No automatic abort.<br>Manual abort by SPS with two cues<br>1. Attitude rate limit is exceeded<br>2. L/V overrate light<br>3. Engine position (ground) |
| Engine Hardover | S-IVB Restart | Loss of control if restarted. Lead thrust would cause vehicle rotation | Crew<br>1. No defined cues<br>Ground<br>1. Engine position<br>2. Loss of hydraulic pressure | No automatic abort.<br>Manual abort by SPS.<br>Do not attempt restart |
| Loss of both actuators for same engine | 0:00-2:30 | Loss of control resulting in structural breakup if engine goes to roll corner | Crew<br>1. Roll rate exceeds 20°/second<br>2. Q-Ball exceeds 3.2 psi<br>Ground<br>1. Engine position<br>2. Loss of hydraulic pressure<br>3. Roll rate | Automatic abort by LES when roll rate limit is exceeded prior to 120 seconds.<br>Manual abort by LES with two cues<br>1. Q-Ball limit exceeded (50 sec-1 min 50 sec)<br>2. Engine position (ground cue) |
| | 2:30-8:54 | No effect | Crew<br>1. No defined cues<br>Ground<br>1. Engine position<br>2. Loss of hydraulic pressure | No abort required |
| Loss of both actuators for same engine | S-IVB Burn Phase | Loss of control | Crew<br>1. Attitude rate exceeds 10°/second<br>2. L/V overrate light<br>Ground<br>1. Engine position<br>2. Loss of hydraulic pressure | No automatic abort.<br>Manual abort by SPS with two cues<br>1. Attitude rate limit is exceeded<br>2. L/V overrate light<br>3. Engine position (ground) |

Figure 9-8 (Sheet 4 of 9)

| TYPICAL INFLIGHT CONTINGENCIES/REACTIONS (CONTINUED) | | | | |
|---|---|---|---|---|
| CONTINGENCY | FLIGHT TIME MIN:SEC | EFFECT | SENSORS OR DISPLAYS | MISSION RULE DATA |
| Loss of Inertial Attitude | 0:00-0:50 | Loss of control resulting in structural failure after 60 seconds | Crew<br>1. Guidance failure light<br>2. Overrate light (used as redundant guidance failure light from 0 to 2 min 5 sec)<br>3. Attitude error exceeds 5°<br>Ground<br>1. Guidance failure light | Abort after 50 seconds |
| | 0:50-1:25 | Loss of control resulting in structural failure during this time period | Crew<br>1. Guidance failure light<br>2. Overrate light (used as redundant guidance failure light from 0 to 2 min 5 sec)<br>3. Attitude error exceeds 5°<br>Ground<br>1. Guidance failure light | Manual abort with two cues<br>1. Guidance failure light<br>2. Overrate light (used as redundant guidance failure light 0-2 min 5 sec) |
| | 1:25-2:05 | | | Manual abort with three cues<br>1. Guidance fail light<br>2. Overrate light<br>3. Attitude error exceeds 5° |
| | 2:05-2:30 | | Same as above except no overrate light | Manual abort on two cues<br>1. Guidance failure light<br>2. Attitude error exceeds 5° |
| Loss of Inertial Attitude | 1:30-2:30 | Loss of guidance with no structural failures | Crew<br>1. Attitude rate exceeds 4°/second<br>2. Guidance failure light<br>3. Q-Ball exceeds 3.2 psi(1 min 30 sec to 1 min 50 sec)<br>4. Attitude error exceeds 5° | Automatic abort by LES when attitude rate limit is exceeded. Manual abort by LES with two cues<br>1. Guidance failure light<br>2. Q-Ball limits exceeded<br>3. Attitude error limit exceeded |
| | 1:30-8:54 | Loss of guidance | Crew<br>1. Guidance failure light<br>Ground<br>1. FDO limits exceeded | No automatic abort. Manual abort by LES prior to tower jettison or by SPS after tower jettison with a single cue<br>1. Guidance failure light<br>2. FDO limits exceeded |

Figure 9-8 (Sheet 5 of 9)

| TYPICAL INFLIGHT CONTINGENCIES/REACTIONS (CONTINUED) | | | | |
|---|---|---|---|---|
| CONTINGENCY | FLIGHT TIME MIN:SEC | EFFECT | SENSORS OR DISPLAYS | MISSION RULE DATA |
| Loss of inertial attitude | S-IVB Burn Phase | Loss of guidance | Crew 1. Guidance failure light Ground 1. FDO limits exceeded | No automatic abort. Manual abort by SPS with single cue 1. Guidance failure light 2. FDO limits exceeded |
| Loss of Inertial Velocity | All Stages | No effect | | No abort required |
| Loss of Attitude Error Commands | All Stages | Loss of control | Crew 1. Attitude rate exceeds 4°/second for S-IC 2. Attitude rate exceeds 10°/second for S-II and S-IVB 3. Attitude error exceeds 5° for S-IC Ground 1. FDO limits exceeded | Automatic abort by LES prior to 2 min 5 sec when attitude rate limit is exceeded. Manual abort by LES prior to tower jettison and SPS after tower jettison with a single cue 1. Attitude error limits exceeded (S-IC only) 2. FDO limits exceeded (all stages) |
| Attitude error command saturated | All stages | Loss of control | Crew 1. Attitude rate exceeds 4°/second in pitch or yaw or 20°/second in roll for S-IC 2. Attitude rate exceeds 10°/second for S-II and S-IVB in pitch or yaw or 20°/second in roll for S-II and S-IVB 3. Attitude error exceeds 5° for S-IC 4. L/V overrate light Ground 1. Guidance failure light 2. FDO limits exceeded | Automatic abort by LES prior to 2 min 5 sec when attitude rate limit is exceeded. Manual abort by LES prior to tower jettison and SPS after tower jettison with two cues 1. L/V overrate light 2. Attitude rate indicator 3. FDO limits exceeded |
| Loss of attitude command signal | All Stages | Loss of guidance | Crew 1. Attitude rate exceeds 4°/second for S-IC in pitch or yaw 2. Attitude rate exceeds 10°/second for S-II and S-IVB in pitch and yaw 3. Attitude error exceeds 5° for S-IC Ground 1. FDO limits exceeded | Automatic abort by LES prior to 2 min 5 sec when attitude rate limit is exceeded. Manual abort by LES prior to tower jettison and SPS after tower jettison with a single cue 1. Attitude rate limits exceeded 2. FDO limits exceeded |

Figure 9-8 (Sheet 6 of 9)

| | | TYPICAL INFLIGHT CONTINGENCIES/REACTIONS (CONTINUED) | | |
|---|---|---|---|---|
| CONTINGENCY | FLIGHT TIME MIN:SEC | EFFECT | SENSORS OR DISPLAYS | MISSION RULE DATA |
| Attitude rate signal saturated | All stages | Loss of control | Crew 1. Attitude rate exceeds 4°/second in pitch or yaw or 20°/second in roll for S-IC 2. Attitude rate exceeds 10°/second for S-II and S-IVB in pitch or yaw or 20°/second in roll for S-II and S-IVB 3. Attitude error exceeds 5° for S-IC 4. L/V overrate light Ground 1. Guidance failure light 2. FDO limits exceeded | Automatic abort by LES prior to 2 min 5 sec when attitude rate limit is exceeded. Manual abort by LES prior to tower jettison and SPS after tower jettison with two cues 1. L/V overrate light 2. Attitude rate indicator 3. FDO limits exceeded |
| Loss of attitude rate signal | All stages | Loss of control | Crew 1. Attitude rate exceeds 4°/second in pitch or yaw or 20°/second in roll for S-IC 2. Attitude rate exceeds 10°/second for S-II and S-IVB in pitch or yaw or 20°/second in roll for S-II and S-IVB 3. Attitude error exceeds 5° for S-IC 4. L/V overrate light Ground 1. Guidance failure light 2. FDO limits exceeded | Automatic abort by LES prior to 2 min 5 sec when attitude rate limit is exceeded. Manual abort by LES prior to tower jettison and SPS after tower jettison with two cues 1. L/V overrate light 2. Attitude rate indicator 3. FDO limits exceeded |
| Lack of S-IC first plane separation | 2:30 | No S-II ignition possible | Lack of S-II thrust | Early S-IVB staging or abort by crew (orbit not possible) |
| Lack of S-II second plane separation (includes partial separation) | 2:57 | Probable subsequent loss of vehicle due to excessive temperature | Second plane separation indicator (redundant indication) | Shutdown prior to overheat and early S-IVB staging required by crew only, limited time between normal jettison time and excessive temperature may preclude ground information backup |

Figure 9-8 (Sheet 7 of 9)

| | | TYPICAL INFLIGHT CONTINGENCIES/REACTIONS (CONTINUED) | | |
|---|---|---|---|---|
| CONTINGENCY | FLIGHT TIME MIN:SEC | EFFECT | SENSORS OR DISPLAYS | MISSION RULE DATA |
| Lack of S-II/ S-IVB plane separation (includes partial) | 8:54 | No S-IVB ignition | Lack of S-IVB thrust | SPS abort required |
| Loss of both APS modules, S-IVB | S-IVB burn phases | Lack of roll control | Crew 1. Roll attitude may be excessive Ground 1. Roll attitude | Manual control by crew with RCS when roll attitude is excessive |
| Loss of one APS S-IVB | S-IVB burn phases | Negligible | Crew 1. No indication | Manual control by crew with RCS when roll attitude is excessive |
| LET fails to jettison | 3:03 | Vehicle may go unstable during S-IVB flight | Crew 1. Visual by crew 2. Attitude rate 3. Vehicle oscillatory Ground 1. Tower off indication | Shutdown S-IVB where vehicle becomes oscillatory. Crew to use emergency procedures for removal of LET if orbit was achieved. |
| Loss of thrust 2 lower adjacent control engines | 2:31.8– 6:20 | Loss of control | Crew 1. Thrust OK lights 2. Attitude rate exceeds 4°/sec 3. Roll attitude error 20° 4. Physiological Ground 1. Thrust OK lights 2. Thrust chamber pressure-zero 3. Longitudinal acceleration 4. Attitude rate 5. FDO limits | No automatic abort. Manual abort by LES on 2 cues prior to tower jettison. Manual abort by SPS on 2 cues after LES jettison and prior to 5 min 50 sec flight time. Early staging to S-IVB on 2 cues after 5 min 50 sec flight time. Cues 1. Thrust OK lights 2. Attitude rate 4°/sec 3. Roll attitude error 20° 4. Physiological 5. FDO limits |
| | 6:20– 8:00 | Failure to get 75 NMI orbit | | |
| | 8:00– 8:41 | No effect | | |
| Loss of thrust 2 upper adjacent control engines | 2:31.8– 6:15 | Loss of control | | |
| | 5:15– 8:41 | No effect | | |
| Loss of thrust 2 side adjacent control engines | 2:31.8– 7:10 | Loss of control | | |
| | 7:10– 8:41 | No effect | | |

Figure 9-8 (Sheet 8 of 9)

| | | TYPICAL INFLIGHT CONTINGENCIES/REACTIONS (CONTINUED) | | |
|---|---|---|---|---|
| CONTINGENCY | FLIGHT TIME MIN:SEC | EFFECT | SENSORS OR DISPLAYS | MISSION RULE DATA |
| Loss of thrust center and one control | 2:31.8- 3:10 | S-IVB depletion before N POI | Crew<br>1. Thrust OK lights<br>2. Attitude rate exceeds 4°/sec<br>3. Roll attitude error 20°<br>4. Physiological<br>Ground<br>1. Thrust OK lights<br>2. Thrust chamber pressure-zero<br>3. Longitudinal acceleration<br>4. Attitude rate<br>5. FDO limits | No automatic abort. Manual abort by LES on 2 cues prior to tower jettison. Manual abort by SPS on 2 cues after LES jettison and prior to 5 min 50 sec flight time. Early staging to S-IVB on 2 cues after 5 min 50 sec flight time.<br>Cues<br>1. Thrust OK lights<br>2. Attitude rate 4°/sec<br>3. Roll attitude error 20°<br>4. Physiological<br>5. FDO limits |
| | 3:10- 3:50 | No effect | | |
| | 3:50- 7:10 | Failure to get 75 NMI orbit | | |
| | 7:10- 8:41 | No effect | | |
| Loss of thrust 2 opposite engines | 2:31.8- 3:10 | S-IVB depletion before N POI | | |
| | 3:10- 3:50 | No effect | | |
| | 3:50- 6:00 | Failure to get 75 NMI orbit | | |
| | 6:00- 8:41 | No effect | | |
| Loss of thrust 2 adjacent control engines | 2:31.8- 7:10 | Possible loss of control during S-II flight | Crew<br>1. Thrust OK lights<br>2. Attitude rate exceeds 4°/second prior to 7 min 55 sec<br>3. Physiological<br>Ground<br>1. Thrust OK lights<br>2. Thrust chamber pressure - zero<br>3. Longitudinal acceleration<br>4. Attitude rate | No automatic abort. Manual abort by LES prior to tower jettison and by SPS after tower jettison with two cues<br>1. Thrust OK lights<br>2. Attitude rate exceeded prior to 5 min<br>3. Physiological |
| | 6:20- 8:00 | Failure to achieve 75 NMI orbit | | Early staging with two cues |
| | 8:00- 8:41 | No effect | | Continue Mission |
| Loss of thrust 2 opposite control or center engine and one control engine | 2:31.8- 3:10 | S-IVB depletion before POI | Crew<br>1. Thrust OK lights<br>2. Attitude rate exceeds 4°/second prior to 7 min 55 sec<br>3. Physiological<br>Ground<br>1. Thrust OK lights<br>2. Thrust chamber pressure - zero<br>3. Longitudinal Accel.<br>4. Attitude rate | Same as 2 min 30 sec- 5 min 50 sec above |
| | 3:10- 3:50 | No effect | | Same as 2 min 30 sec - 5 min 50 sec above |
| | 3:50- 7:10 | Failure to achieve 75 NMI orbit | | Early staging with two cues |
| | 7:10- 8:41 | No effect | | Continue mission |

Figure 9-8 (Sheet 9 of 9)

## VEHICLE FLIGHT CONTROL PARAMETERS

In order to perform flight control monitoring functions, essential data must be collected, transmitted, processed, displayed, and evaluated to determine the space vehicle's capability to start or continue the mission. Representative parameters included in this essential data are briefly described in the following paragraphs.

## PARAMETERS MONITORED BY LCC

The launch vehicle checkout and prelaunch operations monitored by the Launch Control Center (LCC) are briefly discussed in Section VIII of this manual. These operations determine the state of readiness of the launch vehicle, ground support, telemetry, range safety, and other operational support systems. During the final countdown, hundreds of parameters are monitored to ascertain vehicle, system, and component performance capability. Among these parameters are the "redlines". The redline values must be within the predetermined limits or the countdown will be halted. Typical redlines are fuel and oxidizer tank ullage pressure, $GN_2$ and helium storage sphere pressure, hydraulic supply pressures, thrust chamber jacket temperatures, bus voltages, IU guidance computer operations, $H_2$ and $O_2$ concentrations, and S-IVB oxidizer and fuel recirculation pump flow. In addition to the redlines, there are a number of operational support elements such as ALDS, range instrumentation, ground tracking and telemetry stations, ground communications, and other ground support facilities which must be operational at specified times in the countdown.

## PARAMETERS MONITORED BY BOOSTER SYSTEMS GROUP

The Booster Systems Group monitors launch vehicle systems (S-IC, S-II, S-IVB, and IU) and advises the flight director and flight crew of any system anomalies. They are responsible for abort actions due to failure or loss of thrust and overrate conditions; for monitoring and confirming flight events and conditions such as inflight power, stage ignition, holddown release, all engines go, roll and pitch initiate, engine cutoff, attitude control and stage separations; and for digital commanding of LV systems.

Specific responsibilities in the group are allocated as follows:

1. The BSE No. 1 has overall responsibility for the group, for commands to the launch vehicle, and for monitoring and evaluating the S-IC and S-II flight performance. Typical flight control parameters monitored include engine combustion chamber pressure, engine gimbal system supply pressure, fuel and oxidizer tank ullage pressure, helium storage tank pressure, engine actuator (yaw/pitch/roll) position, THRUST OK pressure switches, longitudinal acceleration, vent valve positions, engine ignition/cutoff, and various bus voltages.

2. The BSE No. 2 supports BSE No. 1 in monitoring the S-II flight and assumes responsibility for monitoring the S-IVB burns. Parameters monitored are similar to those monitored by BSE No. 1.

3. The BSE No. 3 monitors the attitude control, electrical, guidance and navigation, and IU systems. Typical parameters monitored include roll/pitch/yaw guidance and gimbal angles, angular rates, ST-124-M3 gimbal temperature and bearing pressure, LVDC temperature, and various bus voltages.

NOTE

The preceding flight controllers are located in the MOCR. The following are located in the vehicle system SSR.

4. The Guidance and Navigation-Digital Systems (GND) engineer monitors the guidance, navigation, and digital (sequential) BSE systems. The GND provides detailed support to BSE No. 3. Typical parameters monitored includes ST-124-M3 accelerometer and gyro pickups (X,Y,Z axes); and fixed position and fixed velocity (X,Y,Z components).

5. The Attitude Control and Stabilization Systems (ACS) engineer monitors the attitude control System, the S-IVB hydraulic and auxiliary propulsion systems, and the emergency detection system. The ACS provides detailed support to BSE's No. 2 and 3. Typical parameters monitored include hydraulic accumulator pressures, hydraulic reservoir piston position, attitude control fuel and oxidizer module temperatures, and rates excessives (pitch/roll/yaw).

6. The Engine Systems Engineer monitors the S-II and S-IVB engine system and $O_2/H_2$ burner and provides detailed support to BSE No. 2. Among the parameters monitored are thrust chamber pressure, engine inlet lox and $LH_2$ pressure, pre-valve position, and $O_2/H_2$ burner chamber dome temperature.

7. The Stage Systems Engineer monitors tank pressurization, repressurization, bulkhead pressure differential, chilldown and provides detailed support to the BSE No. 2. Typical parameters monitored include helium tank pressure, start tank pressure/temperature, and common bulkhead pressure.

8. The Electrical Network and Systems (ENS) engineer monitors the electrical systems (all LV stages), IU environmental control system, and range safety systems (safing at orbital insertion) and provides detailed support to BSE No. 3. Typical parameters monitored include exploding bridgewire voltages, sublimator inlet temperature, $GN_2$ regulator inlet temperature, and various bus voltages and currents.

9. The Command Systems Engineer monitors the commands sent to the launch vehicle and advises BSE No. 1 on their status. In the event of rejection of a command by the onboard computer, he determines the cause of the rejection, i.e., improperly coded command, malfunction of the command system, or malfunction of the computer.

10. The Consumables Engineer monitors status at all times of launch vehicle consumables including all high pressure spheres, APS propellants, and main stage propellants. He advises the BSE No. 2 of mission impact when consumables are depleted beyond predicted limits.

## PARAMETERS MONITORED BY FLIGHT DYNAMICS GROUP

The Flight Dynamics Group monitors and evaluates the powered flight trajectory and makes the abort decisions based on trajectory violations. They are responsible for abort planning, entry time and orbital maneuver determinations, rendezvous planning, inertial alignment correlation, landing point prediction, and digital commanding of the guidance systems.

The MOCR positions of the Flight Dynamics Group include the Flight Dynamics Officer (FDO), the Guidance Officer (GUIDO), and the Retrofire Officer (RETRO). The MOCR positions are given detailed specialized support by the flight dynamics SSR.

The surveillance parameters measured by the ground tracking stations and transmitted to the MCC are computer processed into plotboard and digital displays. The flight dynamics group compares the actual data with pre-mission calculated nominal data and is able to determine mission status. The surveillance parameters include slant range, azimuth and elevation angles, antenna polarization angle, and other data. From these measurements, space vehicle position, velocity, flight path angle, trajectory, ephemeris, etc., may be calculated. Typical plotboard displays generated from the surveillance parameters are altitude versus downrange distance, latitude versus longitude, flight path angle versus inertial velocity, and latitude versus flight path angle.

## PARAMETERS MONITORED BY SPACECRAFT SYSTEMS GROUP

The Spacecraft Systems Group monitors and evaluates the performance of spacecraft electrical, optical, mechanical, and life support systems; maintains and analyzes consumables status; prepares the mission log; coordinates telemetry playback; determines spacecraft weight and center of gravity; and executes digital commanding of spacecraft systems.

The MOCR positions of this group include the Command and Service Module Electrical, Environmental, and Communications Engineer (CSM EECOM), the CSM Guidance, Navigation, and Control Engineer (CSM GNC), the Lunar Module Electrical, Environmental, and Communications Engineer (LM EECOM), and the LM Guidance, Navigation, and Control Engineer (LM GNC). These positions are backed up with detailed support from the vehicle systems SSR.

Typical parameters monitored by this group include fuel cell skin and condenser temperatures, fuel cell current, various battery and bus voltages, launch escape tower and motor discretes, AGCU drift, SPS helium tank pressure, SPS fuel and oxidizer tank pressure, and fuel and oxidizer inlet pressure differential.

## PARAMETERS MONITORED BY LIFE SYSTEMS GROUP

The Life Systems Group is responsible for the well being of the flight crew. The group is headed by the Flight Surgeon in the MOCR. Aeromedical and environmental control specialists in the life systems SSR provide detailed support to the Flight Surgeon. The group monitors the flight crew health status and environmental/biomedical parameters.

## MANNED SPACE FLIGHT NETWORK

The Manned Space Flight Network (MSFN) is a global network of ground stations, ships, and aircraft designed to support manned and unmanned space flight. The network

provides tracking, telemetry, voice and teletype communications, command, recording, and television capabilities. The network is specifically configured to meet the requirements of each mission.

MSFN stations are categorized as lunar support stations (deep-space tracking in excess of 15,000 miles), near-space support stations with Unified S-Band (USB) equipment, and near-space support stations without USB equipment. Figure 9-9 is a matrix listing the stations by designators and capabilities. Figure 9-10 shows the geographical location of each station.

MSFN stations include facilities operated by NASA, the United States Department of Defense (DOD), and the Australian Department of Supply (DOS).

The DOD facilities include the Eastern Test Range (ETR), Western Test Range (WTR), White Sands Missile Range (WMSR), Range Instrumentation Ships (RIS), and Apollo Range Instrumentation Aircraft (ARIA). Recovery forces under DOD are not considered to be part of the MSFN.

## NASA COMMUNICATIONS NETWORKS

The NASA Communications (NASCOM) network is a point-to-point communications systems connecting the MSFN stations to the MCC. NASCOM is managed by the Goddard Space Flight Center, where the primary communications switching center is located. Three smaller NASCOM switching centers are located at London, Honolulu, and Canberra. Patrick AFB, Florida and Wheeler AFB, Hawaii serve as switching centers for the DOD eastern and western test ranges, respectively. The MSFN stations throughout the world are interconnected by landline, undersea cable, radio and communications satellite circuits. These circuits carry teletype, voice, and data in real-time support of the missions. Figure 9-11 depicts a typical NASCOM configuration.

Each MSFN USB land station has a minimum of five voice/data circuits and two teletype circuits. The Apollo insertion and injection ships have a similar capability through the communications satellites.

The Apollo Launch Data System (ALDS) between KSC and MSC is controlled by MSC and is not routed through GSFC. The ALDS consists of wide-band telemetry, voice coordination circuits, and a high speed circuit for the Countdown and Status Transmission System (CASTS). In addition, other circuits are provided for launch coordination, tracking data, simulations, public information, television, and recovery.

## MSFC SUPPORT OF LAUNCH AND FLIGHT OPERATIONS

The Marshall Space Flight Center (MSFC) by means of the Launch Information Exchange Facility (LIEF) and the Huntsville Operations Support (HOSC) provides real-time support of launch vehicle prelaunch, launch, and flight operations. MSFC also provides support via LIEF for postflight data delivery and evaluation.

## LAUNCH INFORMATION EXCHANGE FACILITY

The LIEF encompasses those personnel, communications,

data processing, display, and related facilities used by the MSFC launch vehicle design team to support Apollo-Saturn mission operations. The LIEF operations support organization is shown in figure 9-12.

In-depth real-time support is provided for prelaunch, launch, and flight operations from HOSC consoles manned by engineers who perform detailed system data monitoring and analysis.

System support engineers from MSFC and stage contractors are organized into preselected subsystem problem groups (approximately 160 engineers in 55 groups) to support KSC and MSC in launch vehicle areas which may be the subject of a request for analysis. The capabilities of MSFC laboratories and the System Development Facility (SDF) are also available.

## PRELAUNCH WIND MONITORING

Prelaunch flight wind monitoring analyses and trajectory simulations are jointly performed by MSFC and MSC personnel located at MSFC during the terminal countdown. Beginning at T-24 hours, actual wind data is transmitted periodically from KSC to the HOSC. These data are used by the MSFC/MSC wind monitoring team in vehicle flight digital simulations to verify the capability of the vehicle with these winds. Angle of attack, engine deflections, and structural

# MSFN STATIONS/EQUIPMENT MATRIX

| | Station | USB - 85' Antenna | USB - 30' Ant (Single) | USB - 30' Ant (Dual) | TLM - Ant (AGAVE) | TLM - Ant (Teltrac) | TLM - Ant (TLM-18) | TLM - Ant (TAA-3A) | C-band Radar | C-band Hi-speed Radar Data | FC A/G Comm. | A/G Comm. (Remote) | ALDS | Hi-speed TLM Data | RF Command | RSDP SUM | Gemini Consoles & Displays | Apollo Consoles & Displays | Tone Command | Dig. Comm. System | Downrange Up-link | TLM R&R | TLM D/T | USB - 12' Ant (Single) | FC Manned | Saturn Booster TLM | Voice (SRO) | Voice (SCAMA) | TTY | Apollo Data Processors | Apollo TV Monitor | TLM Flat Broadside Ant. | S-band Tel R&R | RSO Displays |
|---|---|---|---|---|---|---|---|---|---|---|---|---|---|---|---|---|---|---|---|---|---|---|---|---|---|---|---|---|---|---|---|---|---|---|
| Mission Control Center | MCC | | | | | | | | | X | X | X | | | | X | X | X | | X | | | X | | X | | | | X | | | | | |
| MSFN Operations Center | MSFNOC | | | | | | | | | | | X | X | X | | X | | | | X | | | | | | | X | X | X | X | X | | | |
| Central Instrumentation Facility | CIF | | | | | | | | | | | | X | | | | | | | | | | | | | | | | | | | | | |
| | TEL IV | | | | | | | X | | | | | | | | | | | | | | X | X | | | X | | | | | | | X | |
| Ascension Island, U.K. (NASA) | ACN # | | | X | | | | | | | | X | | | | | | | | | | X | X | | | | | X | X | X | X | | X | |
| Antigua, U.K. (NASA) | ANG # | | X | | | | | | | | | X | | X | | | | | | | | X | X | | | X | | X | X | X | X | | X | |
| Antigua (DOD-ETR) | ANT # | | | | | X | X | X | X | X | | X | | X | X | | | | X | | X | X | X | | X | X | X | | X | | | | X | |
| Ascension Island (DOD-ETR) | ASC # | | | | | X | X | | X | | | | | | | | | | X | | X | X | X | | | | X | | X | | | | X | |
| Bermuda, U.K. (NASA) | BDA # | | X | | | X | | | X | X | | X | | X | X | | | | X | | X | X | X | | | | | X | X | X | X | | X | X |
| Canberra, Australia (WRE) | CNB * | X | | | | | | | | | | | | | | | | | | | | | | | | | | X | X | X | X | | X | |
| Vandenberg, California (DOD-WTR) | CAL # | | | | X | | | | X | X | | X | | | | | | | | | | X | X | | | | | X | X | | | | | |
| Cape Kennedy (DOD-ETR) | CNV # | | | | | | X | | X | | | | | | X | | | | X | | X | X | X | | | | X | | X | | | | | |
| Carnarvon, Australia (WRE) | CRO # | | | X | | X | | | X | X | X | | | | X | X | X | X | | X | | X | X | | X | | | X | X | X | X | | X | |
| USNS Coastal Sentry (DOD-ETR) | CBQ # | | | | X | | | | | | X | X | | | X | X | X | | X | X | | X | X | | | | | | | | | | | |
| Canton Island (NASA) | CTN # | | | | X | | | | | | | X | | | | | | | | | | X | X | | | | | X | X | | | | | |
| Canary Islands, Spain (NASA) | CYI # | | X | | X | X | | | X | | X | X | | | X | X | X | | | X | | X | X | | | | | X | X | X | X | | X | |
| Grand Bahama Island (DOD-ETR) | GBI # | | | | | X | | | X | X | | X | | X | X | | | | | | X | X | X | | | X | X | | X | | | | X | |
| Grand Bahamas, U.K. (NASA) | GBM # | | X | | | | | | | | | X | | X | | | | | | | | X | X | | | X | | X | X | X | | | X | |
| Goldstone, California (NASA) | GDS * | X | | | | | | | | | | | | | | | | | | X | | | | | | X | | X | | X | X | | X | |
| Grand Turk Island (DOD-ETR) | GTK # | | | | | X | | | X | X | | | | | | | | | X | | | X | | | | | X | | X | | | | | |
| Guam, U.S. (NASA) | GWM # | | | X | | X | | | | | | X | | | | | | | | X | | | | | | | | X | | X | X | | X | |
| Guaymas, Mexico (NASA) | GYM # | | X | | X | X | | | | | X | X | | | | | X | X | | X | | X | X | | X | | | X | X | X | X | | X | |
| Kokee Park, Hawaii (NASA) | HAW # | | | X | X | X | | | X | | X | X | | | X | X | X | | X | X | | X | X | | | | | X | X | X | X | | X | |
| Madrid, Spain (NASA) | MAD * | X | | | | | | | | | | | | | | | | | | X | | | | | | | | X | | X | X | | X | |
| Merritt Island, Florida (NASA) | MIL # | | | X | | X | | | | | | X | | | | | | X | | X | | X | X | | | X | | X | X | X | | | X | |
| Merritt Island (DOD-ETR) | MLA # | | | | | | | | X | X | | | | | | | | | | | | | | | | | X | | X | | X | | | |
| Patrick AFB, Florida (DOD-ETR) | PAT # | | | | | | | | X | X | | | | | | | | | | | | | | | | | X | | X | | | | | |
| Pretoria, So. Africa (DOD-ETR) | PRE # | | | | | | | | X | | | X | | | | | | | | | | | | | | | X | | X | | | | | |
| USNS Rose Knot (DOD-ETR) | RKV # | | | | X | | | | | | X | X | | | X | X | X | | X | X | | X | X | | | X | | | | | | | | |
| USNS Range Tracker (DOD-WTR) | RTK # | | | | X | | | | X | | | X | | | | | | | | | | X | X | | | | | X | X | | | | | |
| Tananarive, Malagasy Rep. (NASA) | TAN # | | | | X | | | | | | | X | | | | | | | | | | X | X | | | | | X | X | | | | | |
| Corpus Christi, Texas (NASA) | TEX # | | X | | X | X | | | | | X | X | | X | X | X | X | | | X | | X | X | | | | | X | X | X | X | | X | |
| White Sands, N.M. (DOD-WSMR) | WHS # | | | | X | | | | X | X | | | | | | | | | | | | | | | | | | X | X | | | | | |
| Woomera, Australia (DOS-WRE) | WOM # | | | | X | | | | X | | | | | | | | | | | | | | | | | | | | X | | | | | |
| Apollo Range Inst. Aircraft | ARIA 1-8 | | | | | | | | | | | X | | | | | | | | | | X | X | | | | X | | | | | | | |
| USNS Huntsville (DOD - WTR) | Ships: HTV | | | | | | | | X | | | | | | | | | | | | | X | X | X | | | | | X | | | X | X | |
| USNS Mercury (DOD - WTR) | MER | | X | | | | | | X | | | X | | | X | | | X | | X | | X | X | | X | | | | X | | X | X | X | |
| USNS Redstone (DOD - WTR) | RED | | X | | | | | | X | | | X | | | X | | | X | | X | | X | X | | X | | | | X | | X | X | X | |
| USNS Vanguard (DOD - WTR) | VAN | | X | | | | | | X | | | X | | | X | | | X | | X | | X | X | | X | | | | X | | X | X | X | |
| USNS Watertown (DOD - WTR) | WTN | | | | | | | | X | | | | | | | | | | | | | X | X | X | | | | | X | | | X | X | |

* Lunar Support Stations
\# Near Space Support Stations
Supplemental Stations

Figure 9-9

loads are calculated and compared against vehicle limits. Simulations are made on either the IBM 7094 or B5500 computer and results are reported to the Launch Control Center (LCC) within 60 minutes after wind data transmission. At T-2 1/4 hours a go/no-go recommendation is transmitted to KSC by the wind monitoring team. A go/no-go condition is also relayed to the LCC for lox and LH$_2$ loading.

In the event of marginal wind conditions, contingency wind data balloon releases are made by KSC on an hourly basis after T-2 1/2 hours and a go/no-go recommendation is transmitted to KSC for each contingency release. This contingency data is provided MSFC in real-time via CIF/DATA-CORE and trajectory simulations are performed on-line to expedite reporting to KSC.

Ground wind monitoring activities are also performed by MSFC laboratory personnel for developmental tests of displays. Wind anemometer and strain gauge data are received in real-time and bending moments are computed and

compared with similar bending moment displays in CIF.

## LAUNCH AND FLIGHT OPERATIONS SUPPORT

During the prelaunch period, primary support is directed to KSC. Voice communications are also maintained with the Booster Systems Group at MCC and the KSC/MCC support engineers in CIF to coordinate preparations for the flight phase and answer any support request.

At liftoff, primary support transfers from KSC to the MCC. The HOSC engineering consoles provide support, as required to the Booster Systems Group during S-IVB/IU orbital operations, by monitoring detailed instrumentation for system inflight and dynamic trends, by assisting in the detection and isolation of vehicle malfunctions, and for providing advisory contact with vehicle design specialist. This support is normally provided from liftoff through the active launch vehicle post-spacecraft separation phase approximately T + 6 hours, or until LIEF mission support termination.

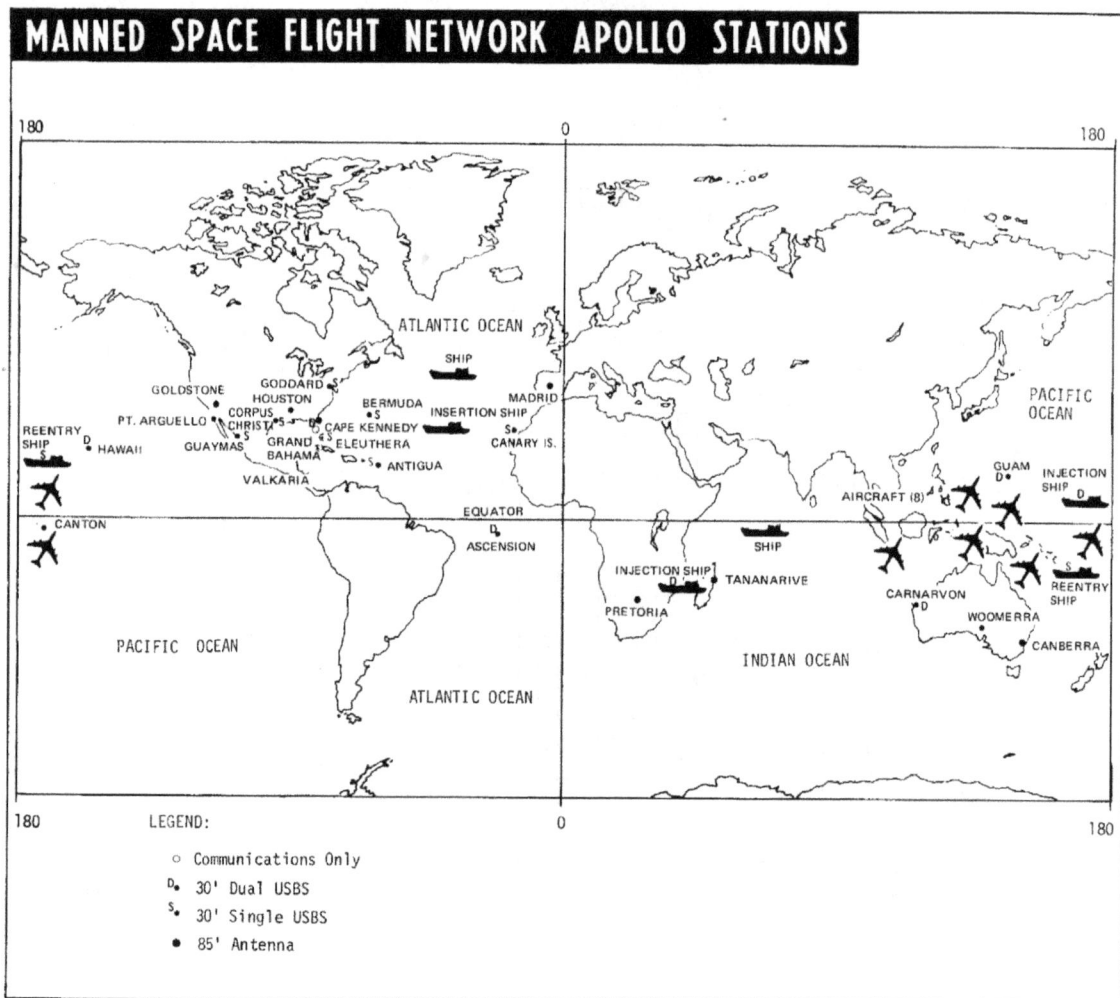

Figure 9-10

# TYPICAL APOLLO COMMUNICATIONS NETWORK CONFIGURATION

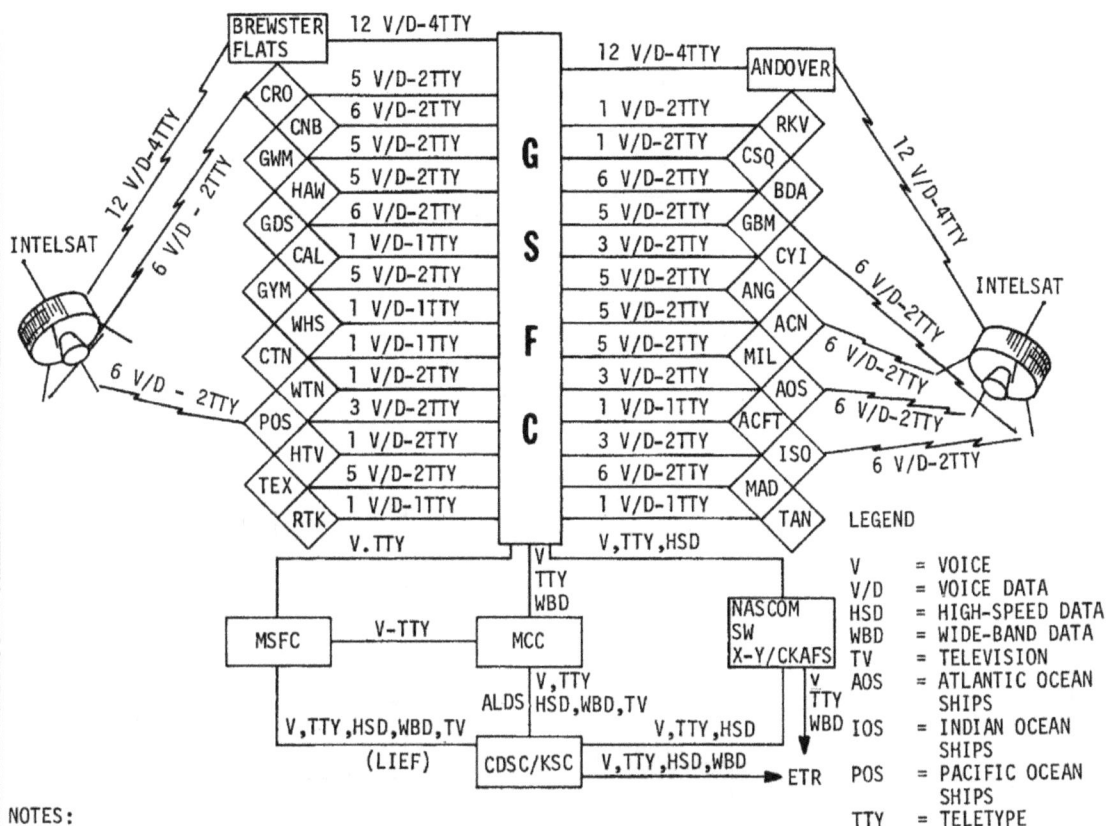

BREWSTER FLATS — 12 V/D-4TTY

12 V/D-4TTY — ANDOVER

CRO — 5 V/D-2TTY

CNB — 6 V/D-2TTY

GWM — 5 V/D-2TTY

HAW — 5 V/D-2TTY

GDS — 6 V/D-2TTY

CAL — 1 V/D-1TTY

GYM — 5 V/D-2TTY

WHS — 1 V/D-1TTY

CTN — 1 V/D-1TTY

WTN — 1 V/D-2TTY

POS — 3 V/D-2TTY

HTV — 1 V/D-2TTY

TEX — 5 V/D-2TTY

RTK — 1 V/D-1TTY

1 V/D-2TTY — RKV

1 V/D-2TTY — CSQ

6 V/D-2TTY — BDA

5 V/D-2TTY — GBM

3 V/D-2TTY — CYI

5 V/D-2TTY — ANG

5 V/D-2TTY — ACN

5 V/D-2TTY — MIL

3 V/D-2TTY — AOS

1 V/D-1TTY — ACFT

3 V/D-2TTY — ISO

6 V/D-2TTY — MAD

1 V/D-1TTY — TAN

GSFC

INTELSAT — 12 V/D-4TTY — 6 V/D-2TTY — 6 V/D-2TTY

INTELSAT — 12 V/D-4TTY — 6 V/D-2TTY — 6 V/D-2TTY — 6 V/D-2TTY — 6 V/D-2TTY

V.TTY    V,TTY,HSD

MSFC — V-TTY — MCC

MCC: V / TTY / WBD

NASCOM SW X-Y/CKAFS

V,TTY HSD,WBD,TV

V,TTY HSD,WBD,TV (LIEF) — CDSC/KSC — V,TTY,HSD,WBD → ETR

V,TTY,HSD

ALDS

V / TTY / WBD

## LEGEND

| | |
|---|---|
| V | = VOICE |
| V/D | = VOICE DATA |
| HSD | = HIGH-SPEED DATA |
| WBD | = WIDE-BAND DATA |
| TV | = TELEVISION |
| AOS | = ATLANTIC OCEAN SHIPS |
| IOS | = INDIAN OCEAN SHIPS |
| POS | = PACIFIC OCEAN SHIPS |
| TTY | = TELETYPE |

## NOTES:

1. NASCOM interface points for ships and aircraft are mission dependent.

2. Additional V/D and TTY circuits between GSFC and centers are augmented as required.

3. Communications circuits as shown exist at this time or will be available for missions as required.

4. Circuit routing is not necessarily as indicated.

5. Voice Data Network to and from the Apollo USB stations will be configured with usage as indicated

    a. Net #1 Mission Conference

    b. Net #2 Biomed Data

    c. Net #3 Network M & O

    d. Net #4 Telemetry Data/Command Data (USB)

    e. Net #5 Tracking Data (USB)

6. One additional voice data circuit will be available to each Apollo station equipped with an 85' antenna and to Bermuda for high-speed tracking data.

7. Apollo stations and ships connected to GSFC via Intelsat can receive at all times, regardless of number of stations or ships. Only two stations or ships can transmit at a given time via each Intelsat.

8. Apollo ship may be assigned to any of these areas.

Figure 9-11

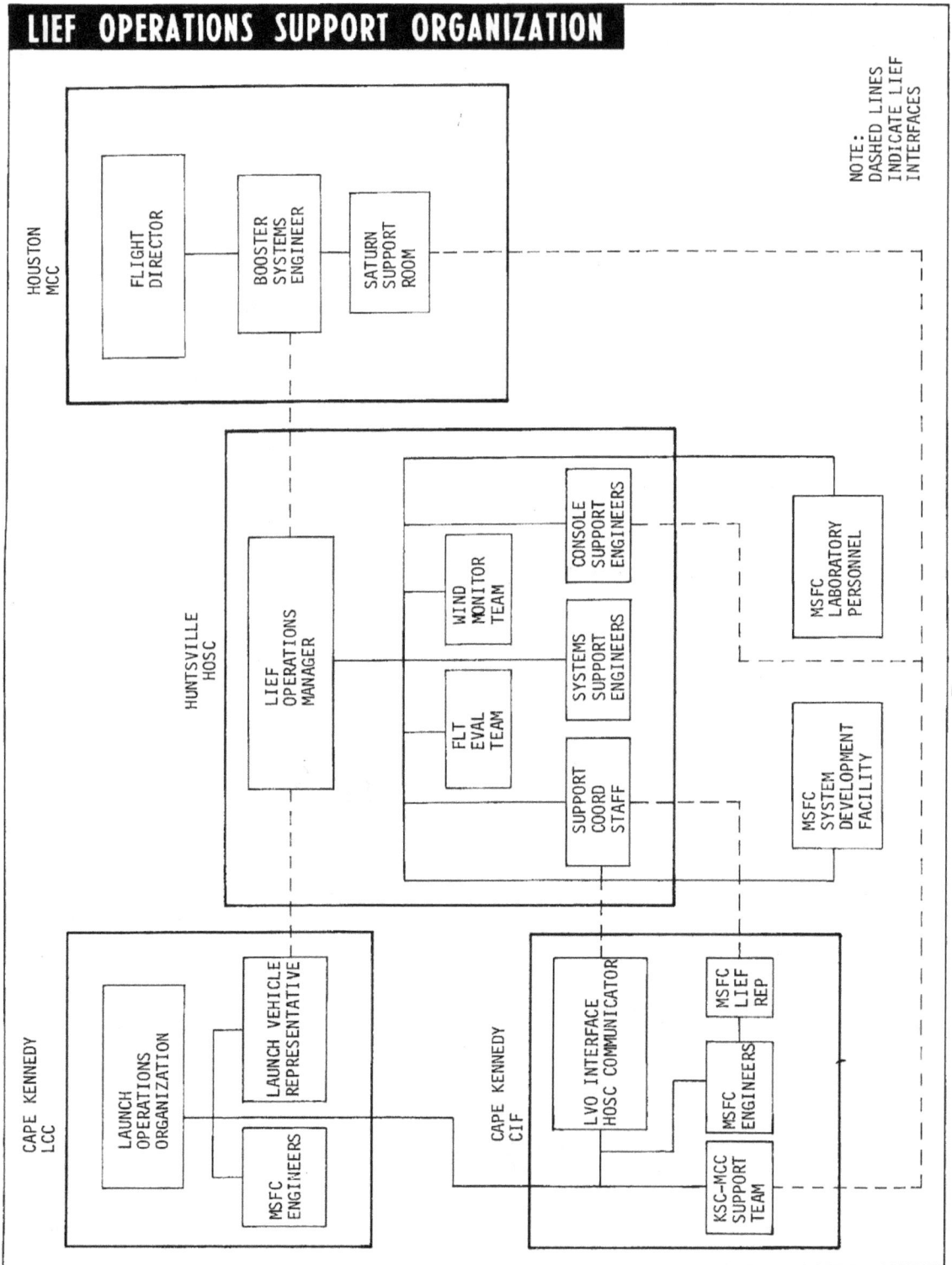

# LIEF OPERATIONS SUPPORT ORGANIZATION

NOTE:
DASHED LINES
INDICATE LIEF
INTERFACES

Figure 9-12

# SECTION X
# MISSION VARIBLES AND CONSTRAINTS

## TABLE OF CONTENTS

## MISSION OBJECTIVES

The mission objectives for the C Prime mission are as follows:

1. Verify the LV capability for a free-return, translunar injection (TLI).

2. Demonstrate the restart capability of the S-IVB stage.

3. Verify the J-2 engine modifications.

4. Confirm S-II and S-IVB J-2 engine environment.

5. Confirm the LV longitudinal oscillation environment during the S-IC burn period.

6. Verify the lox prevalve accumulators (surge chambers) incorporated in the S-IC stage to supress the low frequency longitudinal oscillations (POGO effect).

7. Demonstrate helium heater repressurization system operation.

8. Verify the capability to inject the S-IVB/IU/LTA-B into a lunar slingshot trajectory.

9. Demonstrate the capability to safe the S-IVB stage.

10. Verify the onboard CCS/ground system interface and operation in deep space environment.

## FLIGHT PROFILES

The C Prime mission has two planned profiles, basic and Option 1 as shown in figures 10-1 and 10-2. The attitude maneuvers required during flight are shown in figure 10-3.

In the basic mission (figure 10-1) the launch vehicle places the spacecraft into a 100 nautical mile circular earth orbit. The boost to earth orbit consists of complete burns of the S-IC and S-II stages and a first burn of the S-IVB stage. The CSM is separated from the S-IVB/IU/LTA-B during the first revolution in EPO. The S-IVB is restarted over the Pacific Ocean during the second revolution (first injection opportunity) and the S-IVB/IU/LTA-B is injected into a lunar transfer trajectory. The S-IVB/IU/LTA-B maneuvers to a predetermined attitude and executes retrograde dump of residual propellants (slingshot procedure) to attempt a solar orbit. The S-IVB stage is then "safed" by venting residual gases. The CSM continues in earth orbit to complete spacecraft operations. The CM then separates from the SM and returns to earth.

In the Option 1 mission (figure 10-2) the launch vehicle places the spacecraft into a 100 nautical mile circular orbit. The boost to earth orbit consists of complete burns of the S-IC and S-II stages and a first burn of the S-IVB stage. The launch azimuth will be 72 to 108 degrees (figure 10-4) with a launch window of December 20 to 27, 1968 or January 18 to 24, 1969. The LV/Spacecraft systems are checked out during coast in the parking orbit. The S-IVB stage is restarted over the Pacific Ocean during the second or third revolution (first or second injection opportunity) as shown in figure 10-5. After escape velocity is reached, the S-IVB is cutoff and the S-IVB/IU/Spacecraft is injected into a free-return, translunar trajectory, where the CSM is separated from the S-IVB/IU/LTA-B. The S-IVB/IU/LTA-B maneuvers to a predetermined attitude and executes retrograde dump of residual propellants (slingshot procedure) to attain a solar orbit. The S-IVB is then "safed" by venting residual gases. The CSM continues on the translunar trajectory (64 to 70 hours) and either goes into a lunar orbit or remains on a free-return trajectory. The CSM transearth trajectory takes 75 to 85 hours. On the approach to earth, the CM is separated from the SM and recovered in the Pacific Ocean.

## MISSION CONSTRAINTS

Mission requirements impose the following constraints on the LV/Spacecraft, trajectory profile, and mission planning factors.

1. During Time Base 5, the crew shall be capable of issuing manual attitude commands to the LV. At "hand back," the S-IVB/IU shall maintain the hand back attitude in the mode (local or inertial) which is in the timeline at "hand back."

2. The S-IVB restart attitude will be local horizontal.

3. The crew command S-IVB attitude rate limits are 0.3 degrees per second in pitch and yaw, and 0.5 degrees per second in roll.

4. Vehicle pitch and yaw attitude rates shall not exceed one degree per second.

5. The S-IC center engine cutoff time shall occur prior to attaining 4g acceleration.

6. The S-IVB stage will be fully loaded and have two burn periods.

7. The basic mission will utilize a preselected single day of lunar targeting for the entire launch window.

8. There will be no inflight navigation or target update. All profile options will be obtained by crew action (Option 1 by riding TLI burn).

9. The PU system (S-II and S-IVB stages) will be flown open-loop.

**X**

10. Pacific translunar injection opportunities (S-IVB restart) during second or third revolution in EPO shall be used for Option 1. If both TLI opportunities are inhibited by the crew, no S-IVB restart will be attempted.

11. The flight azimuth will be 72 degrees for the basic

mission profile and 72 to 108 degrees for Option 1 profile.

12. In the basic mission profile, CSM separation occurs in the first revolution in EPO.

13. The maximum vehicle yaw attitude may not exceed 45 degrees (measured from flight azimuth at launch).

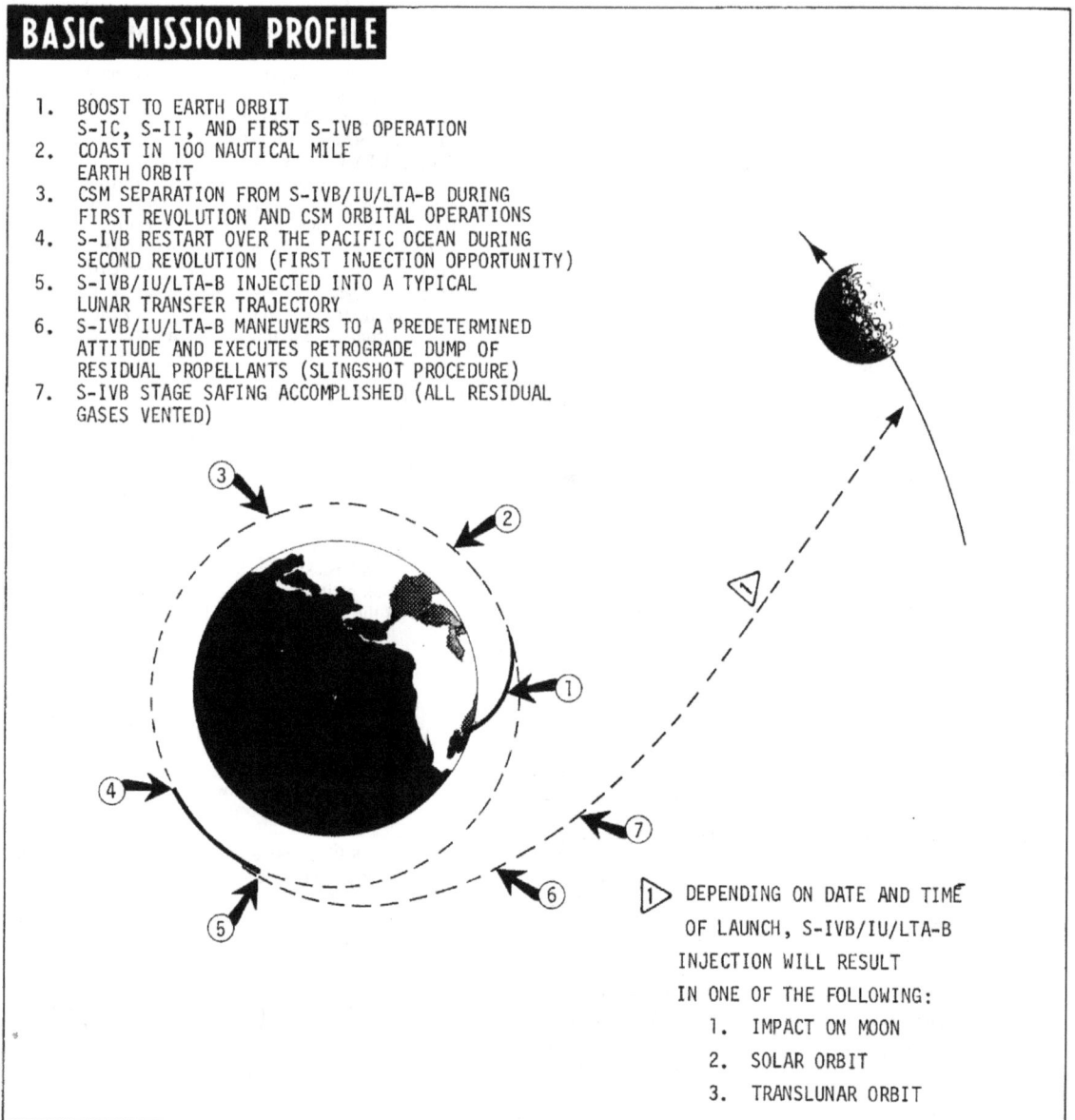

## BASIC MISSION PROFILE

1. BOOST TO EARTH ORBIT
   S-IC, S-II, AND FIRST S-IVB OPERATION
2. COAST IN 100 NAUTICAL MILE
   EARTH ORBIT
3. CSM SEPARATION FROM S-IVB/IU/LTA-B DURING
   FIRST REVOLUTION AND CSM ORBITAL OPERATIONS
4. S-IVB RESTART OVER THE PACIFIC OCEAN DURING
   SECOND REVOLUTION (FIRST INJECTION OPPORTUNITY)
5. S-IVB/IU/LTA-B INJECTED INTO A TYPICAL
   LUNAR TRANSFER TRAJECTORY
6. S-IVB/IU/LTA-B MANEUVERS TO A PREDETERMINED
   ATTITUDE AND EXECUTES RETROGRADE DUMP OF
   RESIDUAL PROPELLANTS (SLINGSHOT PROCEDURE)
7. S-IVB STAGE SAFING ACCOMPLISHED (ALL RESIDUAL
   GASES VENTED)

▷ DEPENDING ON DATE AND TIME
  OF LAUNCH, S-IVB/IU/LTA-B
  INJECTION WILL RESULT
  IN ONE OF THE FOLLOWING:
      1. IMPACT ON MOON
      2. SOLAR ORBIT
      3. TRANSLUNAR ORBIT

Figure 10-1

# OPTION 1 MISSION PROFILE

1. BOOST TO EARTH ORBIT
   S-IC, S-II, AND FIRST S-IVB
   OPERATION
2. COAST IN 100 NAUTICAL MILE
   EARTH ORBIT
3. PERFORM LV/SPACECRAFT SYSTEMS
   CHECKOUT DURING COAST IN PARKING
   ORBIT
4. S-IVB RESTART OVER THE PACIFIC
   OCEAN DURING THE SECOND OR THIRD
   REVOLUTION (FIRST OR SECOND
   INJECTION OPPORTUNITY)
5. S-IVB SECOND BURN CUTOFF AND S-IVB/
   IU/SPACECRAFT INJECTED INTO A FREE
   RETURN TRANSLUNAR TRAJECTORY
6. CSM SEPARATED FROM THE S-IVB/IU/LTA-B
7. S-IVB/IU/LTA-B MANEUVERS TO A
   PREDETERMINED ATTITUDE AND EXECUTES
   RETROGRADE DUMP OF RESIDUAL
   PROPELLANTS (SLINGSHOT PROCEDURE)

8. S-IVB/IU/LTA-B TRAJECTORY
   (TRAJECTORY IS SUCH THAT THE
   INFLUENCE OF THE MOON'S
   GRAVITATIONAL FIELD INCREASES
   THE VELOCITY OF THE S-IVB/IU/LTA-B
   SUFFICIENTLY TO PLACE IT IN SOLAR
   ORBIT.)
9. S-IVB STAGE SAFING ACCOMPLISHED
   (ALL RESIDUAL GASES VENTED)
10. CSM TRANSLUNAR TRAJECTORY (64
    TO 70 HOURS)
11. OPTIONAL LUNAR MISSION: EITHER
    LUNAR ORBIT OR FREE RETURN
    TRAJECTORY
12. CSM TRANSEARTH TRAJECTORY
    (75 TO 85 HOURS)
13. CM SEPARATES FROM SM
14. CM RECOVERY (PACIFIC OCEAN)

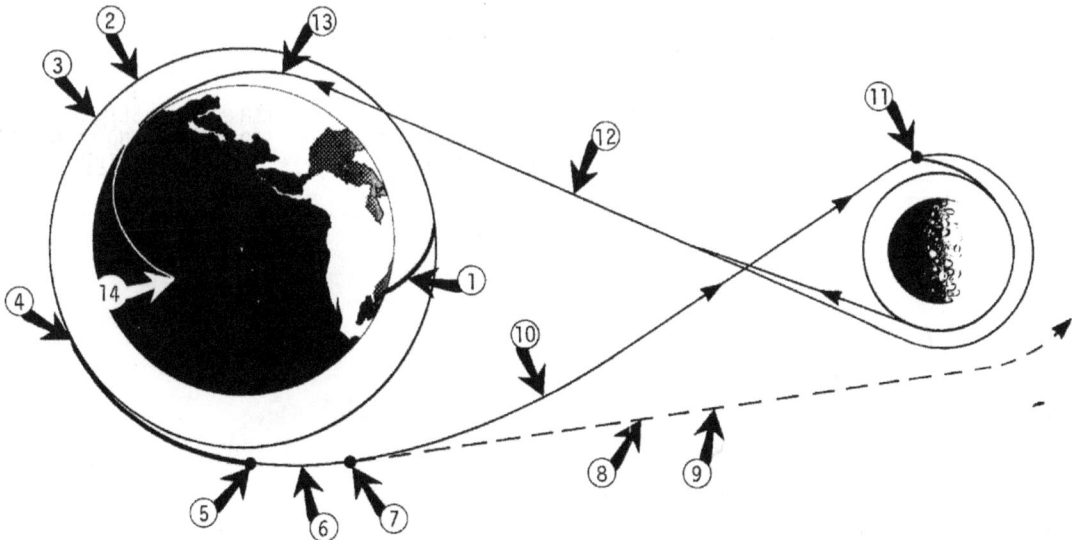

Figure 10-2

# ATTITUDE MANEUVER REQUIREMENTS

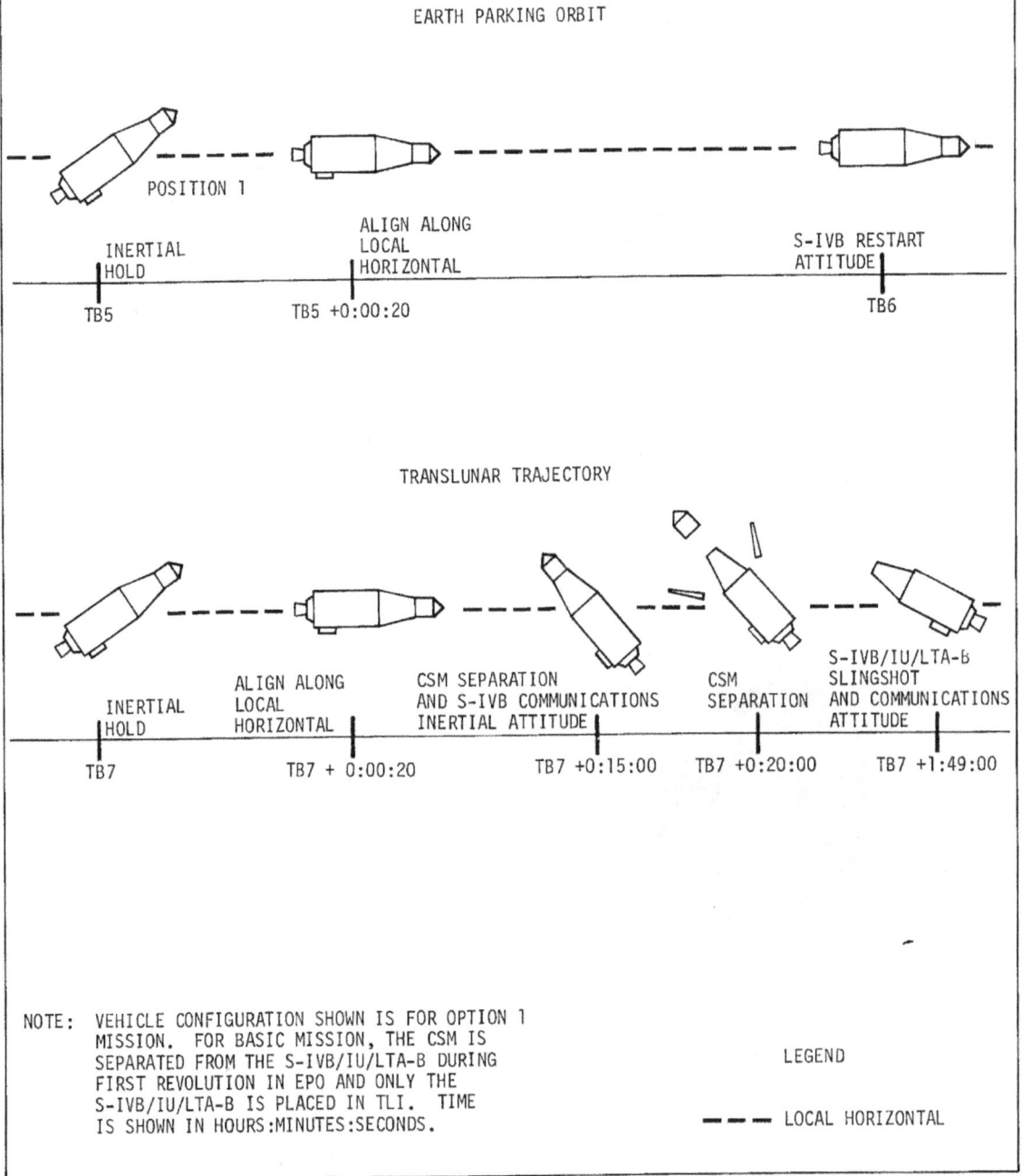

EARTH PARKING ORBIT

POSITION 1

INERTIAL
HOLD

TB5

ALIGN ALONG
LOCAL
HORIZONTAL

TB5 +0:00:20

S-IVB RESTART
ATTITUDE

TB6

TRANSLUNAR TRAJECTORY

INERTIAL
HOLD

TB7

ALIGN ALONG
LOCAL
HORIZONTAL

TB7 + 0:00:20

CSM SEPARATION
AND S-IVB COMMUNICATIONS
INERTIAL ATTITUDE

TB7 +0:15:00

CSM
SEPARATION

TB7 +0:20:00

S-IVB/IU/LTA-B
SLINGSHOT
AND COMMUNICATIONS
ATTITUDE

TB7 +1:49:00

NOTE:  VEHICLE CONFIGURATION SHOWN IS FOR OPTION 1
MISSION.  FOR BASIC MISSION, THE CSM IS
SEPARATED FROM THE S-IVB/IU/LTA-B DURING
FIRST REVOLUTION IN EPO AND ONLY THE
S-IVB/IU/LTA-B IS PLACED IN TLI.  TIME
IS SHOWN IN HOURS:MINUTES:SECONDS.

LEGEND

– – –  LOCAL HORIZONTAL

Figure 10-3

# GROUND PROJECTIONS FOR POWERED FLIGHT INTO EARTH PARKING ORBIT

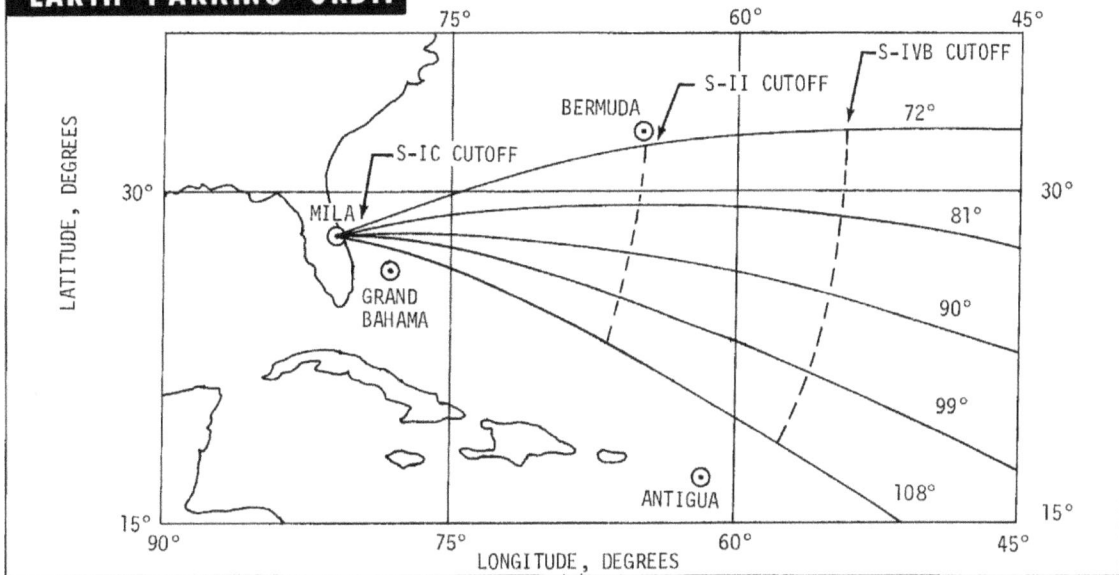

Figure 10-4

# GROUND PROJECTION OF TRANSLUNAR INJECTION BURNS, DECEMBER 21, 1968

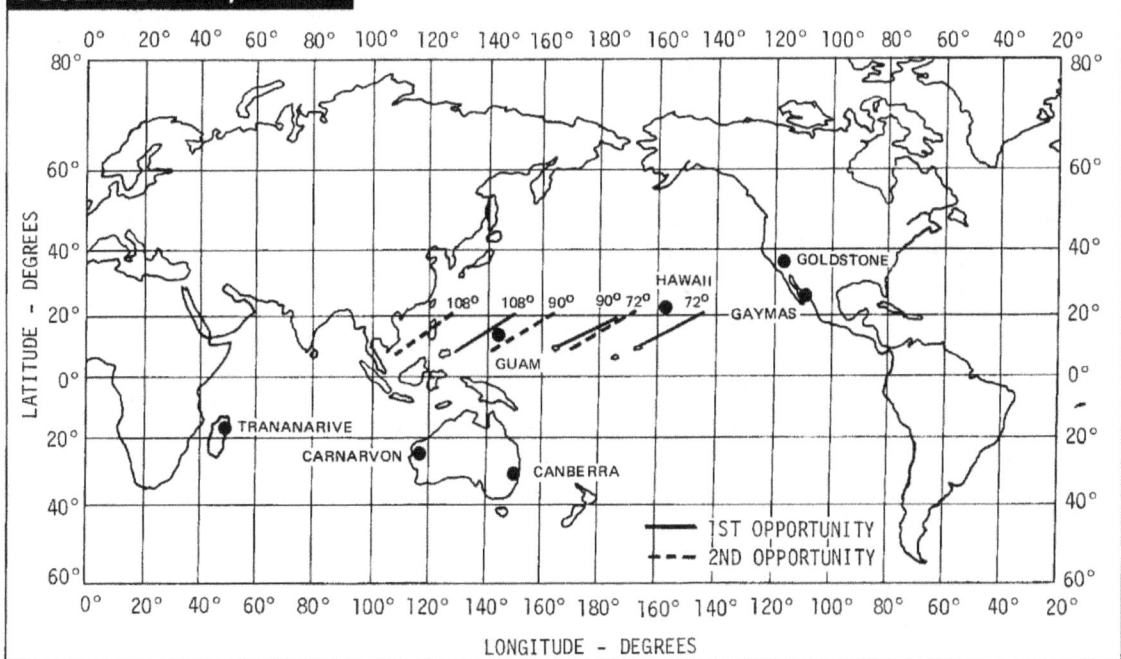

Figure 10-5

# APPENDIX A
# ABBREVIATIONS AND ACRONYMS

**A**

A .................................................Astronaut
ac .......................................... Alternating Current
ACCEL .........................................Acceleration
ACE ............................ Acceptance Checkout Equipment
ACM ..............................Actuation Control Module
ACN ............... NASA MSFN Station, Ascension Island
ACS ......... Attitude Control and Stabilization Systems
                      Engineer (Booster Systems)
ACT'R ..........................................Actuator
AF ............................................. Air Force
AFB ......................................... Air Force Base
AFD ............................ Assistant Flight Director
AFETR ....................... Air Force Eastern Test Range
AGAVE ................. Automatic Gimbaled Antenna
                      Vectoring Equipment
AGC ..................... Apollo Guidance Computer
A/G COMM ........... Air to Ground and Ground to Air
                      Communications
AGCU ................. Apollo Guidance Control Unit
ALDS ...................Apollo Launch Data System
Alt ...........................................Altitude
AM ..............................Amplitude Modulation
amp ................................ Ampere or Amplifier
AMR ........................ Atlantic Missile Range
ANG ...................NASA MSFN Station, Antigua
Ant ............................................Antenna
ANT ...............DOD-ETR MSFN Station, Antigua
AOA ............................. Angle of Attack
AOS .............. Acquisition of Spacecraft (by a site)
AOS ........................... Atlantic Ocean Ships
APS ..................... Auxiliary Propulsion System
ARIA .......... Apollo Range Instrumentation Aircraft
ASC ........DOD-ETR MSFN Station, Ascension Island
ASD ..................... Abort Summary Document
ASI ....................Augmented Spark Igniter
ATT .............................................Attitude
AUTO ................................ Automatic

**B**

BDA ....................NASA MSFN Station, Bermuda
BIOMED ..............................Biomedical
BMAG ................... Body Mounted Attitude Gyro
BEF ...................... Blunt End Forward
BSE ...................... Booster Systems Engineer

**C**

CAL..DOD-WTR MSFN Station, Vandenberg AFB, California
calips ................. Calibrationable Pressure Switch
CAPCOM ................... Spacecraft Communicator
CASTS .......Countdown and Status Transmission System
CBQ ...... DOD-ETR MSFN Station, USNS Coastal Sentry
CCATS . Command, Communications and Telemetry System
CCF ................Converter/Compressor Facility
CCS ..............Command Communications System
CCW .................................Counterclockwise
CDDT ............... Countdown Demonstration Test

CDF ...................... Confined Detonating Fuse
CDP ............. Command Data Processor, MSFN Site
CDR ...................... Critical Design Review
CDR ...................... Spacecraft Commander
CECO ...................... Center Engine Cutoff
CG ...................................Center of Gravity
CIF ......... Central Instrumentation Facility
                  (Located at Kennedy Space Center)
CIU ...................... Computer Interface Unit
CKAFS ............... Cape Kennedy Air Force Stations
CM ...................... Command Module
CMC ...................Command Module Computer
CMD ...................................Command
CNB ........ WRE MSFN Deep-Space Station, Canberra
CNV .......... DOD-ETR MSFN Station, Cape Kennedy
CO ...................................Checkout
COFW ...................Certificate of Flight Worthiness
COM ...................................Common
cps ...................... Cycles per Second (Hertz)
CRO .......... WRE MSFN Station, Carnarvon
CRT .......................Cathode Ray Tube
CSM .........Command and Service Module
C-T ...................Crawler-Transporter
CTN ........... NASA MSFN Station, Canton Island
CW ...................................Clockwise
CW ...................................Continuous Wave
CYI ...........NASA MSFN Station, Canary Islands
CZR ...................High Speed Metric Camera

**D**

DATA-CORE ..........CIF Telemetry Conversion System
db .................................................Decibel
dbm .......... Decibels Referenced to One Milliwatt
dc .......................................... Direct Current
DCR ...................... Destruct Command Receiver
DCR ...................... Design Certification Review
DCS .......................Digital Computer System
DDAS ...................Digital Data Acquisition System
DEE .......................Digital Event Evaluator
deg ....................................... Degree
Dig. Comm...........................Digital Command
DOD ...................... Department of Defense
DOS .......................Department of Supply (Australia)
DRK .......................Display Request Keyboard
DSKY ........ Display and Keyboard (Spacecraft Guidance
                      and Control)
DTS ...................... Data Transmission System

**E**

EAO ...................... Experiment Activities Officer
EBW .......................Exploding Bridgewire
ECA .......................Electrical Control Assembly
ECS .......................Environmental Control System
ECU .......................Environmental Control Unit
EDS .......................Emergency Detection System
EECOM ... Electrical, Environmental, and Communications
                      Engineer
ELS ...................... Earth Landing System

**A**

| | |
|---|---|
| EMS | Entry Monitor System |
| ENG | Engine |
| ENS | Electrical Network and Systems Engineer (Booster Systems) |
| EPO | Earth Parking Orbit |
| ESE | Electrical Support Equipment |
| ETR | Eastern Test Range |

**F**

| | |
|---|---|
| F | Fahrenheit |
| F | Force |
| FACI | First Article Configuration Inspection |
| FAO | Flight Activities Officer |
| FC | Flight Controller |
| FCO | Flight Control Office |
| FCSM | Flight Combustion Stability Monitor |
| FD | Flight Director |
| FDAI | Flight Director Attitude Indicator |
| FDK | Forced Display Module |
| FDO | Flight Dynamics Officer |
| FET | Flight Evaluation Team |
| FEWG | Flight Evaluation Working Group |
| FGR | Flight Geometry Reserves |
| FLSC | Flexible Linear Shaped Charge |
| FLTR | Filter |
| FM | Frequency Modulation |
| FPR | Flight Performance Reserves |
| FRT | Flight Readiness Test |
| FT | Foot or Feet |
| FTS | Flight Termination System |

**G**

| | |
|---|---|
| G | Acceleration of Gravity |
| GAL | Gallon(s) |
| GBI | DOD-ETR MSFN Station, Grand Bahama Island |
| GBM | NASA MSFN Station, Grand Bahama Island |
| GCC | Ground Control Computer |
| GDC | Gyro Display Coupler |
| GDS | NASA MSFN Deep Space Station, Goldstone California |
| GET | Ground Elapsed Time |
| GG | Gas Generator |
| $GH_2$ | Gaseous Hydrogen |
| GHz | Gigihertz (One Billion Hertz) |
| GLOTRAC | Global Tracking System |
| $GN_2$ | Gaseous Nitrogen |
| G&N | Guidance and Navigation |
| GNC | Guidance, Navigation and Control Engineer |
| GND | Guidance, Navigation and Digital Systems Engineer (Booster Systems) |
| GOX | Gaseous Oxygen |
| gpm | Gallons per Minute |
| GRR | Guidance Reference Release |
| GSE | Ground Support Equipment |
| GSE-ECU | Ground Support Equipment-Environmental Control Unit |
| GSFC | Goddard Space Flight Center |
| GTK | DOD-ETR MSFN Station, Grand Turk Island |
| GUID | Guidance |
| GUIDO | Guidance Officer |
| G-V | Gravity versus Velocity |
| GWM | NASA MSFN Station, Guam |

**H**

| | |
|---|---|
| $H_2$ | Hydrogen |

| | |
|---|---|
| HAW | NASA MSFN Station, Hawaii |
| HD | Highly Desirable Mission Rule Item |
| He | Helium |
| HF | High Frequency (3-30 MHz) |
| HOSC | Huntsville Operations Support Center |
| HSD | High-Speed Data |
| HTV | DOD-WTR MSFN Station, USNS Huntsville |
| Hz | Hertz (one cycle per second) |

**I**

| | |
|---|---|
| IA | Input Axis |
| IBM | International Business Machines Corporation |
| ICD | Interface Control Document |
| ICO | Inboard Cutoff |
| IECO | Inboard Engine Cutoff |
| IF | Intermediate Frequency |
| IFV | Igniter Fuel Valve |
| IGM | Iterative Guidance Mode |
| IMP | Impulse |
| IMU | Inertial Measurement Unit |
| IMV | Ignition Monitor Valve |
| in | Inch |
| INTELSAT | Communications Satellite |
| IO | Industrial Operations |
| IOS | Indian Ocean Ships |
| IP | Impact Predictor (Located at Kennedy Space Center) |
| IRIG | Inter-Range Instrumentation Group |
| IU | Instrument Unit |

**J-K**

| | |
|---|---|
| kbps | Kilobits per Second |
| kHz | Kilohertz (One Thousand Hertz) |
| KM | Kilometer |
| KOH | Potassium Hydroxide |
| KSC | Kennedy Space Center |

**L**

| | |
|---|---|
| lb(s) | Pound(s) |
| LCC | Launch Control Center |
| LCR | LIEF Control Room |
| LE | Launch Escape |
| LEM | Lunar Excursion Module |
| LES | Launch Escape System |
| LET | Launch Escape Tower |
| LOS | Loss of Spacecraft (by a site) |
| $LH_2$ | Liquid Hydrogen |
| LIEF | Launch Information Exchange Facility |
| LM | Lunar Module |
| LO | Liftoff |
| lox | Liquid Oxygen |
| LSC | Linear Shaped Charge |
| LSD | Low-Speed Data |
| LTA-B | Lunar Test Article, Model B |
| LV | Launch Vehicle |
| LVDA | Launch Vehicle Data Adapter |
| LVDC | Launch Vehicle Digital Computer |

**M**

| | |
|---|---|
| M | Mandatory Mission Rule Item |
| M | Mass |
| MA | Apollo Program Office (Symbol) |

MAD .....NASA MSFN Deep Space Station, Madrid, Spain
MAP . . . . . . . . . . . . . . . . . . . Message Acceptance Pulse
MAX . . . . . . . . . . . . . . . . . . . . . . . . . . . . . . .Maximum
MCC . . . . . . . . . . . . . . . . . . . . . . Mission Control Center
MCP . . . . . . . . . . . . . . . . . . Mission Control Programmer
MCR . . . . . . . . . . . . . . . . . . . . . . . Main Conference Room
MD . . . . . . . . . . . . . . . . . . . . . . . . . . . . .Mission Director
MDC . . . . . . . . . . . . . . . . . . . . . . .Main Display Console
MDC . . . . . . . . . . . . . . . . McDonnell-Douglas Corporation
MDF . . . . . . . . . . . . . . . . . . . . . . . . Mild Detonating Fuse
MED . . . . . . . . . . . . . . . . . . . . . . . . . . . . . . . . . . Medium
MER . . . . . . . . . . DOD-WTR MSFN Station, USNS Mercury
MESC . . . . . . . . . . . Master Event Sequence Controller
MFCO . . . . . . . . . . . . . . . . . . . . . . Manual Fuel Cutoff
MHz . . . . . . . . . . . . . . . . . .Megahertz (One Million Hertz)
MIL . . . . . . .NASA MSFN Station, Merritt Island, Florida
MILA . . . . . . . . . . . . . . . Merritt Island Launch Area
MIN . . . . . . . . . . . . . . . . . . . . . . . . . . . . . . . . . Minimum
Min . . . . . . . . . . . . . . . . . . . . . . . . . . . . . . . . . . Minute
ML . . . . . . . . . . . . . . . . . . . . . . . . . . . . Mobile Launcher
MLA . . . . DOD-ETR MSFN Station, Merritt Island, Florida
MMH . . . . . . . . . . . . . . . . . . . . . Monomethyl Hydrazine
M&O . . . . . . . . . . . . . . . . . . Maintenance and Operations
MOC . . . . . . . . . . . . . . . . . . Mission Operations Computer
MOCR . . . . . . . . . . .Mission Operations Control Room
MOD . . . . . . . . . . . . . . . . . . . . . . . . . . . . . . . . . .Model
MOD . . . . . . . . . . . . . . . . . . . . . . . . . . . . . .Modification
MR . . . . . . . . . . . . . . . . . . . . . . . . . . . . . . .Mixture Ratio
MSC . . . . . . . . . . . . . . . . . . . . . . Manned Spacecraft Center
m/sec . . . . . . . . . . . . . . . . . . Millisecond (1/1000 Second)
MSFC . . . . . . . . . . . . . . . . . . Marshall Space Flight Center
MSFN . . . . . . . . . . . . . . . . . Manned Space Flight Network
MSFNOC . Manned Space Flight Network Operations Center
MSK . . . . . . . . . . . . . . . . . . . . . .Manual Select Keyboard
MSR . . . . . . . . . . . . . . . . . . . . . . . .Mission Support Room
MSS . . . . . . . . . . . . . . . . . . . . . . Mobile Service Structure
MTF . . . . . . . . . . . . . . . . . . . . . Mississippi Test Facility
MTVC . . . . . . . . . . . . . . . .Manual Thrust Vector Control
MUX . . . . . . . . . . . . . . . . . . . . . . . . . . . . .Multiplexer

**N**

N/A . . . . . . . . . . . . . . . . . . . . . . . . . . . . . . .Not Applicable
NASA . . . . . National Aeronautics and Space Administration
NASCOM . . . . . . . . . . . . . .NASA Communications Network
No . . . . . . . . . . . . . . . . . . . . . . . . . . . . . . . . . . . Number
NPSH . . . . . . . . . . . . . . . . . . . . Net Positive Suction Head
NMI . . . . . . . . . . . . . . . . . . . . . . . . . . . . Nautical Mile

**O**

OA . . . . . . . . . . . . . . . . . . . . . . . . . . . . . . . .Output Axis
OAT . . . . . . . . . . . . . . . . . . . . . . .Overall Acceptance Test
OBECO . . . . . . . . . . . . . . . . . . .Outboard Engine Cutoff
OCO . . . . . . . . . . . . . . . . . . . . . . . . . . Outboard Cutoff
ODOP . . . . . . . . . . . . . . . . . . . . . . . . . . . . Offset Doppler
OECO . . . . . . . . . . . . . . . . . . . . .Outboard Engine Cutoff
O2 . . . . . . . . . . . . . . . . . . . . . . . . . . . . . . . . . .Oxygen
O&P . . . . . . . . . . . . . . . . . . . . Operations and Procedures
OSC . . . . . . . . . . . . . . . . . . . . . . . . . . . . . . . Oscillator
OSR . . . . . . . . . . . . . . . . . . . . Operations Support Room
OTV . . . . . . . . . . . . . . . . . Operational Television System
OX . . . . . . . . . . . . . . . . . . . . . . . . . . . . . . . . . .Oxidizer

**P**

P . . . . . . . . . . . . . . . . . . . . . . . . . . . . . . . . . . . .Pitch

PAM . . . . . . . . . . . . . . . . . . . . Pulse Amplitude Modulation
PAT . . . . . . DOD-ETR MSFN Station, Patrick AFB, Florida
PBI . . . . . . . . . . . . . . . . . . . . . . . . Push-Button Indicator
PC . . . . . . . . . . . . . . . . . . . . . . . . . . . . . . . Pitch Control
PCM . . . . . . . . . . . . . . . . . . . . . . .Pulse Code Modulation
pct . . . . . . . . . . . . . . . . . . . . . . . . . . . . . . . . . .Percent
PDFRR . . . . . . . Program Director's Flight Readiness Review
PDR . . . . . . . . . . . . . . . . . . . . . .Preliminary Design Review
PDS . . . . . . . . . . . . . . . . . . . . .Propellant Dispersion System
PEA . . . . . . . . . . . . . . . . . . . . .Platform Electronics Assembly
PETN . . . . . . . . . . . . . . . . . .Pentaerythrite Tetranitrate
PMPFR . . . . . . . . . . . .Program Manager's Preflight Review
POGO . . . . . . . . . Undesirable Launch Vehicle Longitudinal
                                                    Oscillations
POS . . . . . . . . . . . . . . . . . . . . . . . . . . . Pacific Ocean Ships
POS . . . . . . . . . . . . . . . . . . . . . . . . . . . . . . . . Position
POT . . . . . . . . . . . . . . . . . . . . . . . . . . . . Potentiometer
pps . . . . . . . . . . . . . . . . . . . . . . . . . .Pulses per Second
PRE . . . . .DOD-ETR MSFN Station, Pretoria, So. Africa
PRF . . . . . . . . . . . . . . . . . . . . Pulse Repetition Frequency
PRN . . . . . . . . . . . . . . . . . . . . . . . . Psuedo-Random Noise
PRPLNT . . . . . . . . . . . . . . . . . . . . . . . . . . . .Propellant
psi . . . . . . . . . . . . . . . . . . . . . . . . Pounds per Square Inch
psia . . . . . . . . . . . . . . . . .Pounds per Square Inch Absolute
psid . . . . . . . . . . . . . . Pounds per Square Inch Differential
psig . . . . . . . . . . . . . . . . . . . .Pounds per Square Inch Gauge
PTCR . . . . . . . . . . . . . . . . . . Pad Terminal Connection Room
PTCS . . . . . . . . . Propellant Tanking Computer System
PTL . . . . . . . . . . . . . . . . . . . . . . . . . . Prepare to Launch
PU . . . . . . . . . . . . . . . . . . . . . . . . . .Propellant Utilization

**Q**

QLDS . . . . . . . . . . . . . . . . . . . . . Quick Look Data Station

**R**

R . . . . . . . . . . . . . . . . . . . . . . . . . . . . . . . . . . . . Roll
RACS . . . . . . . . . . . . Remote Automatic Calibration System
RASM . . . . . . . . . . . . . . . .Remote Analog Submultiplexer
RCR . . . . . . . . . . . . . . . . . . . . . . . . Recovery Control Room
RCS . . . . . . . . . . . . . . . . . . . . . Reaction Control System
RDM . . . . . . . . . . . . . . . . . . . Remote Digital Multiplexer
RDX . . . . . . . . . . . . . . . Cyclotrimethylene-trinitramine
R&DO . . . . . . . . . . Research and Development Operations
RDSM . . . . . . . . . . . . . . . . Remote Digital Submultiplexer
RED . . . . . . . . . .DOD-WTR MSFN Station, USNS Redstone
RETRO . . . . . . . . . . . . . . . . . . . . . . . . . Retrofire Officer
RF . . . . . . . . . . . . . . . . . . . . . . . . . . . . . Radio Frequency
RI . . . . . . . . . . . . . . . . . . . . . . . . . . . . Radio Interference
RIS . . . . . . . . . . . . . . . . . . .Range Instrumentation Ship
RKV . . . . . . . . DOD-ETR MSFN Station, USNS Rose Knot
RMS . . . . . . . . . . . . . . . . . . . . . . . . Root Mean-Square
RNG . . . . . . . . . . . . . . . . . . . . . . . . . . . . . . . . . .Range
RP-1 . . . . . . . . . . . . . . . . . . . . . . . . . . Rocket Propellant
R&QA . . . . . . . . . . . . . . .Reliability & Quality Assurance
R&R . . . . . . . . . . . . . . . . . . . . . . . . . Receive and Record
RSCR . . . . . . . . . . . . . . .Range Safety Command Receiver
RSDP . . . . . . . . . . . . . . . . . . Remote Site Data Processor
RSO . . . . . . . . . . . . . . . . . . . . . . . Range Safety Officer
RSS . . . . . . . . . . . . . . . . . . . . . . . . . . .Root Sum Square
RTC . . . . . . . . . . . . . . . . . . . . . . . . . . Real Time Command
RTCC . . . . . . . . . . . . . . . . .Real Time Computer Complex
R/T CMD . . . . . . . . . . . . . . . .Real-Time Command Controller
RTK . . . . . . DOD-WTR MSFN Station, USNS Range Tracker

## S

S&A . . . . . . . . . . . . . . . . . . . . . . . Safety and Arming
SACTO . . . . . . . . . . . . . . . . . . .Sacramento Test Operations
SC . . . . . . . . . . . . . . . . . . . . . . . . . . . . . . . . . . Spacecraft
SCAMA . . . . . . . . .Switching, Conferencing, and Monitoring
                                                    Arrangement
SCO . . . . . . . . . . . . . . . . . . . . . . . .Subcarrier Oscillator
SCS . . . . . . . . . . . . . . . . . . . . . . . Stability Control System
SDF . . . . . . . . . . . . . . . . . . .Systems Development Facility
sec . . . . . . . . . . . . . . . . . . . . . . . . . . . . . . . . . . . .Second
SECO . . . . . . . . . . . . . . . . . . . . . . Single Engine Cutoff
SEP . . . . . . . . . . . . . . . . . . . . . . . . . . . . . . Separation
SIT . . . . . . . . . . . . . . . . . . . . . . . .Systems Interface Test
SLA . . . . . . . . . . . . . . . . . . . . . Spacecraft-LM Adapter
SLV . . . . . . . . . . . . . . . . . . . . . . .Saturn Launch Vehicle
SM . . . . . . . . . . . . . . . . . . . . . . . . . . . . . .Service Module
SMEK . . . . . . . . . . . Summary Message Enable Keyboard
S/N . . . . . . . . . . . . . . . . . . . . . . . . . . . Serial Number
SPS . . . . . . . . . . . . . . . . . . . . . Service Propulsion System
SRA . . . . . . . . . . . . . . . . . . . . . . . . . .Spin Reference Axis
SRM . . . . . . . . . . . . . . . . . . . . . . Status Report Module
SRO . . . . . . . . . . . . . . Superintendent of Range Operations
s/s . . . . . . . . . . . . . . . . . . . . . . . . . . Samples per Second
SSB . . . . . . . . . . . . . . . . . . . . . . . . . Single Sideband
SSR . . . . . . . . . . . . . . . . . . . . . . . . .Staff Support Room
ST-124-M3 . . . . . . . . . . . . . . . . . . Saturn V Stable Platform
STDV . . . . . . . . . . . . . . . . . . . Start Tank Discharge Valve
SW . . . . . . . . . . . . . . . . . . . . . . . . . . . . . . . .Switch(ing)
SYNC . . . . . . . . . . . . . . . . . . . . . . . . . . . . . Synchronize

## T

TAN . . . . . . . . . NASA MSFN Station, Tananarive, Malagasy
TBD . . . . . . . . . . . . . . . . . . . . . . . . . To Be Determined
TCS . . . . . . . . . . . . . . . . . . Thermal Conditioning System
TEL IV . . . . . . .AFETR Telemetry Station at Cape Kennedy
Teltrac . . . . . . . . . . . . . . . . . . . . . . . Telemetry Tracking
TEX . . . . . . . . NASA MSFN Station, Corpus Christi, Texas
TLI . . . . . . . . . . . . . . . . . . . . . . . .Translunar Injection
TLM . . . . . . . . . . . . . . . . . . . . . . . . . . . . .Telemetry
TM . . . . . . . . . . . . . . . . . . . . . . . . . . . . . .Telemetry
TSM . . . . . . . . . . . . . . . . . . . . . . Tail Service Mast
TTY . . . . . . . . . . . . . . . . . . . . . . . . . . . . . .Teletype
TV . . . . . . . . . . . . . . . . . . . . . . . . . . . . . Television
TVC . . . . . . . . . . . . . . . . . . . . . Thrust Vector Control
TWR JETT . . . . . . . . . . . . . . . . . . . . . . . . Tower Jettison

## U

UDL . . . . . . . . . . . . . . . . . . . . . . . . . . .Up-Data-Link

UHF . . . . . . . . . . . . Ultra High Frequency (300-3000 MHz)
USB . . . . . . . . . . . . . . . . . . . . . . . . . . .Unified S-Band
USBS . . . . . . . . . . . . . . . . . . . . .Unified S-Band Station
USNS . . . . . . . . . . . . . . . . . . . . .United States Navy Ship

## V

V . . . . . . . . . . . . . . . . . . . . . . . . . . . . . . . . . . . .Velocity
V . . . . . . . . . . . . . . . . . . . . . . . . . . . . . . . . . . . . . Voice
v . . . . . . . . . . . . . . . . . . . . . . . . . . . . . . . . . . . . . Volts
v-a . . . . . . . . . . . . . . . . . . . . . . . . . . . . . . . .Volt-ampere
VAB . . . . . . . . . . . . . . . . . . . . .Vehicle Assembly Building
VAN . . . . . . . . . .DOD-WTR MSFN Station, USNS Vanguard
V/D . . . . . . . . . . . . . . . . . . . . . . . . . .Voice and Data
vdc . . . . . . . . . . . . . . . . . . . . . . . . Volts, Direct Current
VHF . . . . . . . . . . . . . Very High Frequency (30-300 MHz)
vswr . . . . . . . . . . . . . . . . . . . .Voltage Standing Wave Ratio

## W

WBD . . . . . . . . . . . . . . . . . . . . . . . . . . . . Wide Band Data
W/G . . . . . . . . . . . . . . . . . . . . . . . . . . . . . . . Water/Glycol
WHS . . . . . .DOD MSFN Station, White Sands Missile Range,
                                                    N. Mexico
WOM . . . . . . . . . .WRE MSFN Station, Woomera, Australia
WRE . . . . . . . . .Weapons Research Establishment, Australian
                                          Department of Supply
WSMR . . . . . . . . . . . . . . . . . . . White Sands Missile Range
WTN . . . . . . . . .DOD-WTR MSFN Station, USNS Watertown
WTR . . . . . . . . . . . . . . . . . . . . . . . . . . . Western Test Range

## X-Z

XLUNAR . . . . . . . . . . . . . . . . . . . . . . . . . . Translunar
XTAL . . . . . . . . . . . . . . . . . . . . . . . . . . . . . . . . .Crystal
Y . . . . . . . . . . . . . . . . . . . . . . . . . . . . . . . . . . . . . . Yaw

## SYMBOLS

$\beta$ . . . . . . . . . . . . . . . . . . Thrust Vector Angular Deflection
$\Delta P$ . . . . . . . . . . . . . . . . . . . . . . .Differential Pressure
$\Delta V$ . . . . . . . . . . . . . . . . . . . . . . . Velocity Increment
$\Delta\rho$ . . . . . . . . . . . . . . . . . . .Atmospheric Density Increment
$\mu$ sec . . . . . . . . . . . . . . . . . . . . . . . . . . . .Microsecond
$\phi$ . . . . . . . . . . . . . . . . . . . . . . . . . . .Vehicle Attitude
$\dot{\phi}$ . . . . . . . . . . . . . .Vehicle Attitude Angular Change Rate
$\psi$ . . . . . . . . . . . . . . . . . . . . . . Attitude Error Command
$\chi$ . . . . . . . . . . . . . . . . . . . . . . . . . . . .Desired Attitude

# APPENDIX B
# BIBLIOGRAPHY

Apollo Design Certification Review; April 21, 1966; Apollo Program Directive No. 7.

Apollo Flight Readiness Reviews, Part I; November 8, 1965; Apollo Program Directive No. 8.

Apollo Mission Rules; April 12, 1967; NASA Management Instruction 8020.9.

Apollo/Saturn V Facility Description; October 1, 1966; K-V-012.

Apollo Test Requirements; March 1967; NHB 8080.1.

Astrionics System Handbook, Saturn Launch Vehicle; August 15, 1966; Not numbered.

Automatic Terminal Countdown Sequence Interlocks SA-502; February 16, 1968; 40M50547.

Battery Assembly; May 19, 1965; 40M30780.

Battery, S-IVB Aft Bus No. 2; Rev. J, February 13, 1967; 1A68317.

Battery, S-IVB Forward Bus No. 1 and Aft Bus No. 1; Rev. L; March 29, 1967; 1A59741.

Battery, S-IVB Forward Bus No. 2; Rev. K; February 3, 1967; 1A68316.

Battery Technical Manual; June 1, 1967; IBM 66-966-0024.

C-Band Radar Transponder Set SST-135C; March 1, 1967; IBM 66-966-0008.

Cable Interconnection Diagram, Instrument Unit, S-IU-503; June 27, 1966; 7909181.

Channel Assignments, S-IU-503; Rev. R; April 27, 1967; 7910974.

Checklist, Technical, Saturn V, Revision 15, July 1, 1968; Not numbered.

Clearance Envelope, LEM/S-IVB/IU, Physical; Rev. B; Not dated; 13M50123.

Command Decoder Technical Manual; March 1, 1967; IBM 66-966-0019.

Components Clearance Envelope S-IVB to IU; May 27, 1966; 13M50117.

Composite Mechanical Schematic; Rev. J; September 1, 1968; 10M30531.

Design Certification Review Report S-IC Stage, AS-503, Volume 4; April 4, 1968; Not numbered.

Design Certification Review Report, S-II-3 Stage; February 26, 1968; SD68-125.

Design Certification Review Report, S-IVB Stage, AS-503, Volume 6; February 21, 1968; DAC-56644.

Dynamic Launch Decision Criteria Feasibility Study Report Saturn V; January 31, 1967; D5-15685.

Electrical Interface, Saturn V Launch Vehicle and Apollo Block II CSM; January 19, 1968; 40M37521.

Electrical Interface, S-IC/S-II, Saturn V Vehicle; October 27, 1965; 40M30596.

Electrical Interface, S-II/S-IVB, Saturn V Vehicle; November 2, 1965; 40M30593.

Electrical Interface, S-IVB/Instrument Unit, Saturn V Vehicle; Rev. D; December 1, 1967; 40M30597.

Emergency Detection and Procedures (Draft Copy) Received from MSFC November 7, 1967; Not numbered.

Emergency Detection and Procedures, Addendum, (Draft copy); Received from MSFC November 17, 1967; Not numbered.

Emergency Detection System Analysis, Saturn V Launch Vehicle, SA-503; Rev. C; September 20, 1968; D5-15555-3C.

Emergency Detection System Description, Apollo-Saturn V; July 1, 1967; IBM 66-966-0022.

Engine Actuation, S-II, FISR; June 30, 1966; D-515657.

Engine Compartment Conditioning Subsystem Report, Saturn S-II; November 15, 1963; SID 62-142.

Environmental Control System, Operational Description of IU and S-IVB; Rev. B; April 28, 1965; 20M97007.

Environmental Control System Saturn V Vehicle/Launch Complex 39, FISR; June 30, 1966; D5-15662.

F-1 Engine Allocation Chart, S-IC; Rev. F; June 13, 1967; 60B37303.

Flight Control Data Requirements, Saturn Launch Vehicle, AS-503; January 15, 1968; Not numbered.

Flight Control Data Requirements, Saturn Launch Vehicle, Standard; August 1, 1967; Not numbered.

Flight Control, Introduction to; June 16, 1967; SCD No. T.01.

Flight Control System, Saturn S-II; February 1, 1965; SID 62-140.

**B**

Flight Loads Analysis, Saturn V Launch Vehicle; AS-503 (Manned); August 15, 1968; D5-15531-3A

Flight Operations Planning and Preparation for Manned Orbital Missions; (John H. Boynton and Christopher C. Kraft, Jr.) AIAA Paper No. 66-904

Flight Performance Handbook for Powered Flight Operations; (Space Technology Laboratories, Inc.) Not dated; Not numbered.

Flight Sequence Program, Definition of Saturn SA-503; Rev. C; September 26, 1968; 40M33623.

Fluids Requirements, Saturn V Vehicle Instrument Unit, ICD; Rev. A; March 29, 1967; 13M50099.

Fluids Requirements, Saturn V Launch Vehicle, S-II Stage; Rev. E; March 9, 1967; 13M50097.

Fluids Requirements, Saturn ICD, S-IVB Stage; Rev. B; February 3, 1966; 13M50098.

Frequency Plan, Saturn Apollo, Instrumentation and Communications; Rev. B; May 9, 1966; 13M60003.

Fuel Tank Pressurization System, Advanced Functional Schematic; Rev. L; November 21, 1967; 1B62829.

Gaseous Helium System, Functional Interface Fluid Requirements; Rev. A; February 28, 1967; 65ICD9401.

Gaseous Nitrogen System, Functional Interface Fluid Requirements; Rev. A; February 28, 1967; 65ICD9402.

Ground Safety Plan, Apollo/Saturn IB; April 1, 1966; K-IB-023.

Ground Support Equipment Fact Booklet, Saturn V Launch Vehicle; Chg. August 25, 1967; MSFC-MAN-100.

Hazardous Gas Detection System; June 19, 1967; MSFC-MAN-234.

Hold Limit and Event Constraint Analysis, SA-504; May 17, 1968; SPAR 68-2.

Inboard Profile, Saturn V; Rev. M; February 25, 1966; 10M03369.

Instrumentation Program and Components, Instrument Unit, S-IU-5-3; Rev. A; February 23, 1966; 6009048.

Instrumentation Subsystem Report; July 1, 1967; SID 62-136

Instrumentation System Description; October 1, 1966; IBM 65-966-0021.

Instrumentation System Description, AS-503; September 1, 1965; D5-15338-5.

Instrument Unit Assembly; Rev. G; June 18, 1968; 10Z22501-1.

Instrument Unit Environmental Control System Description, Saturn IB/V; March 15, 1966; IBM 66-966-0009.

Instrument Unit-Saturn V, Functional Interlock Requirement; Rev. A; May 16, 1966; 10M30554.

Instrument Unit System Description and Component Data (S-IU-201-212/501-515), Saturn IB/V; Chg. June 6, 1967; IBM 66-966-0006.

Instrument Unit System Equipment, AS-500 Series, Functional Block Diagram; October 1, 1966; IBM 66-157-0003.

Instrument Unit to Spacecraft Physical Requirements; August 1, 1967; 13M50303.

J-2 Rocket Engine Data Manual; Chg. 4; July 10, 1968; R-3825-1.

Launch Strategy Study, Saturn V; June 14, 1968; SPAR 68-6

LIEF Operations Plan, AS-501; August 23, 1967; Not Numbered.

Lox Tank Pressurization System, Advanced Functional Schematic; Rev. C; August 1, 1966; 1B62828.

Manned Spacecraft: Engineering Design & Operation (Purser, Faget & Smith); Not numbered.

Measuring Rack and Measuring Rack Selector (S-IU-204 thru 212/501 thru 515); October 1, 1966; IBM 66-966-0017.

Mission Implementation Plan, Mission C Prime, AS-503/Apollo-8; September 17, 1968; I-V-8010.

Mission Program Plan, Saturn Instrument Unit; July 5, 1966; IBM 66-966-0023.

Mobile Launcher, Vehicle Holddown Arms and Service Platforms, Apollo Saturn V, Operations and Maintenance; September 15, 1967; TM 487-MD.

Model Specification for Saturn V S-II Stage; April 1, 1964; SID 61-361.

Navigation, Guidance, and Control System Description; November 1, 1966; IBM 66-966-0003.

Network Operations Document, All Flights, Goddard Space Flight Center; December 15, 1967; Not numbered.

Onboard EDS, Design Criteria, Saturn V; Rev. C; May 5, 1966; 13M65001.

Operational Data Acquisition System Design Requirements Drawing, Saturn S-IVB; Rev. F; May 23, 1966; 1B38225.

Operational Trajectory Analysis-Option 1, December Launch Opportunity; September 24, 1968; MSFC Memo R-AERO-FMT-199-68.

Operational Trajectory, Preliminary, AS-503 C Prime Mission (December Launch Window); October 8, 1968; Boeing Memo 5-9600-H-122.

Ordnance Systems, Saturn V Launch Vehicle; April 1, 1968; SE 005-003-2H.

Performance Prediction, S-IC Stage, for the Flight of the Apollo/Saturn 503 Vehicle; October 1968; Boeing Report D5-13600-3.

POGO, Joint MSFC/MSC Plan for Apollo/Saturn V Space Vehicle Oscillation Program, Preliminary; July 18, 1968; Not numbered.

POGO Solution for AS-503, Certification of; August 5, 1968; MPR SAT V-68-1.

Prelaunch Checkout and Launch Operations Requirements, Saturn V Launch Vehicle; July 18, 1967; I-V-TR-501-10.

Program Operations Plan, MSFC, Apollo/Saturn; January 31, 1966; Not numbered.

Program Orientation, Saturn S-IVB; August 1964; SM-46688.

Program Schedules, Status and Analysis (MSFC), Issue 25; August 1, 1968; I-V-P.

Propulsion Control System, Saturn S-II, Description (Preliminary); Rev. B; May 31, 1967; 20M97011.

Propulsion Performance Prediction Update for the Translunar Injection Mission, Saturn V SA-503 C Prime, (S-II and S-IVB Stages); October 15, 1968; Boeing Memo 5-9350-H-150.

Propulsion Systems Handbook (Preliminary); August 4, 1967; D5-15660.

Protective Hood and Tower Installation, Tail Service Mast; January 26, 1968; 76K 04274.

Range Safety; September 10, 1965; AFETRR 127-9.

Range Safety Analysis (Preliminary), Saturn V, AS-503; November 10, 1966; D5-15523-3.

Range Safety Manual; September 10, 1965; AFETRM 127-1.

Recycle to Next Launch Window. Sequence of Events - Time Bars - Saturn V Vehicle - KSC; October 19, 1967; D5-16004-913.

Redlines Digest, Countdown Observer, AS-503; June 6, 1968; I-V-RLD-503

Reference Trajectory, AS-504 Launch Vehicle; May 15, 1968; D5-15481-1.

S-IC Flight Readiness Report; Not dated; Not numbered.

S-IVB Stage Project Review, Saturn, (AS-204 thru AS-515); September 12, 1967; Not numbered.

S-IVB to Instrument Unit Functional Requirements, ICD; November 18, 1967; 13M07002.

S-IVB to Instrument Unit Physical Requirements; April 14, 1965; 13M50302.

S-IVB-V Stage End Item Test Plan; Rev. B; April 13, 1967; 1B66684.

S-IVB-V-503 Stage End Item Test Plan; Rev. B; December 29, 1966; 1B66170.

Saturn Data Summary Handbook; Douglas Logistics Support Department; October 1, 1965; Not numbered.

Saturn S-II General Manual; August 1, 1963; SID 63-1.

Secure Range Safety Command System for Saturn; November 9, 1964; NASA TM X-53162.

Sequence and Flow of Hardware Development and Key Inspection, Review, and Certification Checkpoints; August 30, 1966; Apollo Program Directive No. 6A.

Sequence of Events Time Bar; Assembly, Checkout, and Launch; Saturn V SA-504 Vehicle, KSC; Rev. C; June 13, 1968; D5-16004-911.

Sequence of Events Time Line Analysis; Assembly, Checkout, and Launch; Saturn V SA-504 Vehicle, KSC Pad; Rev. C; June 17, 1968; D5-16004-910.

Spacecraft (Block II) Description, Apollo Operations Handbook; March 1, 1967 (Revised July 29, 1968); SM2A-03-Block II-(1).

Specification Addendum, Saturn V Program, SA-503 (MSFC); July 12, 1966; RS02W-1000003.

Specification Baseline, Saturn V Program (SA-504 through SA-515); September 23, 1966 (Revised September 15, 1967); RS02W-1000004A.

Stage Flight Measurements, S-IC-3; April 15, 1967; MSFC-MAN-042-3.

Stage System Description S-IC-3; April 14, 1967; MSFC-MAN-040-3.

Start Times, F-1 Engine, S-IC Stage; Rev. E; August 2, 1967; 60B37301.

Switch Selector, Model II Saturn IB/V Instrument Unit; Chg. March 1, 1967; IBM 66-966-0001.

Systems and Performance Briefing, MSFC, April 27-28, 1967; Not numbered. Technical Checklist (MSFC), Saturn V, Rev. 15; July 1, 1968; Not numbered.

Telemetry System Manual, Saturn V, Vol. 2; Rev. A; January 26, 1968; 50M71535.

Test and Checkout Requirements Document, Launch Vehicle, SA-502 through SA-510; September 29, 1967; I-V-TR-502-10.

Test Plan, S-IVB Stage End Item; October 19, 1967; 1B66684.

Test Specifications and Criteria, KSC Prelaunch Checkout and Launch Operations, S-IVB-V; Rev. A; August 18, 1966; 1B66162.

Thermal Control, Subsystem Report for Saturn S-II; April 15, 1965; SID 62-137.

Timer, 30-Second, Multiple Engine Cutoff Enable; May 26, 1967; Rev. C; 7915671-1.

# ALPHABETICAL INDEX

index

www.ingramcontent.com/pod-product-compliance
Lightning Source LLC
Chambersburg PA
CBHW080327270326
41927CB00014B/3127